T0258585

Biodiesel Production

Biodiesel Production

Edited by **Kurt Marcel**

LANRYE
INTERNATIONAL

New Jersey

Published by Clanrye International,
55 Van Reypen Street,
Jersey City, NJ 07306, USA
www.clanryeinternational.com

Biodiesel Production
Edited by Kurt Marcel

International Standard Book Number: 978-1-63240-078-9 (Hardback)

This book contains information obtained from authentic and highly regarded sources. Copyright for all individual chapters remain with the respective authors as indicated. A wide variety of references are listed. Permission and sources are indicated; for detailed attributions, please refer to the permissions page. Reasonable efforts have been made to publish reliable data and information, but the authors, editors and publisher cannot assume any responsibility for the validity of all materials or the consequences of their use.

The publisher's policy is to use permanent paper from mills that operate a sustainable forestry policy. Furthermore, the publisher ensures that the text paper and cover boards used have met acceptable environmental accreditation standards.

Trademark Notice: Registered trademark of products or corporate names are used only for explanation and identification without intent to infringe.

Printed in the United States of America.

Contents

Preface

The world is advancing at a fast pace like never before. Therefore, the need is to keep up with the latest developments. This book was an idea that came to fruition when the specialists in the area realized the need to coordinate together and document essential themes in the subject. That's when I was requested to be the editor. Editing this book has been an honour as it brings together diverse authors researching on different streams of the field. The book collates essential materials contributed by veterans in the area which can be utilized by students and researchers alike.

This book highlights the latest developments and growing trends in the field of biodiesel production. The beginning few chapters, deal with feedstocks required for the production of biodiesel. It also encompasses issues related to the use of inexpensive inedible raw materials and production of biomass feedstock with properties that may help it to produce biodiesel. Waste vegetable oils, animal fats, algae etc. are some sources of oil to produce biodiesels. This book would be a rich source of information to the scientists, students and professionals who are engaged in biodiesel production.

Each chapter is a sole-standing publication that reflects each author's interpretation. Thus, the book displays a multi-facetted picture of our current understanding of application, resources and aspects of the field. I would like to thank the contributors of this book and my family for their endless support.

Editor

Part 1

Feedstocks for Biodiesel Production

Non Edible Oils:
Raw Materials for Sustainable Biodiesel

C.L. Bianchi et al.[*]
Università degli Studi di Milano,
Dipartimento di Chimica Fisica ed Elettrochimica, Milano,
Italy

1. Introduction

In EU directive 2003/30/EC biodiesel is defined as "methyl ester produced from vegetable or animal oil, of diesel quality, to be used as biofuel". The more recent EU directive 2009/28/EC has set the targets of achieving, by 2020, a 20% share of energy from renewable energy sources in the EU's overall energy consumption and a 10% share of energy from renewable sources in each member State's transport energy consumption. In this context special consideration is paid to the role played by the development of a sustainable and responsible biofuels production, with no impact on food chain.

Nowadays most biodiesel is produced through triglycerides transesterification of edible oils with methanol, in the presence of an alkaline catalyst (Lotero et al., 2005). The so obtained product has low viscosity and is a biofuel (fatty methyl ester) that can replace petroleum-based diesel fuel with no need of engine modifications (Suwannakarn et al., 2005). Furthermore, if compared to fossil fuel, the formed ester fuels are non-toxic, safe to handle, and biodegradable (Krawczyk, 1996). Glycerine is also obtained as by-product as shown in Fig. 1.

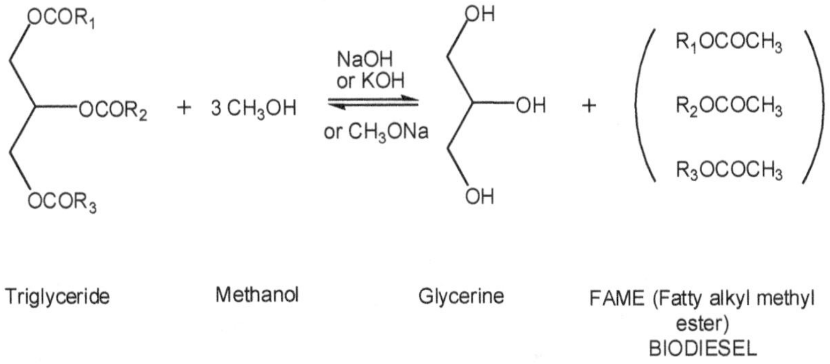

| Triglyceride | Methanol | Glycerine | FAME (Fatty alkyl methyl ester) BIODIESEL |

Fig. 1. Transesterification of a trygliceride.

[*] C. Pirola[1], D.C. Boffito[1], A. Di Fronzo[1], G. Carvoli[1], D. Barnabè[2], R. Bucchi[2] and A. Rispoli[2]
1 *Università degli Studi di Milano – Dipartimento di Chimica Fisica ed Elettrochimica, Milano, Italy,*
2 *Agri2000 Soc. Coop., Bologna, Italy.*

Refined, low acidity oilseeds (e.g. those derived from sunflower, soybean, rapeseed, tobacco etc.) may be easily converted into biodiesel, but their exploitation significantly raises the production costs, resulting in a biofuel that is not competitive with the petroleum-based diesel (Loreto et al., 2005). Presumably, as the market increases and technology is improved, costs will be driven down. In any case, the raw materials constitute a large portion of the manufacturing cost of biodiesel (up to 80%) (Bender, 1999).

Current oilseeds production systems raise environmental concerns because lands are intensively cultivated requiring high fertilizer and water inputs. These practices, aiming to increase yield, must be reduced or carefully regulated to prevent emissions of greenhouse gases or other environmental impacts. To do this, improved agronomic practices as the use of mixed species or crop rotation undoubtedly play a key role in mitigating negative impacts and enhancing biodiversity. A deep understanding of the microbial diversity of soils, its impacts on nutrient uptake and therefore on yield is crucial for sustainable cropping systems (The Royal Society, 2008).

Energy crops for industrial destination may represent a strategic opportunity in land use and income generation. However, in addition to the environmental aspects, economical concerns exist regarding the subtraction of lands for food cultivation. In a high market tension, it could have major impact on food/feed prices, increasing inequality, especially in developing countries. In addition, increased demand for food can result in the slow-down in biodiesel production due to reduced raw material availability. This was noticed in 2007 with industrial plants exploiting only 50% of their production capacity (Carvoli et al., 2008).

For all these reasons, it is highly desirable to produce biodiesel from crops specifically selected for their high productivity and characterized by low input requirements, or from low-cost feedstock such as waste cooking oil (WCO), animal fats and greases (Canakci et al., 2005; Zhang et. al, 2003).

While edible crops available for biodiesel production are restricted to few species (mainly palm/ soybean in the U.S. and palm/ rapeseed in the E.U.), the intent of using dedicated alternative feedstock opens a wide choice for new species that may be more suitable for specific conditions resulting on high yields.

The high WCO potential is recognized also by the EU directive 2009/28/EC, where waste vegetable or animal oil biodiesel is reported to save about the 88% of greenhouse emissions, a quite high value if compared to biodiesel from common vegetable oils, whose greenhouse emission savings range from 36 to 62%. The main issue posed by such a raw material is the need of its standardization, especially with regard to acidity decrease. Several methods have been proposed to solve this problem. Among them it is worth mentioning, besides the cited alkali refining method, addition of excess catalyst (Ono & Yoshiharu, 1979), extraction with a solvent (Rao et al., 2009), distillation refining process (Xingzhong et al., 2008) and pre-esterification method (Loreto et al., 2005; Pirola et al., 2010; Bianchi et. al, 2010; Parodi and Martini, 2008). This last seems to be the most attractive approach and has recently received much attention.

In the following paragraphs, the authors expose how it is possible to exploit waste materials or oils derived from crops not addressed to the food as potential raw materials for biodiesel production. Both the agronomic and chemical aspects deriving from the experimental work of the authors will be displayed.

2. Agronomical aspects

The authors present here preliminary results of a three years study about the feasibility of using new oilseed species for biodiesel production in Italy[1]. The intent is to propose an innovative agronomic solution that may affect the energy balance and the ability to achieve a high level of sustainability in the oilseeds production.

2.1 Non edible oil crops in the Mediterranean basin

A considerable amount of studies are available on mainstream and alternative crops for biodiesel feedstock. The authors made a selection of the most promising crops to be introduced in the Mediterranean zone, taking into account that currently the Mediterranean basin comprises not only temperate climate but also slightly-arid lands. Some of these are being effectively tested under the mentioned project as part of a unique rotation program.

Among oil crops the Brassicaceae family has an outstanding position. Rapeseed (*Brassica napus*) is the third largest oil crop with 12% of the world plant oil market with best yields when cultivated in cold-temperate regions (Carlsson, 2009). Yet, the large biodiversity of Brassicaceae reveal incipient species, among which *Brassica juncea*, *Brassica nigra*, *Brassica rapa*, *Brassica carinata*, *Sinapis alba*, *Camelina sativa*, *Eruca sativa* ssp. *oleifera*, etc. Besides the potential as raw material for biodiesel, their high content of glucosinolates (GSL) make them able to recover soils made marginal by soil-borne pests as nematodes (e.g. galling nematodes from the *Meloidogyne* genus and cist nematodes from *Heterodera* and *Globodera* genera) (Romero et al., 2009; Curto & Lazzeri, 2006). Many researchers also report weed-suppressive effects of Brassicaceae (e.g. Al-Khatib, 1997; Krishnan, 1998) as well as filtering-buffering effects against heavy metals pollution (Palmer et al., 2001).

On the other side an unexpected source of oilseed seems to arise from the tobacco culture. In anticipation of changes in tobacco market, selections of new varieties destined for energy production are coming out. Tobacco, as drought resistant species, seems a good option to face the shift of some previously fertile into arid lands caused by climate change.

2.1.1 *Brassica carinata*

The recent interest in *B. carinata* (also known as Ethiopian or Abyssinian mustard) is mainly a result of its high resistance to biotic and abiotic stresses such as drought tolerance. *Brassica carinata,* is an annual crop noted to be highly resistant to many rapeseed pests: blackleg (*Leptosphaeria maculans)*, white rust (*Albugo candida)*, *Sclerotinia* sp. and *Phyllotreta cruciferae* (Pan, 2009). According to Razon (2009), *B.carinata*, together with *E. sativa* ssp. *oleifera*, is the most promising oilseed for biodiesel purpose in temperate zones, not just for the yield but also for its adaptability to hard pedo-climatic conditions. It may be used in a crop rotation system with cereals and on low nutrient soils. Best results are achieved sowing on autumn (IENICA, 2004). Harvesting may be done with same equipment used for rapeseed with the advantage that *B. carinata* shows a good resistance to the dehiscence of mature siliquae. The vegetable oil obtained from *B.carinata* is characterized by the presence of erucic acid, making it unsuitable for human consumption. On the other hand, its physico-chemical properties meet the European

[1] SUSBIOFUEL project ("Studio di fattibilità per la produzione di biocarburanti da semi oleosi di nuove specie e da sottoprodotti o materiali di scarto" – D.M. 27800/7303/09), financially supported by the Ministry of Agricultural, Food and Forestry Policies – Italy.

specifications defined for biodiesel destination by the normative EN 14214:2002. Beyond its oil production capabilities, it was pointed out that the *B. carinata*'s lignocellulosic biomass can also be used to generate power and especially heat (Gasola et al., 2007), revealing an even greater potential.

2.1.2 *Brassica juncea*

Brassica juncea (also known as wild mustard or Indian mustard) varieties are grown for edible leaves or for condiment mustard only in some countries, while its use as an oilseed crop is increasingly growing. Canadian plant breeders have developed *B. juncea* cultivars with canola characteristics (Potts et al., 1999). As a result, canola varieties of *B. napus* and canola-type *B. juncea* have similar compositional characteristics. The key differences between *B. napus* and canola-type *B. juncea* lie in their agronomic characteristics. *Brassica juncea* tolerates high temperatures and drought better than *B. napus*, and thus it is better suited for the warmer, drier climates as the Upper Plains of the U.S. or the Mediterranean area. Green manure of *B.juncea* is a current practice in some countries (e.g. Italy and U.S.) making use of the GSL-Myrosinase system as a natural biofumigant. At the same time, this practice supplies organic matter to soil. To make the most of its biocidal activity against soil-borne pests and diseases, the mulching and incorporation to soil must be done at flowering time (Curto & Lazzeri, 2006).

2.1.3 *Nicotiana tabacum*

The tobacco (*N. tabacum*) is an annual herbaceous plant belonging to the Solanaceae family, widespread in North and South America, commonly grown for the collection of leaves. The seeds are very small (up to 10,000/g) and contain 36 to 39% of oil having a high percentage of linoleic acid (Giannelos et al., 2002). Currently, the common varieties directed to leaf production reach the modest order of 1 to 1.2 t seeds/ha (Patel, 1998, as cited in Usta, 2005) as a result of selection to reduce the amount of seed produced. Recently researchers were able to over express, through genetic engineering, genes responsible for the oil production in the leaves (Andrianov et al., 2010). However, the seeds potential for oil production is much higher. In this sense, another recent outcome on tobacco improvement is a variety that can at least triple seed (up to 5 t/ha) and oil production. The energy tobacco varieties exist both in the non GMO and the GMO version for resistance factors against herbicides and insects (Fogher, 2008). Its high oil yield makes it very competitive in front of mainstream oil crops as rapeseed, sunflower and soybean. The remaining meal revealed to be relevant for combustion or to be used as a protein source for livestock. Tests with pigs demonstrated its palatability to animals, a good conversion rate and therefore its equivalence to the soybean meal (Fogher, 2002). In addition, the presence of consolidate agricultural practices and know-how make clear the advantage of using a well-known species as tobacco as alternative feedstock for biodiesel. The research on Energy Tobacco has also found new economies for the transplant management as well as direct sowing techniques are currently under test. Combine-harvesters for the harvest of the whole inflorescences are available.

2.1.4 *Ricinus communis*

Ricinus communis (castor bean) is an oilseed crop that belongs to the Euphorbiaceae family, which includes other energy crops as cassava (*Manihot esculenta*), rubber tree (*Hevea brasiliensis*) and physic nut (*Jatropha curcas*). Among non-edible oils, the one extracted from castor bean is the most used for a wide variety of industrial purposes. Its oil is primarily of economic interest having cosmetic, medical and chemical applications. The presence of a high proportion of ricin oleic acid makes it suitable for the production of high-quality lubricants (Sanzone & Sortino, 2010). The use of castor oil is particularly supported in Brazil, with attempts to extract the ethyl esters using ethanol from sugarcane fermentation (although less reactive than methanol), making it a complete natural and renewable product (Pinto et al., 2005). Albeit the actual productivity is not very high, between 600 and 1,000 kg seeds/ha year, this value could triplicate with genetic improvement (Holanda, 2004). With the recent report on the draft genome sequence of castor bean revealing some key genes involved in oil synthesis (Chan et al., 2010), this possibility becomes even more palpable. In addition to this, the ease with which it can be cultivated in unfavorable environments contributes to its appeal as a raw material for sustainable biodiesel. In agreement to this, a two years field experiment conducted in south Italy using local ecotypes yielded around 2.3 t/ha of seeds, with up to 38% oil content, a quite high number for the dry conditions of the region (Sanzone & Sortino, 2010). The main limitation is the hand harvest, the current practice in the biggest producer countries as India, Brazil and China. However mechanization of harvesting is recently available for the collection of dwarf hybrid plants (Clixoo, 2010).

2.1.5 *Cynara cardunculus*
Among the species of interest for the production of biodiesel, the cardoon or artichoke thistle (*C. cardunculus*) is an important resource to be exploited, particularly in light of its adaptability to different soils. *Cinara cardunculus* is a perennial herbaceous species belonging to the family Asteraceae. Its deep root system allows the plant to extract water and nutrients from very deep soil zones revealing a plant with a small demand for fertilization and extremely resistant to drought. This characteristic makes it suitable to be grown on dry marginal or abandoned lands in the Mediterranean basin. Production reaches 30-35 t/ha per year, with about 2 tons of seeds; the seeds contain up to 25 % oil, with a similar composition to sunflower oil (Pasqualino, 2006). Recently, studies have been conducted within the EU project "Biocard - Global Process to Improve *C. cardunculus*". In the framework of this project, a research on the harvesting procedures, i.e. a crucial point of the cultivation of the thistle has also been conducted. As an example, a combine prototype designed to separate and thresh the capitula and to drop the biomass proved to be feasible, with a good cost/working capacity relation (Pari et al., 2008).

2.2 A new proposal for biodiesel production
The rationale of this proposal consists in the use of non-edible crops on soils no longer suitable for food production due to infestation by nematodes. The authors tested the possibility to rescue marginal soil fertility in consequence of the cultivation and the green manure of a naturally biocidal crop (*B. juncea* and *B. carinata*). Thanks to this practice the soil could be quickly good enough to produce oilseeds with satisfying yields for industrial destination. Furthermore a reduction in inputs of fertilizers is also expected due to preservation of organic matter content of soil. This practice offers the possibility to rescue soils availability for food production. Indeed, after some cycles of this rotation, the pest

control and the progressive increase of organic matter should make the soil eligible again for quality productions.

2.2.1 Experimental details

The agronomic rotation was tested under a wide range of situations. Three field trial locations were chosen taking into account Italy's wide latitudinal distribution[2]. Experimental design was thought to produce oilseed from N. tabacum and from traditional oilseed crops (sunflower, soybean, and rapeseed), used as comparison to validate the methodology. Each field was divided into two parts and B.juncea was sown only in one half of the field. To maximize the biofumigant effect, green manuring with B.juncea biomass was carried out when the crop reached flowering. After this, sowing of soybean, sunflower and rapeseed as well as the transplant of tobacco plantlets took place in both parts of the field. In order to make the proposal as flexible as possible, four different fertilization treatments were used: low input (30 kg/ha of chemical fertilizer[3]), medium input (90 kg/ha of chemical fertilizer), high input (140 kg/ha of chemical fertilizer) or organic input (10000 kg/ha of poultry manure). Untreated plots were set up as control. All field tests were conducted under Good Experimental Practices (GEP).

To evaluate the effect of the green manure of *B.juncea* on nematode infection, countings of *Meloidogyne* spp. were carried out on soil samples taken from both sides of the field while effects on yield of crops grown in succession were monitored recording the fresh weight per hectare (kg/ha) of plant biomass from both sides of the field. Since the green manure of *B.juncea* supplies organic matter to soil, possibly increasing also its sulphur content, it's relevant to ensure that crops grown after this agronomical practice are not enriched in sulphur and therefore less suitable for biodiesel production[4].To check this, sulphur quantification in sunflower seeds and oil were done. Seed samples were taken from the unfertilized plots of both sides of the field, and sulphur content detected by ICP-MS (Inductively Coupled Plasma Mass Spectrometry).

2.2.2 Results and discussion on agronomical aspects

Research on alternative biofuel aims to face the increasing demand for energy requirements by means of a more sustainable energy supply. From this point of view, greenhouse gases saving is expected from biofuels.

The first year of experimentation makes clear that plants grown in succession of *B. juncea* resulted in higher biomass. This could be due either to the increase in the organic matter content or to the pest control. Indeed, counting of nematodes revealed a strong effect of the green manure of *B.juncea* on nematode control. The average number of larvae found was almost four times lower in the presence of the biofumigant crop. The use of *B. juncea* as green manure does not influence the sulphur content in sunflower seeds and oil, suggesting no sulphur accumulation occurs in succeeding crops.

In order to assess the chemical properties of *B. juncea* oil for biodiesel destination, the authors quantified the total sulphur, nitrogen and phosphorus content in oil from commercial seeds of *B. juncea*. In table 1 data of the quantifications are reported.

---

[2] Altedo (BO), Vaccolino (FE) and Santa Margherita di Savoia (FG).
[3] Urea (Nitrogen 46%)
[4] The contents of this element in the final product must be under 10 ppm (UNI EN 14214 - Automotive fuels. Fatty acid methyl esters (FAME) for diesel engines. Requirements and test methods).

Element	Unit	Value	Standard Test Method
sulphur	mg/kg	112	UNI EN 20846:2005
nitrogen	% (mass)	0,35	ASTM D5291-09
phosphorus	mg/kg	< 4	UNI EN 14107:2003

Table 1. Nitrogen, sulphur and phosphor content in *B. juncea* oil.

In table 2 the mean percentage increasing of biomass of *B. napus*, *H. annus*, *G. max*, and *N. tabacum* produced after green manuring of *B. juncea* is summarized.

Crop	Unit	Biomass increasing
N. tabacum	%	21
B. napus	%	15
H. annus	%	26
G. max	%	28

Table 2. Increasing of biomass of oilseed crops produced after green manuring of B. *juncea*.

3. Chemical aspects: Standardization of the raw materials and biodiesel production

3.1 Oil characterization

Oil characterization before proceeding with the standardization of the raw material is a very important issue. Some properties remain in fact unchanged from the starting material to the finished biodiesel, or they are anyway predetermined. It is so important to check that the values of such chemical and physical oil properties are in range with those required by the standard regulations (see Table 3). The experimental procedures to get the values of such properties are also standardized and are indicated in the regulations. The following are parameters for starting oil that can affect the quality of the final biodiesel.

- **Sulfur and phosphorous content:**

High sulphur and phosphorous content in the fuels cause greater engine wear and in particular shorten the life of the catalyst. Biodiesel derived from soybean, rapeseed, sunflower and tobacco oils are known to contain virtually no sulphur (Radich, 2004; Zhiyuan et al., 2008).

The authors have nevertheless found that the oil obtained from *B.juncea* seeds may contain high concentrations of sulphur due to the presence in the plant's tissues of glucosinolates, the molecules responsible for the biofumigation effect.

- **Linoleic acid methyl ester, iodine value and viscosity**

Soybean, sunflower, peanut and rapeseed oils contain a high proportion of linoleic fatty acids, so affecting the properties of the derived ester with a low melting point and cetane number. Quantitative determination of linoleic acid methyl ester is accomplished by gas chromatography with the use of an internal standard after the substrate has been transesterificated and allows also the quantification of the other acid methyl esters (Environment Australia, 2003). The super-critical chromatography is another useful analytical technique, suitable for the direct analysis of the oils.

Specification	Units	limits		Method
		Min	Max	
Ester content	% (m/m)	96.5		EN 14103
Density 15°C	kg/m³	860	900	EN ISO 3675 EN ISO 12185
Viscosity 40°C	mm²/s	3.50	5.00	EN ISO 3104
Sulphur	mg/kg	-	10.0	preEN ISO 20846 preEN ISO 20884
Carbon residue (10% dist.residue)	% (m/m)	-	0.30	EN ISO 10370
Cetane number		51.0	-	EN ISO 5165
Sulphated ash	% (m/m)	-	0.02	ISO 3987
Water	mg/kg	-	500	EN ISO 12937
Total contamination	mg/kg	-	24	EN 12662
Cu corrosion max			-	EN ISO 2160
Oxidation stability, 110°C	h (hours)	6.0	-	EN 14112
Acid value	mg KOH/g	-	0.5	EN 14104
Iodine value	gr I₂/100 gr	-	120	EN 14111
Linoleic acid ME	% (m/m)	-	12.0	EN 14103
Methanol	% (m/m)	-	0.20	EN 14110
Monoglyceride	% (m/m)	-	0.80	EN 14105
Diglyceride	% (m/m)	-	0.20	EN 14105
Triglyceride	% (m/m)	-	0.20	EN 14105
Free glycerol	% (m/m)	-	0.02	EN 14105
Total glycerol	% (m/m)	-	0.25	EN 14105
Gp I metals (Na+K)	mg/kg	-	5.0	EN 14108 EN14109
Gp II metals (Ca+Mg)	mg/kg	-	5.0	EN14538
Phosphorous	mg/kg	-	5.0	EN 14538

Table 3. European Standard specifications for biodiesel (automotive fuels).

An indicative fatty acid methyl esters composition of the raw oils typically used for biodiesel production and of the ones adopted by the authors, is given in Table 4 (Velasco et al., 1998; Tyson, 2002; Winayanuwattikun at al. 2008, Zheng & Hanna, 1996).

Oil	Comon Name	Fatty acid composition, wt%
Arachis hypogea	Peanut	11.9 (16:0), 3.0 (18:0), 40.0 (18:1), 40.7 (18:2), 1.2 (20:0), 3.2 (22:0)
Brassica juncea	Indian mustard	3.6 (16:0), 1.1 (18:0), 13.9 (18:1), 21.5 (18:2), 13.7 (18:3), 8.7 (20:1), 33.5 (22:1)
Brassica napus	Canola	4.7 (16:0), 0.1 (16:1), 1.6 (18:0), 66.0 (18:1), 21.2 (18:2), 5.2 (18:3), 0.9 (20:0), 0.3 (22:0)
Carthamus tinctorius	Safflower	0.1 (14:0), 6.4 (16:0), 2.2 (18:0), 14.1 (18:1), 76.6 (18:2), 0.2 (18:3), 0.2 (20:0) 0.2 (22:0)
Elaeis guineensis	Palm	0.5 (12:0), 1.0 (14:0), 38.7 (16:0), 3.3 (18:0), 45.5 (18:1), 10.8 (18:2), 0.1 (18:3), 0.1 (20:0)
Glycine max	Soybean	10.7 (16:0), 3.0 (18:0), 24.0 (18:1), 56.6 (18:2), 5.3 (18:3), 0.2 (20:0), 0.2 (22:0)
Helianthus annus	Sunflower	6.6 (16:0), 3.1 (18:0), 22.4 (18:1), 66.2 (18:2), 1.0 (18:3), 0.3 (20:0), 0.4 (22:0)
Jatropha curcas	Physic nut	0.1 (12:0), 0.2 (14:0), 14.8 (16:0), 0.8 (16:1), 4.2 (18:0), 41.0 (18:1), 38.6 (18:2), 0.3 (18:3)
Nicotiana tabacum	Tobacco	6.6 (16:0), 3.1 (18:0), 22.4 (18:1), 66.2 (18:2), 1.0 (18:3), 0.3 (20:0), 0.4 (22:0)
Lard	-	4.8 (14:0), 28.4 (16:0), 4.7 (16:1) 14.8 (18:0), 44.6 (18:1), 2.7 (18:2)
Yellow grease	-	1.0 (14:0), 23.0 (16:0), 1.0 (16:1) 10.0 (18:0), 50.0 (18:1), 15.0 (18:2)
Brown grease	-	1.7 (14:0), 23.0 (16:0), 3.1 (16:1) 12.5 (18:0), 42.5 (18:1), 12.2 (18:2), 0.8 (18:3)

Table 4. Indicative acidic composition of some raw materials for biodiesel production.

- **Iodine value, viscosity and density**

The iodine value (IV) is an index of the number of double bonds in biodiesel, and therefore is a parameter that quantifies the degree of unsaturation of biodiesel. Both EN and ASTM standard methods measure the IV by addition of an iodine/chlorine reagent. Biodiesel viscosity is directly correlated to the IV of biodiesel for biodiesel with iodine numbers of between 107 and 150 (Environment Australia, 2003).

One of the main reasons for processing vegetable oils for use in engines is to reduce the viscosity thereby improving fuel flow characteristics. High viscosities can cause injector spray pattern problems that lead to excessive coking and oil dilution. These problems are associated with reduced engine life. Nevertheless, the necessary characteristics depend also on the end use; the engines for the production of energetic power in fact allow the use of fuels with higher viscosity (i.e. from palm oil).

Density dictates the energy content of fuel where high densities indicate more thermal energy for the same amount of fuel and therefore better fuel economy.

The authors have already published the results of the measurement of the IV obtained for some oils selected as potential raw materials for BD production (Pirola et al., 2011). In Table 5 the values of IV, viscosity and density found by the authors for waste cooking oil and its mixture with raw rapeseed oil are shown, demonstrating that the properties of the feedstock can be improved by the use of blends of different oils. The values reported in the Table 5 evidences that with the dilution with rapeseed oil it is possible to decrease the viscosity of WCO but increasing the number of IV. Nevertheless also in the case of most diluted sample the IV value is lower than those of rapeseed oil.

Oil	Iodine value (gI$_2$/100g oil)	Viscosity (mm^2/s 40 °C)	Density (kg/m^3 15° C)
WCO	54	82.2	918
WCO:rapeseed oil 1:1	85	52.8	914
WCO:rapeseed oil 3:1	100	40.5	926
Rapeseed	115	n.d.	n.d.

Table 5. IV, viscosities and densities of some potential raw materials for biodiesel production.

It has to be taken into account that after the transesterification process the IV of the feedstock remain unchanged, the viscosity is reduced from 10 to 15 times, whereas density has been found to remain almost the same or to be reduced in some cases (Zheng & Hanna, 1996).

3.2 Oil standardization: Free fatty acids esterification reaction

As already mentioned in the introduction paragraph, the use of raw, non edible oils poses the problem of standardization before the transesterification process, especially with regard to acidity decrease. In fact oils, besides triglycerides contain also free fatty acids (FFA). These lasts are able to react with the alkaline catalyst used for the transesterification reaction yielding soaps which prevent the contact between the reagents. A FFA content lower than 0.5% wt is also required by the EN 14214.

Among the different deacidification methods listed in the introduction, the authors have recently paid attention to the pre-esterification process (Loreto et al., 2005; Pirola et al., 2010; Bianchi et al., 2010). This method is particularly convenient as it is not only able to lower the acidity content of the oils but also provides methyl esters already at this stage, so increasing the final yield in biodiesel. A scheme of the FFA esterification reaction is given in Fig.2.

$$RCOOH + CH_3OH \overset{\text{acid catalysis}}{\rightleftharpoons} RCOOCH_3 + H_2O$$

Fig. 2. Scheme of the Free Fatty Acid Esterification Reaction.

The use of heterogeneous catalysts (Sharma & Singh, 2011) is usually preferred to the use of homogeneous ones (Alsalme et al., 2008) as it prevents neutralization and separation costs, besides being not corrosive, so avoiding the use of expensive construction materials. Another important advantage is that the recovered catalysts can be potentially used for a long time and/or multiple reaction cycles.

In the recent years the authors have deepened the study of the pre-esterification process investigating the effect of the use of different kinds of oils, different types of reactors and catalysts and different operating conditions (Pirola et al., 2010; Bianchi et al., 2010; Pirola et al. 2011)

In the following paragraphs, the most relevant aspects of the experimental work and the results obtained by the authors for what concerns the pre-esterification process are reported.

3.2.1 Experimental details

A remarkable aspect of the proposed process is represented by the mild operative conditions, i.e. low temperature (between 303 and 338 K) and atmospheric pressure. Moreover, the adopted working temperature is the same of the following transesterification reaction and of the methanol recovery by distillation. Each single reaction has been carried out for six hours withdrawing samples from the reactor at pre-established times and analysing them through titration with KOH 0.1 M. The percentage of FFA content per weight was calculated as otherwise reported (Marchetti & Errazu, 2007, Pirola et al. 2010).

All the esterification experiments have been conducted using a slurry reactor as the one already described elsewhere (Bianchi et al., 2010). A slurry reactor is the simplest type of catalytic reactor, in which the catalyst is suspended in the mass of the regents thanks to the agitation.

Much attention has been paid by the authors to the use of acid ion exchange resins. Amberlyst ®46 (named A46 in this chapter), i.e. a commercial product by Dow Advanced Materials, and D5081, a catalyst at the laboratory development stage by Purolite® have been successfully applied in this reaction. The main features of the employed catalysts are reported in Tab. 6.

Resin	Matrix	Functional Group	Ionic form	Acid capacity (meq H^+/g)	Max. operating Temp (°C)
D5081	Styrene-divinylbenzene	$R-SO_3^-$	H^+	1,0	130
A46	Styrene-divinylbenzene	$R-SO_3^-$	H^+	0,60	120

Table 6. Main features of the ion exchange resins adopted as catalysts in the FFA esterification reaction.

The acid capacity of the catalysts, corresponding to the number of the active sites per gram of catalyst was also experimentally determined by the authors by ion exchange with a NaCl-saturated solution and successive titration with NaOH (López et al., 2007). The values were found to be in agreement with the ones provided by technical sheets.

A distinguishing feature of A46 and D5081 is represented by the location of the active acid sites: these catalysts are in fact sulphonated only on their surface and not inside the pores. Consequently, A46 and D5081 are characterized by a smaller number of acid sites per gram if compared to other Amberlysts®, which are also internally sulphonated (Bianchi et al., 2010).

3.2.2 Deacidification results

In Fig. 3 the results from the esterification reaction performed on different raw oils are shown.

From the graph it can be noticed that in almost all the cases it is possible to obtain a FFA concentration lower than 0.5% wt after 6 hours of reaction. The differences in the acidic composition seem not to affect the final yield of the reaction. What seems to influence the FFA conversion is the refinement degree of the oil. Waste cooking oil (WCO) is in fact more hardly processable with the esterification in comparison to refined oils, probably due to its higher viscosity which results in limitations to the mass transfer of the reagents towards catalysts. Indeed, the required acidity limit is not achieved within 6 hours of reaction. Adding rapeseed oil, less viscous, to the WCO in different ratios it is possible to increase the final FFA conversion and reaching a FFA content lower than 0.5% wt. The blend of a raw oil characterized by high viscosity with a less viscous one is also effective in shortening the time to reach the plateau of conversion, as displayed in Fig. 4.

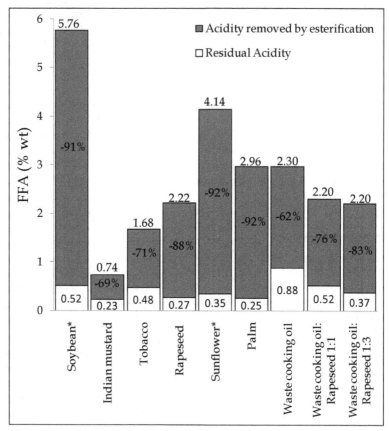

Fig. 3. Acidity removed by esterification (6 hr) and residual acidity of different oils used as raw material: slurry reactor, T=338K, catalyst: Amberlyst® 46 weight ratio methanol/oil= 16:100, weight ratio catalyst/oil=1:10; *commercial, refined oils with the addition of pure oleic acid.

In Fig. 4 the conversion curves concerning the recycles of the use of the catalyst A46 in the case of WCO are also shown. The catalyst does not show a drastic drop in its activity notwithstanding the used substrate is not refined. This decrease in the catalytic performance might be ascribable to the catalyst's settling in the reaction environment (Pirola et al., 2011) or to the presence of cations inside the oil. This aspect is still under investigation.

It is convenient to use an excess of methanol respect the stoichiometric amount in order to shift the equilibrium towards the product. Nevertheless, when adding methanol a double phase system is formed (the maximum solubility of methanol in oil is in the interval 6- 8%) and therefore it is not convenient to increase further this parameter.

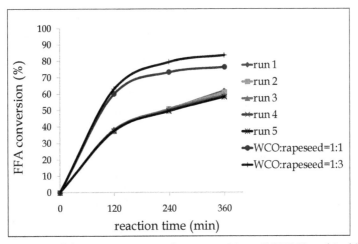

Fig. 4. FFA conversion (%) vs reaction time of waste cooking oil (WCO) and its blends with rapeseed oils: slurry reactor, T=338K, catalyst: Amberlyst® 46 weight ratio methanol /oil= 16:100, weight ratio catalyst/oil=1:10.

The lifetime of the catalyst is a very important issue from an industrial standpoint. The authors have already performed a deep study on the ion exchange resins endurance in the FFA esterification reaction (Pirola et al., 2010). The most important outcome of this study is that resins like A46 (Dow Advanced Materials) and D5081 (Purolite), which are functionalized only on their surface are very stable in the reaction conditions and can guarantee long operating times without being replaced.

A comparison between these two resins is displayed in Fig. 5.

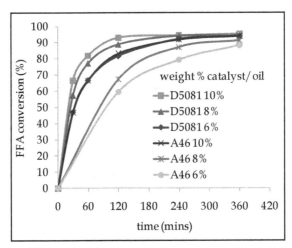

Fig. 5. FFA conversion (%) vs reaction time for different amounts of catalysts A46 and D5081, rapeseed oil with initial acidity=5%, slurry reactor, weight ratio methanol/oil= 16:100, T=338K. Dots are experimentally obtained. Continue lines are simulated (see paragraph 3.2.3)

As can be seen from the graph, catalyst D5081 shows better results than A46 at lower catalyst's loading. This can be easily explained by the higher number of acid sites located on its surface. In particular, the use of a ratio of 10 %wt of catalyst D5081 vs. oil allows reaching the maximum conversion in 2 hours. From the graph can be seen how the curves for 6% of D5081 and 10% wt catalyst/oil of A46 perfectly overlap. This outcome suggested that a fixed amount of acid active sites per gram of FFA was required to reach the maximum of conversion in 4 hours. Based on the experimental data obtained, this amount was found to be equal to 1.2 meq of H^+.

3.2.3 Simulation of the catalytic results

The considered reaction system turns out to be an highly non-ideal system, being formed by a mixture of oil, methylester, methanol, FFA and water. Indeed, activity coefficients instead of concentrations are used not only for the phase and chemical equilibria calculations, but also for the kinetic expressions. Modified UNIFAC model was used adopting the parameters available in literature and published by Gmehling et al., 2002 (Pirola et al., 2011).

A pseudohomogeneous model was used for describing the kinetic behavior of the reaction (Pöpken et al., 2000). The adopted model is displayed in the following equation:

$$r = \frac{1}{m_{cat}} \frac{1}{\upsilon_i} \frac{dn_i}{dt} = k_1 a_{FFA} a_{methanol} - k_{-1} a_{methylester} a_{water}$$

where:
r= reaction rate
m_{cat}= dry mass of catalyst, gr
υ_i= stoichiometric coefficients of component i
n_1= moles of component i
t = reaction time
k_1= kinetic constant of direct reaction
k_{-1}= kinetic constant of indirect reaction
a_i= activity of component i

The temperature dependence of the rate constant is expressed by the Arrhenius law:

$$k_i = k_i^0 \exp\left(\frac{-E_{A,i}}{RT}\right)$$

where k_i^0 and $E_{A,i}$ are the pre-exponential factor and the activation energy of the reaction i, respectively (i=1 for the direct reaction, i=-1 for the indirect reaction), T is the absolute temperature and R the Universal Gas Constant. The adopted parameters set is the same reported by Steinigeweg (Steinigeweg & Gmehling, 2003).

All the simulations were carried using Batch Reactor of PRO II by Simsci – Esscor. The model turned out to be able to reproduce qualitatively the behavior of different systems, characterized by different catalyst type and content.

In the previous Figure 5, continue lines represent simulated behaviors using the same parameters, but considering a different catalyst mass due to different catalyst acidity and concentration.

3.3 Oil transformation: The transesterification reaction

The transesterification reaction has been performed by the authors on the rapeseed and *B.juncea* (Indian mustard) oilseeds deacidified with the esterification process described in the previous paragraph.

Sodium Methoxide (MeONa) was employed as catalyst. MeONa is known to be the most active catalyst for triglycerides transesterification reaction, but it requires the total absence of water (Schuchardt, 1996). For this reason, the unreacted methanol and the reaction water were evaporated from the deacidified oils before processing them with the transesterification reaction.

The employed experimental setup was the same employed for the slurry esterification.

Being the transesterification an equilibrium reaction, it was performed in two steps, removing the formed glycerine after the first step. The adopted conditions were the following:

- 1st step: weight ratio methanol/oil=20:100, weight ratio MeONa/oil=1:100, 233 K, 1,5 h;
- 2nd step: weight ratio methanol/oil=5:100, weight ratio MeONa/oil=0.5:100, 233 K, 1 h.

The total ester content is a measure of the completeness of the transesterification reaction. Many are the factors affecting ester yield in the transesterification reaction: molar ratios of glycerides to alcohol, type of catalyst(s) used, reaction conditions, water content, FFA concentration, etc.

The European prEN14214 biodiesel standard sets a minimum limit for ester content of >96.5% mass, whereas the US ASTM D 6751 biodiesel standard does not set a specification for ester content.

Mono- and di-glycerides as well as tri-glycerides can remain in the final product in small quantities. Most are generally reacted or concentrated in the glycerine phase and separated from the ester.

Both in the case of rapeseed oil and *B.juncea* oilseed, the final yield in methylester was higher than 98%.

The analyses of methyl esters and unreacted mono-, di- and triglycerides are accomplished through gas chromatography.

The detailed requirements for biodiesel according to both EN 14214 and US ASTM D 6751 are listed in paragraph 1.

In the US a standard for biodiesel (ASTM D 6751 – Standard Specification for Biodiesel Fuel (B100) does not include the same number of parameters as prEN 14214 but the parameters that coincide have similar limits. The US specification covers sulfur biodiesel (B100) content much higher if compared to the one of European Standard. For use as a blend component with diesel fuel oils defined by ASTM D 975 Grades 1-D, 2-D, and low sulfur 1-D and 2-D. (Environment Australia, 2003).

4. Conclusion

The use of the oilseed deriving from alternative crops or waste oils as a feedstock for biodiesel production represents a very convenient way in order to lower the production costs of this biofuel.

From the agronomic point of view the authors verified that the green manure of *B.juncea* resulted in nematode infestation drastically decreased and improved soil quality, reflected in higher yield of crops in agronomic succession. In the first year of experimentation *B. juncea* was preferred to *B.carinata* because of its suitability to spring planting (starting period of the project SUSBIOFUEL). Further work will be necessary to improve the setting up of the agronomic proposal. Winter sowing of *B.carinata* will be done in the next years and alternative promising patented variety of tobacco (selected for seed production)[5] are currently under test. The authors are also evaluating the proposed rotation in comparison with commercial pellets[6] of defatted Brassicaceae meal. In addition, more outcomes are attended: yield grains[7], evaluation of the weed control potential of *B. juncea* and survival rate of transplanted *N.tabacum* plantlets following the green manuring or not.

The flexibility of Brassicaceae (efficient green manure and/or oil crop) allows using these species with a dual aim according to the situation, thus increasing the sustainability of the system. On the other hand new tobacco varieties promise yields above the best rape harvests around Europe. Under this light tobacco is a really interesting alternative oil crop especially in countries like Italy where it has been cultivated since a long time and Good Agricultural Practices (GAP) for this crop have long been known: all points in favour to the conversion of tobacco cultivation toward oil seeds production. To give a more comprehensive evaluation of innovations introduced in the whole biodiesel production chain, the authors aim to develop a method able to assess biodiesel sustainability.

The authors are aware that their proposal alone does not solve the overall sustainability problem of biodiesel production, but it contributes significantly to a wider portfolio of land-use strategy, stimulating the call for innovations both in technology and emissions reduction measures. Food production from marginal soils would worsen soil depletion and nematodes infestation. The restoring of soil fertility avoiding the chemicals usage, and in the mean time the generation of income from vegetable oils, assure the ethical, economical and environmental sustainability of the solution. Policy strategies will be needed to increasingly shift abandoned or low biodiversity value marginal lands to this kind of ecologically-friendly practices.

From the chemical point of view, the high concentration of FFA contained in these raw materials (waste or alternative crops) leading to the formation of soaps during the final transesterification step can be easily overcome by performing a pre-esterification reaction. This treatment allows lowering the acid content of the raw material below the limit required by the biodiesel standard, so avoiding also the formation of soaps during the transesterification stage. The FFA esterification is also helpful in increasing the final yield in biodiesel as it produces methyl esters.

Oilseeds of Brassica juncea, Nicotiana tabacum, rapeseed, palm, soybean and sunflower have been successfully deacidified with esterification reaction. Waste cooking oil (WCO) itself does not represent a good potential raw material for biodiesel production due to its properties which hardly match the required standards. Nevertheless it is possible to exploit this kind of feedstock by its use in blends with other oils characterized by a lower viscosity. The authors have successfully deacidified blend of WCO and rapessed oil, also obtaining an increase of the reaction rate.

Two acid ion exchange resins have been selected as catalysts: Amberlyst®46 (Dow Advanced Materials) and Purolite® D5081 (Purolite). Both these resins gave satisfactory results in the studied reaction. D5081 resulted to me more active than A46, being able to give the maximum of conversion in shorter times than A46, other conditions being equal.

---

[5] Kindly supplied by Sunchem Holding S.r.l.

[6] Biofence by Triumph Italia S.p.a.

[7] This kind of data is necessary to express results in terms of functional unit as required by a life cycle thinking approach.

A process simulation of the FFA esterification, able to predict the reaction progress through a thermodynamic and kinetic analysis was successfully performed using the software PRO II (SimSci). A pseudohomogeneous model was used for describing the kinetic behaviour of the reaction, using a modified UNIFAC model for the calculation of the activity coefficients (used not only for the phase and chemical equilibria calculations, but also for the kinetic expressions). The data obtained from the use of this model showed to be in a very good correlation with the experimental results

5. Acknowledgment

The authors gratefully acknowledge the financial support by Italian Ministero delle Politiche Agricole, Alimentari e Forestali (project SUSBIOFUEL – D.M. 27800/7303/09).

6. References

Al-Khatib, K., Libbey, C. & Boydston, R. (1997). Weed Suppression with Brassica Green Manure Crops in Green Pea. *Weed Science*, Vol. 45, No. 3, pp. 439-445, ISSN 1939-7291.

Alsalme, A. Kozhevnikova, E.F. & Kozhevnikov, I. (2008) Heteropoly acids as catalysts for liquid-phase esterification and transesterification. *Applied Catalysis A*, Vol. 349, pp.170-176.

Andrianov, V., Borisjuk, N., Pogrebnyak, N., Brinker, A., Dixon, J., Spitsin, S., Flynn, J., Matyszczuk, P., Andryszak, K., Laurelli, M., Golovkin, M. & Koprowski, H. (2010). Tobacco as a Production Platform for Biofuel: Overexpression of *Arabidopsis DGAT* and *LEC2* genes Increases Accumulation and Shifts the Composition of Lipids in Green Biomass. *Plant Biotechnology Journal*, Vol. 8, (April 2010), pp. 277–287, ISSN 1467-7644.

Bender, M. (1999). Economic Feasibility Review for Community-Scale Farmer Cooperatives for Biodiesel. *Bioresource Technology*, Vol. 70, (October 1999), pp. 81-87, ISSN 0960-8524.

Bianchi, C.L., Boffito, D.C., Pirola, C. & Ragaini, V. (2010) Low temperature de-acidification process of animal fat as a pre-step to biodiesel production. *Catalysis Letters*, Vol. 134, (November, 2009), pp. 179-183.

Canakci, M. & Sanli, H. (2008) Biodiesel production from various feedstocks and their effects on the fuel properties. *Journal of Industrial Microbiology Biotechnology*, Vol. 35, pp. 431-441.

Carlsson, A. S. (2009). Plant Oils as Feedstock Alternatives to Petroleum - A Short Survey of Potential Oil Crop Platforms. *Biochimie*, Vol. 91, (June 2009), pp. 665–670, ISSN 0300-9084.

Carvoli, G., Ragaini, V. & Soave, C. (2008) Il futuro è dietro l'angolo. *La Chimica e l'Industria*, Vol. 1, pp. 38-41.

Chan, A. P., Crabtree,J., Zhao, Q., Lorenzi, H., Orvis, J., Puiu, D., Melake-Berhan, A., Jones, K. M., Redman, J., Chen, G., Cahoon, E. B., Gedil, M., Stanke, M., Haas, B. J., Wortman, J. R., Fraser-Liggett, C. M., Ravel, J. & Rabinowicz, P. D. (2010). Draft Genome Sequence of the Oilseed Species. *Ricinus communis*. *Nature Biotechnology*, Vol. 28, (August 2010), pp. 951-956, ISSN 1087-0156.

Clixoo (2010). Preview of Comprehensive Castor Oil Report, In: *Castor Oil Industry Reference & Resources*, 30.05.2011, Available from http://www.castoroil.in

Curto, G. & Lazzeri, L. (2006). Brassicacee, un Baluardo Sotterraneo Contro i Nematodi. *Agricoltura*, Vol. 34, No. 5, (May 2006), pp.110-112.

Environment Australia (2003). National Standards for Biodiesel – Discussion Paper, In *Setting National Fuels Quality Standards*, 0 642 54908 7 pp. 1-196, Retrieved from <www.ea.gov.au/atmosphere/transport/biodiesel/index.html> .

Fogher, C. (2008). Sviluppo di un Ideotipo di Tabacco per la Produzione di Seme da Usare a Fini Energetici. Presented at the workshop *"La Filiera del Tabacco in Campania: Ristrutturazione e/o Riconversione. La ricerca come motore per l'innovazione tecnologica, la sostenibilità e la competitività della filiera"*, Portici, Italy, February, 2008

Gasol, C. M., Gabarrella, X., Antonc, A., Rigolad, M., Carrascoe, J., Ciriae, P., Solanoe, M. L. & Rieradeva, J. (2007). Life Cycle Assessment of a *Brassica carinata* Bioenergy Cropping System in Southern Europe. *Biomass and Bioenergy*, Vol. 31, No. 8, (August 2007), pp. 543-555, ISSN 0961-9534.

Giannelos, P. N., Zannikos, F., Stournas, S., Lois, E. & Anastopoulos, G. (2002). Tobacco Seed Oil as an Alternative Diesel Fuel: Physical And Chemical Properties. *Industrial Crops and Products*, Vol.16, (July 2002), pp. 1-9, ISSN 0926-6690.

Holanda, A. (2004). *Biodiesel e Inclusão Social*, Câmara dos Deputados, Coordenação de Publicações, Brasília, Brazil, Retrieved from http://www2.camara.gov.br/a-camara/altosestudos/pdf/biodiesel-e-inclusao-social/biodiesel-e-inclusao-social

IENICA (2004). Agronomy Guide, Generic guidelines on the agronomy of selected industrial crops. *IENICA - Interactive European Network for Industrial Crops and their Applications*, Retrieved from http://www.ienica.net/agronomyguide/agronomyguide05.pdf

Krawczyk, T. (1996). Biodiesel - Alternative Fuel Makes Inroads but Hurdles Remain. *INFORM*, Vol. 7, No. 8, (August 1996), pp. 801-829.

Krishnan, G., Holshauser, D. L. & Nissen, S. J. (1998). Weed Control in Soybean (*Glycine max*) with Green Manure Crops. *Weed Technology*, Vol. 12, No. 1, (Jan.-Mar. 1998), pp. 97-102.

López, D.E., Suwannakarn, K., Bruce, D.A., & Goodwin, J.G. (2007) Esterification and transesterification on tungstated zirconia: Effect of calcination temperature. *Journal of Catalysis*, Vol. 247, pp. 43-50.

Lotero, E., Liu, Y., Lopez, D.E., Suwannakara, K. Bruce, D.A. & Goodwin J.G. (2005). Synthesis of Biodiesel Via Acid Catalysis. *Industrial& Engineering Chemistry Research*, Vol. 44 (14), (January 2005) pp. 5353-5363.

Palmer, C.E., Warwick, S. & Keller, W. (2001). Brassicaceae (Cruciferae) Family, Plant Biotechnology, and Phytoremediation. *International Journal of Phytoremediation*, Vol. 3, No. 3, pp. 245-287, ISSN: 1549-7879.

Pan, X. (2009). A Two Year Agronomic Evaluation of Camelina sativa and Brassica carinata in NS, PEI and SK (Master's thesis). Dalhousie University, Halifax, Canada, Retrieved from Digital Repository Unimib database of Dalhousie University at http://dalspace.library.dal.ca/handle/10222/12370

Pari, L., Fedrizzi, M. & Gallucci, F. (2008). *Cynara cardunculus* Exploitation for Energy Applications: Development of a Combine Head for Theshing and Concurrent

Residues Collecting and Utilization. *Proceedings of 16th European Biomass Conference & Exibition*, ISBN 978-88-89407-58-1, Valencia, Spain, 2-6 June 2008

Parodi, A., Marini, L. (2008) Process for the production of biodiesel. Patent WO 2008/007231.

Pasqualino, J.C. (2006). *Cynara cardunculus as an Alternative Crop for Biodiesel Production* (Ph.D. Thesis). Universitat Rovira I Virgili, Tarragona, Spain, Retrieved from http://tdx.cat/bitstream/handle/10803/8545/PhDThesisJPasqualino.pdf?sequenc e=1

Pinto A. C., Guarierio L. L. N., Rezende M. J. C., Ribeiro N. M., Torres E. A., Lopes W. A., Pereira, P. A. de P. & de Andrade J. B. (2005). Biodiesel: an Overview. *Journal of the Brazilian Chemical Society*, Vol. 16, No.6b, pp. 1313-1330, ISSN 0103-5053

Pirola, C., Bianchi, C.L., Boffito, D.C., Carvoli, G. & Ragaini, V. (2010) Vegetable oil deacidification by Amberlyst : study of catalyst lifetime and a suitable reactor configuration. *Industrial & Engineering Chemistry Research*, Vol. 49 (2010), pp. 4601-4606.

Pirola, C. Boffito, D.C. Carvoli, G., Di Fronzo, A. Ragaini, V. & Bianchi, C.L. (2011) Soybean oil de-acidification as a first step towards biodiesel production. In *Soybean/Book 2*, ISBN 978-953-307-533-4

Pöpken, T. Götze, L. Gmehling, J. (2000). *Reaction Kinetics and Chemical Equilibrium of Homogeneously and Heterogeneously Catalyzed Acetic Acid Esterification with Methanol and Methyl Acetate Hydrolysis*. Industrial Engineering Chemistry Research, Vol. 39, pp. 2601-2611.

Potts, D. A., Rakow, G. W. & Males, D. R. (1999). Canola-Quality *Brassica juncea*, a New Oilseed Crop for the Canadian Prairies. *Proceedings of the 10th International Rapeseed Congress*, Canberra, Australia, September 1999

Radich, A. (1998). Biodiesel Performance, Costs, and Use. Energy Information Administration Available from: http://www.eia.doe.gov/oiaf/analysispaper/biodiesel/pdf/biodiesel.pdf

Rao, K.V., Krishnasamy, S., Penumarthy, V. (2009) WO 2009047793

Razon, L. F. (2009). Alternative Crops for Biodiesel Feedstock. *CAB Reviews: Perspectives in Agriculture, Veterinary Science, Nutrition and Natural Resources*, Vol. 4, No. 56, (October 2009), pp. 1-15, ISSN 1749-8848

Romero, E., Barrau, C. & Romero, F. (2009). Plant Metabolites Derived from *Brassica* spp. Tissues as Biofumigant to Control Soil Borne Fungi Pathogens. *Proceedings of the 7th International Symposium on Chemical and non- Chemical Soil and Substrate Disinfestation*, ISBN 9789066056237, Leuven, Belgium, September 2009.

Sanzone, E. & Sortino, O. (2010). Ricino, Buone Prospettive Negli Ambienti Caldo-Aridi. *Terra e Vita*, No. 7, pp. 28-29, ISSN 0040-3776.

Sharma, Y.S. & Singh, B. (2011). Advancements in solid acid catalysts for ecofriendly and economically viable synthesis of biodiesel. *Biofuels, Bioproducts & Biorefining*, Vol. 5, pp. 69-92.

Steinigeweg, S. & Gmehling, J. (2003). Esterification of a Fatty Acid by Reactive Distillation. *Industrial Engineering Chemistry Research*, Vol. 42, pp. 3612-3619

Suwannakarn, K., Loreto, E., Goodwin J.G.Jr. & Lu, C. (2008). Stability of sulfated zirconia and the nature of the catalytically active species in the transesterification of tryglicerides. *Journal of Catalysis*, Vol 255, pp. 279-286.

T. Ono, K. Yoshiharu, US Patent 4,164, 506, 1979.

The Royal Society (2008). Sustainable Biofuels: Prospects and Challenges. *The Clyvedon Press Ltd*, ISBN 978 0 85403 662 2, Retrieved from http://royalsociety.org/Sustainable-biofuels-prospects-and-challenges/

Tyson, K.S. (2002) Brown grease feedstocks for biodiesel, In: *National Renewable Energy Laboratory*, January 3, 2008 available from http://www.nrbp.org/pdfs/pub32.pdfS.

Usta, N. (2005). Use of Tobacco Seed Oil Methyl Ester in Turbocharged Indirect Injection Diesel Engine. *Biomass & Bioenergy*, Vol. 28, (January 2005), pp. 77-86, ISSN 0961-9534.

Velasco, L., Goffman, F.D. & Becker H.C. (1998). Variability for the fatty acid composition of the seed oil in a germplasm collection of the genus Brassica. *Genteic Resources and Crop Evolution*, Vol. 45, pp. 371-382.

Winayanuwattikun, P., Kaewpiboon, C., Piriyakananon, K., Tantong, S., Thakernkarnkit, W., Chulalaksananukul, W. & Yongvanich, T. (2008). Potential plant oil feedstock for lipase-catalyzed biodiesel production in Thailand. *Biomass and Bioenergy*, Vol. 32, pp. 1279-1286, ISSN 0961-9534.

Xingzhong, Y., Jia, L., Guanming, Z., Jingang, S., Jingyi, T. & Guohe, H. (2009). Optimization of conversion of waste rapeseed oil with high FFA to biodiesel using response surface methodology. *Renewable Energy*, Vol. 33, pp. 1678-1684.

Zhang, Y., Dube, M.A., McLean, D.D. & Kates, M. (2003). Biodiesel production from waste cooking oil: 1. Process design and technological assessment. *Bioresource Technology*, Vol. 90, pp. 229-240.

Zheng, D. & Hanna, M.A. (1996) Preparation and properties of methyl esters of beef tallow. *Bioresource Technology*, Vol. 57, pp. 137-142, ISSN 0960-8524.

Zhiyuan, H., Piqiang, T., Xiaoyu & Y. Diming, L. (2008). Life cycle energy, environment and economic assessment of soybean-based biodiesel as an alternative automotive fuel in China. *Energy*, Vol. 33, pp. 1654-1658, ISSN 0360-5442.

2

Animal Fat Wastes for Biodiesel Production

Vivian Feddern et al.[*]
Embrapa Swine and Poultry,
Brazil

1. Introduction

Our society is highly dependent on petroleum for its activities. However, petroleum is a finite source and causes several environmental problems such as rising carbon dioxide levels in the atmosphere. About 90% is used as an energy source for transportation, heat and electricity generation, being the remaining sources used as feedstocks in the chemical industry (Carlsson, 2009). As demands for energy are increasing and fossil fuels are limited, research is directed towards alternative renewable fuels (Bhatti et al., 2008). High petroleum prices and the scarcity of known petroleum reserves demand the study of other sources of energy. In this context, agroindustrial wastes (animal fats, wood, manure) play an important role as energetic materials. Oils and fats are basically triacylglycerols (TAG) composed of three long-chain fatty acids. These triacylglycerols have higher viscosity and therefore cannot be used as fuel in common diesel engines. In order to reduce viscosity, triacylglycerols are converted into esters by transesterification reaction. By this means, three smaller molecules of ester and one molecule of glycerin are obtained from one molecule of fat or oil. Glycerin is removed as by-product and esters are known as biodiesel (Fazal et al., 2011).

Biodiesel fuels are attracting increasing attention worldwide as a blending component or a direct replacement for diesel fuel in vehicle engines. Biodiesel consists of a mixture of fatty acid (chain length C_{14}-C_{22}) alkyl esters, derived from a renewable lipid feedstock, such as vegetable oil or animal fat. In the case when methanol or ethanol are used as reactants, it will be a mixture of fatty acid methyl esters (FAME) or fatty acid ethyl esters (FAEE), respectively. However, methanol is commonly and widely used in biodiesel production due to its low cost and availability. Other alcohols such as isopropanol and butyl may also be used. A key quality factor for the primary alcohol is the water content, which interferes with the transesterification reactions and can result in poor yields and high level of soap, free fatty acids (FFA) and TAG in the final fuel (Demirbas, 2009a; Lam et al., 2010).

Biodiesel is a low-emission diesel substitute fuel made from renewable resources and waste lipid. The most common way to produce biodiesel is through transesterification, especially

[*] Anildo Cunha Junior, Marina Celant De Prá, Paulo Giovanni de Abreu, Jonas Irineu dos Santos Filho, Martha Mayumi Higarashi, Mauro Sulenta and Arlei Coldebella
Embrapa Swine and Poultry, Brazil.

alkali-catalyzed transesterification (Leung et al, 2010). The most commonly used catalysts for converting TAG to biodiesel are sodium hydroxide, potassium hydroxide and sodium methoxide. The alkaline catalysts are highly hygroscopic and form chemical water when dissolved in the alcohol reactant. They also absorb water from the air during storage. Acid catalysts include sulfuric and phosphoric acids, being more related to directly esterification of FFA, although they are considered to be slow for industrial processing (Demirbas, 2009a). When the raw materials (oils or fats) have a high percentage of FFA or water, the alkali catalyst will react with the FFA to form soaps (Leung et al, 2010).

An alternative fuel to petrodiesel must be technically feasible, economically competitive, environmentally acceptable and easy available (Demirbas, 2009a). FAME from vegetable oils and animal fats have shown promise as biodiesel, due to improved viscosity, volatility and combustion behaviour relative to triacylglycerols, and can be used in conventional diesel engines without significant modifications (Bhatti et al., 2008). The advantages of biodiesel over diesel fuel are its portability, ready availability, renewability, higher combustion efficiency, lower sulphur and aromatic content, higher cetane number, higher biodegradability, better emission profile, safer handling, besides being non-toxic (Lapuerta et al., 2008; Demirbas, 2009a, Balat & Balat, 2010). Besides the superb lubricating property of biodiesel and its similarities in physicochemical properties to diesel, makes it an excellent fuel for compression ignition engines, revealing its potentials and practical usability for the replacement of petrodiesel in the nearest future (Atadashi et al., 2010). Moreover, biodiesel offers advantages regarding the engine wear, cost, and availability. When burned, biodiesel produces pollutants that are less detrimental to human health (Fazal et al., 2011).

Biodiesel has superior emission profile than diesel, substantially reducing emissions of unburned hydrocarbons, carbon monoxide, sulfates, polycyclic aromatic hydrocarbons, nitrated polycyclic aromatic hydrocarbons, and particulate matter (Lapuerta et al., 2008). Diesel blends containing up to 20% biodiesel can be used in nearly all diesel-powered equipment, and higher level blends and pure biodiesel can be used in many engines with little or no modification. Lower-level blends are compatible with most storage and distribution equipments, but special handling is required for higher-level blends (Demirbas, 2009a).

Usage of biodiesel will allow a balance to be sought between agriculture, economic development and environment (Demirbas, 2009a). Lower cost feedstocks are needed since biodiesel from food-grade oils is not economically competitive with petroleum-based diesel fuel. Main animal fat sources are beef tallow, lard, poultry fat and fish oils. Yellow greases can be mixtures of vegetable oils and animal fats. The FFA content affects the type of biodiesel process used and the yield of fuel from that process. Other contamination present can affect the extent of feedstock preparation necessary to use a given reaction chemistry (Demirbas, 2009a). Tallow is beef fat produced by slaughterhouse, while lard is hog fat and chicken fat refers to poultry. Brown grease comes from restaurant grease traps, sewage plants, and "black grease" (sludge). The brown one is gelatinous at room temperature and has low overall oil content. Yellow and brown grease as well as tallow can be converted into biodiesel, although the costs of processing are higher and the per-gallon biodiesel yield is lower. According to the USDA, the United States produces over 1.4 billion gallons of used cooking oil and animal fat each year. In fact, around 74% of the inedible tallow and grease produced goes to animal feed, while the remainder is used to make soaps, lubricants and other products such as Biodiesel (Tickell, 2006).

Soybean oil is the major feedstock for biodiesel in the USA and in other parts of the world. Rapeseed oil is the major source of oil in Europe and it contributes about 85% of the oil for world biodiesel production, followed by sunflower seed oil, soybean oil and palm oil. Some sources for vegetable oil extraction to be use in biodiesel production are: castor berry, palm pulp, palm kernel oil, babassu kernel, sunflower seeds, coconut kernel, cotton seed, peanut grain, canola seed (Leung et al., 2010). According to European Biodiesel Board (EBB, 2008), European production of biodiesel reached 5.7 million tons compared to US production of 1.7 billion liters in 2007. Germany is the largest producer of biodiesel among EU countries, accounting for about half of the total European biodiesel production. In the East Asian countries, palm oil is the major feedstock for biodiesel, being the annual average production expected to be about 31.4 million tons/year over the period 2006-2010 (Shrestha & Gerpen, 2010).

In 2010, about 2.4 billion liters of biodiesel were produced in Brazil, corresponding to 14% of the global participation. The country has a wide variety of feedstocks to be used in the production of oil and fatty acids. However, it is important to find new sources that don't compete with food chains. Therefore, it is necessary to invest in finding residual oils and other products (Pacheco, 2006). Sustainable alternatives for biodiesel production are being researched with the use of enzymes, which allow for mild reaction conditions and easier recovery of glycerol, preventing the drawbacks of the chemical synthesis (Rodrigues & Ayub, 2011).

2. Meat production around the world

In the last years, meat production has increased significantly. World meat production reached 237.7 million tons in 2010, from which 42.7%, 33.4%, 23.9% corresponds to respectively pork, poultry and beef (USDA, 2010). Consequently, a larger amount of residues from animal processing-plants has been generated in countries with intensive livestock production. Within agroindustrial residues, lipid sources may be used as feedstock to biodiesel supply, helping to solve inappropriate environmental disposal, besides contributing to energy demand.

Animal protein consumption in the world is a great well-being indicator of corporations (excluding those who decide for several reasons do not consume animal protein). As can be seen in Figure 1, consumption growth is directly related to the population income level and tends to rise as income rises, because in rich countries energy consumption is 3,470 kcal, while in poorest countries this value is 2,660 kcal (FAO, 2010). In Brazil, the studies done by Hoffmann (2000), Schlindwein (2006) and Pintos-Payeras (2009) also demonstrate income great importance in meat consumption. The percentage of meat in the diet is approximately twice the richer countries. In Brazil, it is noted that meat national consumption is already, in proportional terms, similar to consumption in rich countries, although in absolute terms the energy consumption of 3,060 kcal was below the same period.

These data show that income growth in peripheral countries will have a pronounced impact on meat consumption. Thus, per capita income growth in underdeveloped countries (China, India, Brazil and Russia) in the last three decades will be determinant of consumption growth and meat production. The possibility of per capita income growth in Africa will undoubtedly be a propeller of meat consumption in the near future. The surprising economic growth in China over the past 10 years and its impact on animal protein production has turned this country into a major propeller of meat dynamics in the world. In this period the gross national per capita income has grown at rates of 13.44%, and following the same line, meat consumption grew 2.3% annually (12.7 million tons in the period). In

India, increase in per capita income of 9.23% over the same period was responsible for an increase in meat consumption in the same order of 6.68% per year (2.2 million tons). In Brazil, although less intense, we can see economic growth of 5.22% from 2000 to 2009. However, it must be taken into account that the best income distribution of the Brazilian economy lead to higher meat consumption by the poorest population (+3.64% per year meaning 4.8 million tons). A similar phenomenon was observed in Russia where meat consumption grew 5.07% per year between 2000 and 2009 (3 million tons).

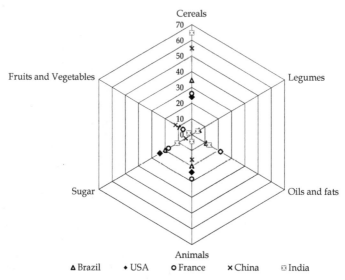

Fig. 1. Sources of per capita energy consumption from the diet as a percentage of the average period 2001-2003.
Source: FAO (2010).

The gradual effect shown in Figure 1 is a major driver of food economy, domestic and global levels, through increasing the middle class and the adoption by families who arrive there, the consumer behavior of those already there (Homem de Mello, 1990). As a result of this graduation, there would be huge demand in animal protein, legumes, fruits and vegetables. This dynamic evolution of food demand, as economic growth explains meat production growth in the world in the last two decades. Although it has been reported a large percentage increase in major meat consumption, the absolute volume consumed in India is still very low (3.8 kg) including beef, pork and poultry, when compared to 89.69, 49.9 and 57.2 kg in Brazil, China and Russia, respectively. Thus, in this country is to be expected that the continuous increase in per capita income by more than a decade will boost meat consumption to a level closer to developing countries. Worldwide, source of animal protein (except milk) most produced and consumed is pork with 29.86% (Figure 2), followed by chicken meat (22.97%), eggs (18.05%) and beef (17.56%). These four groups of sources account for 88.44% of animal protein total consumption in the world.

In a second group of sources, the following four are responsible for more than 7.00% of animal protein consumption. This group comprises the consumption of sheep (2.39%), turkey (1.77%), eggs of other birds (1.42%) and goat (1.42%). Rounding out the meat group: duck (1.09%), buffalo (0.97%), goose and guinea fowl (0.69%), rabbits (0.53%), hunt (0.49%), other meat (0.36%), horse (0.29%), camel (0.10%), other birds (0.03%) and ostrich (0.004%). At the same group, deserves attention the consumption of buffalo that occurs almost entirely in India, where beef consumption is forbidden by Hindu religion (83% of India population). India and Pakistan are also important producers of sheep and goats. Similarly horse consumption is concentrated in Asian countries. Rabbits are produced mostly in China, Venezuela and Italy. China also concentrates the production of geese, goats, ducks and sheep.

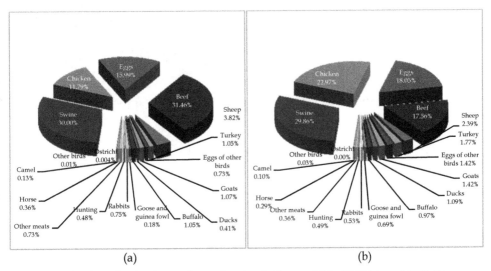

Fig. 2. Distribution of animal protein production in the world in 1975 (a) and 2008 (b)
Source: FAO (2010) adapted by the authors.

Globally, a great dynamic in animal protein production can be noted. Even being less expressive regarding total production, the highest growth rates in animal protein production are focused on meat of birds. Among the most important sources, chicken may be highlighted showing an annually growth of 4.1% over the past 10 years. The negative highlight is related to beef which presented one of the lowest growth rates (1.1%). In an intermediate form, pork production increased in the order of 2.52%. Since 1975, year after year, poultry industry is consolidating itself as one of the most important animal protein sources for the population. According to data from the United States Department of Agriculture (USDA), world production of broilers grew consistently over the past 35 years, from 10.6 million tons in 1975 to 71 million tons by the end of the first decade of this century. Brazil has a different dynamic for meat production. Unlike the rest of the world, the main animal protein is chicken (41.31%), beef (36.49%), pork (12.19%), and eggs (7.38), which represent 97.37% of total produced in the country (Figure 3).

As a result of these factors one should expect a continued growth in production and consumption of meat, mainly chicken, followed by pork and beef (Figure 4).

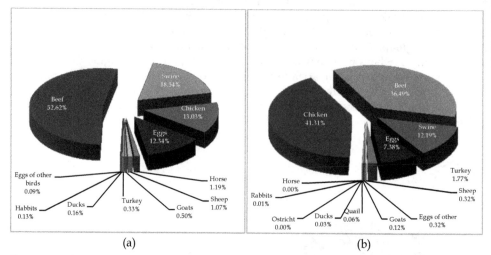

Fig. 3. Distribution of animal protein production in Brazil in 1975 (a) and 2008 (b)
Source: FAO (2010) adapted by the authors.

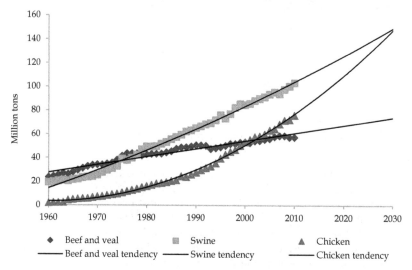

Fig. 4. Tendency of main animal protein production in the World
Source: USDA (2010) adapted by the authors.

3. Characterization and generation of animal fat wastes

Oils and fats are found in living organisms, consisting essentially of fatty acid esters and glycerin mixtures, and are known as triacylglycerols (commonly called triglycerides), which are hydrolyzed during extraction processes and storage, releasing fatty acids and glycerin.

Moreover, the use of oils or fats as fuel for internal combustion engines and their derivatives have been proposed for this intention over the past 100 years, when Rudolf Diesel applied in their assays crude petroleum and peanut oil. However, the problems of petroleum supply on the world market, generated by armed conflict that began in the 30s, led to the search for viable solutions for replacing fossil fuel.

The use of greases and animal fats eliminates the need to dispose them, besides contributing to the supply of biodiesel (Janaun & Ellis, 2010). Animal fats are highly viscous and mostly in solid form at ambient temperature because of their high content of saturated fatty acids. The high viscous fuels lead to poor atomization and result in incomplete combustion. The consequences are the increased emissions of pollutants and particulate in the exhaust gas (Kerihuel et al., 2006). Animal fats are readily available because slaughter industries are generally well managed for product control and handling procedures. However, there's a biosafety issue related to animal fats that could come from the contaminated animals. The future research to ensure biodiesel quality from animal waste (cradle to grave) has been highlighted (Janaun & Ellis, 2010). Biodiesel made from used cooking oil or from animal fat is less resistant to cold weather than biodiesel made from virgin soybean oil or most other virgin oils. As additives are developed specifically for the biodiesel industry, even this distinction could soon disappear (Tickell, 2006).

Black greases are defined loosely as greases resulting from sewage or other unconventional oil sources. It has a low conversion factor to biodiesel due to its high content in FFA. Brown greases are generally defined as a combination of greases and trappings from the slaughter industry. Yellow greases comprehend the oils and greases produced in the fast food industry and collected by the rendering industry (Tickell, 2006).

In Brazilian meat chain, most animal fats are generated in slaughterhouses and rendering plants. Products of rendering industry usually have lower market value. Materials that for aesthetic or sanitary reasons are not suitable for human food are are intended as feedstocks for rendering processes. Among these materials, there are fatty trimmings, bones, and offal, as well as entire carcasses of animals condemned at slaughterhouses, and those that have died on farms (deadstock) or in transit. The raw materials are collected in slaughterhouses, butcherhouses and supermarkets by trucks that take them to rendering plants. There are some industrial-scale slaughterhouses that process the residues within their own facilities. Once in the rendering plants the residues are chopped and heated in a steam-jacketed vessel to drive off the moisture and simultaneously release the fat from the fat cells using this so called "dry" method. The internal temperature of reactor reaches 1200 °C under 5-6 kg/cm^2 of pressure during 2 h per batch.

In Brazil, does not exists an animal fat classification, only a general designation based on the animal from which the fat originates, such as chicken fat or fish oil, tallow and lard. The greases produced in Brazil are generally described as follows:

a. Tallow: extracted from residues of bovine slaughter and it can be filtered or not since it has guaranteed that the product contains minimum 90% total fatty acids, unsaponifiable impurities maximum 1.5% and no FFA or fat degradation products;

b. Lard: extracted from swine slaughter residues, being its specification and quality guarantees the same as for tallow;

c. Chicken fat: extracted from broiler slaughter residues and it can be filtered or not since it has guaranteed that the product contains minimum 90% total fatty acids, maximum 3% unsaponifiable impurities, without FFA or fat degradation products;

d. Animal fat mix: extracted from slaughter residues of mammals or birds. It can be filtered or not since it has guaranteed that the product contains total fatty acids minimum 90%, maximum 2% unsaponifiable impurities, without FFA or products of fat degradation unless the ones generated even with good production practices implemented.

The animal species from which the fat originates must be specified. Additions of antioxidants must be informed in any of these products. The main difference between animal fat and vegetable oil is their fatty acid composition. Vegetable oils have high content of unsaturated fatty acids, mainly oleic and linoleic acid, while animal fat composition has higher proportion of saturated fatty acids (Table 1).

Oil or Fat	12:0	14:0	16:0	16:1	18:0	18:1	18:2	18:3	20:4	≥ 20
Chicken	0.1	1-1.3	17-20.7	5.4	6-12	42.7	20.7	0.7-1.3	0.1	1.6
Lard	0.1	1-2	23.6-30	2.8	12-18	40-50	7-13	0-1	1.7	1.3
Tallow	0.1	3-6	23.3-32	4.4	19-25	37-43	2-3	0.6-0.9	0.2	1.8
Fish	0.2	6.1	14.3	10.0	3.0	15.1	1.4	0.7	0.7	56.5
Butter	-	7-10	24-26	-	10-13	1-2.5	2-5	-	-	-
Soybean	-	0.1	6-10.2	-	2-5	20-30	50-60	-	-	-
Rapeseed	0.2	0.1	3.9	0.2	1.7	60.0	18.8	9.5	-	4.0
Corn	-	1-2	8-12	0.1	2-5	19-49	34-62	0.7	-	2.0
Olive	-	-	9-10	-	2-3	73-84	10-12	Traces	-	-
Cotton	-	-	20-25	-	1-2	23-35	40-50	Traces	-	-

Table 1. Average fatty acid composition of some vegetable oil and animal fat (Pearl, 2002; Rostagno et al., 2011)

Traditionally in Brazil, cleaning and toilet products industries use part of animal fat residues to produce soaps and waxes while other parts are employed in the production of lubricants and leather preservatives. Nevertheless, in Brazil, the beginning of National Program of Biodiesel Production and Use (Law #11.097) has rapidly changed this scenario and between October/2008 and March/2009 biodiesel plants consumed 43% of total tallow, which corresponds to approximately 15% of whole biodiesel produced. Although Brazil is also a major producer of chicken and swine meat, fats from these species are still not being used for biodiesel production. According to UBABEF (2009) data, Brazil produced around 23 million tons of meat, from which 3, 9 and 11 million correspond to swine, cattle and poultry, respectively. Considering the amount of residues 45% (wt/wt) cattle and 25% (wt/wt) swine and poultry contain approximately 15% fat. Thus, feedstock potential amount is 607,500, 412,500 and 112,500 tons for cattle, poultry and swine, respectively.

Wastes from slaughterhouses are constituted by non-edible by-products and wastewater which pass through flocculation and flotation process. Non-edible animal by-products are sent for rendering plants where flours are processed and good-quality fats besides acid fats are originated. Good-quality fats are destined for drugs and cosmetics, while acidic fats (which don't attend industry acid requirements and flotation stage) have low or no commercial value, being their promising target energy or biodiesel production. Wastewater undergoes flocculation and flotation process with the aid of coagulants, being separated into solid (flotation stage) and liquids (liquid phase). The first one is destined to rendering plants, while the second one goes to treatment lagoons, as is shown in Figure 5. Animal fats are classified in three categories (low, medium and high-grade quality fat) according to the risk level, following Regulation (EC) 1774/2002 of the European Parliament and of the

Council of 3 October 2002. High-grade quality fat has below 2% of FFA, which are mainly used for drugs and cosmetics, besides pet food. Medium-grade fat presents 3-5% FFA, while low-grade fat has above 5% FFA and are destined to biofuel production.

In Brazil, according to the National Petroleum Agency (ANP 2011), raw materials of animal origin used for biodiesel production account for 14.82% of authorized nominal capacity. This value is still low compared to the raw materials of vegetable origin that account for 84.45% of biodiesel production. However, volume of animal feedstocks tends to grow since Brazil has one of the largest animal herds in the world. Brazil is currently the second largest cattle producer (over 9.1 million tons), the fourth largest pork producer (more than 3.1 million tons), and the third largest chicken meat producer, with more than 11.4 million tons. In this context, lipid by-products from slaughterhouses should become attractive, especially for economical and environmental reasons.

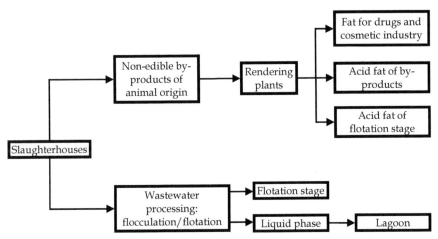

Fig. 5. Scheme of fat processing in slaughterhouses.

4. Biodiesel production from animal fat wastes: technical challenges

The feedstocks issue is the critical point affecting the economic feasibility of biodiesel production, since accounts around 80% of the biofuel total cost. In this context, several efforts have been carried out in order to reduce biodiesel prices, essentially by altering lipid sources (Zhang et al., 2003a, 2003b; Canakci, 2007; Canakci & Sanli, 2008; Wang, 2009; Janaun & Ellis, 2010; Martins et al., 2011). Nowadays, edible vegetable oils are the major starting materials for biodiesel preparation. In consequence, prospection for novel feedstocks has been primarily attributed to investigations involving oleaginous species for inedible oil extraction (Nass et al., 2007). In recent, alternatively lipid residues as waste frying oil and inedible animal fats have also receiving considerable attention from biofuel sector. To take advantage of these low cost and low quality resources, a convenient action would be to reuse residues in order to integrate sustainable energy supply and waste management in food processing facilities.

To get a better understanding of challenges involved on biodiesel synthesis from animal fat wastes, a brief review regarding to fundamental reactions of carboxylic acids and esters is presented (Formo, 1954; Carey & Sundberg, 1983; Morrison & Boyd, 1996). As illustrated in figure 6a, carboxylic acids originate their salts (soaps) by treatment with aqueous alkaline solutions (hydroxides or carbonates). Additionally, carbonyl group confers an interesting synthetic versatility to carboxylic acids since they can be converted into derivatives by nucleophilic substitution. In fact, esters are directly obtained reacting carboxylic acids using alcohols as acyl-acceptors under acidic conditions, being this process usually referred as Fisher esterification (Figure 6b). The strategy frequently employed to shift equilibrium to the right includes the use of large amounts of alcohol and water removal from the reactional medium.

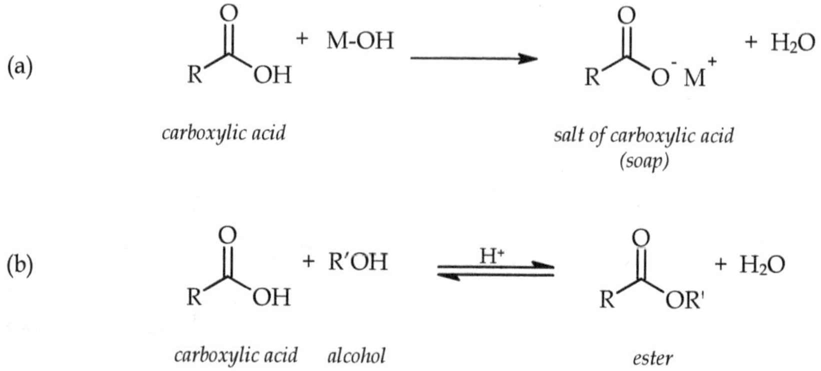

Fig. 6. Reactions of carboxylic acids: (a) acid-base neutralization (where M is Na^+ or K^+); (b) acid-catalyzed esterification

Esters are carboxylic acid derivatives that can be hydrolyzed either in acid or basic medium. The alkali-catalyzed process is essentially irreversible (Figure 7a). On the other hand, hydrolysis in acidic solution is an equilibrium reaction, being dependent on the relative alcohol and water concentrations (Figure 7b).

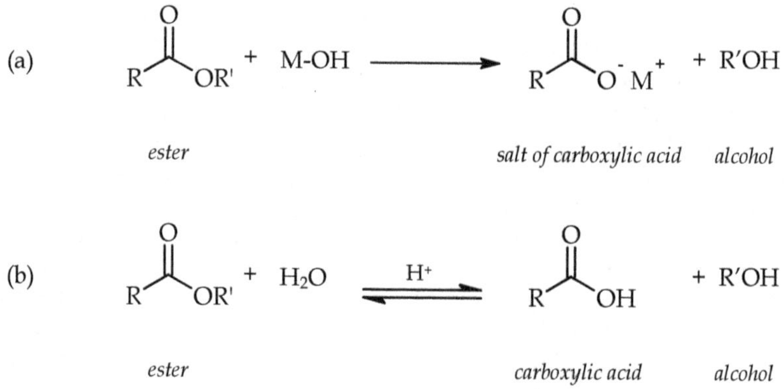

Fig. 7. Ester hydrolysis: (a) alkali-catalyzed (where M is Na^+ or K^+); (b) acid-catalyzed

According to Xu (2003), interesterification is a general term for the reactions between an ester and a fatty acid (acidolysis), an alcohol (alcoholysis), or another ester (transesterification). Esters are converted into another by alkoxy group exchanger as exemplified in figure 8. The displacement of $-OR'$ molecular unit is carried out when the original ester react with an alcohol to provide a new carboxylic acid derivative. Alcoholysis is usually denominated by most authors as transesterification, general term that will be used from now on to describe biodiesel production reaction. The transesterification is an equilibrium process and addition of an excessive amount of alcohol can be used in favor of products' synthesis.

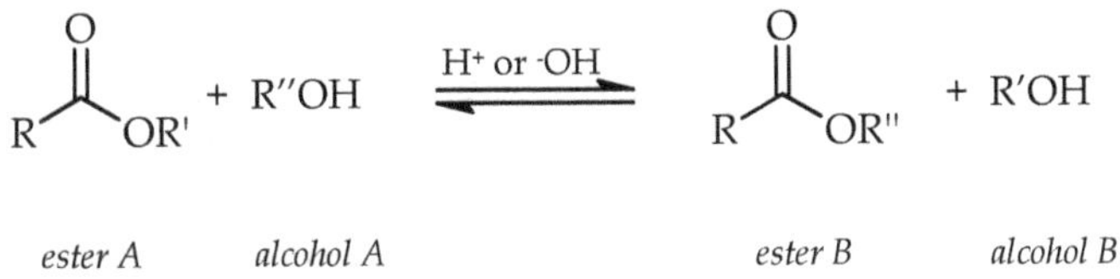

ester A   alcohol A   ester B   alcohol B

Fig. 8. Transesterification type alcoholysis

Oils and fats are complex lipids derived respectively from vegetable and animal sources. Their compositions are primarily based on triacylglycerols (TAG), which molecules consist of a glycerol backbone attached by ester bonds to three long-chain carboxylic acids (fatty acids). Reactions of ester linkages of oils and fats were recognized a long time by their technological importance (Formo, 1954). Nowadays, non-hydrolytic ester reactions (esterification and interesterification) play a fundamental role in the applied chemistry. For instance, biodiesel is a mixture of fatty acid mono-alkyl esters readily produced from TAG transesterification by using a short chain alcohol, as showed in figure 9.

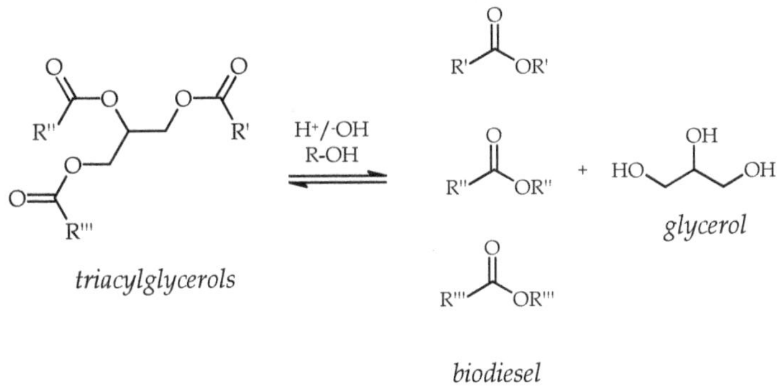

Fig. 9. Overall scheme of the TAG transesterification for biodiesel production (where R is $-CH_3$ or $-CH_3CH_2$)

Transesterification of TAG is a process of three consecutives and reversible acid- or basic-catalyzed reactions. Diacylglycerols (DAG) and monoacylglycerols (MAG) are intermediates. The stoichiometry of the overall reaction requires a molar ratio of 1:3 (TAG:alcohol) to give 3 mol of ester and 1 mol of glycerol. Its course involves stepwise conversions of TAG to DAG to MAG to glycerol (GL) (Figure 10).

Reaction 1 $$\text{TAG + alcohol} \overset{\text{catalyst}}{\rightleftharpoons} \text{DAG + RCOOR}'$$

Reaction 2 $$\text{DAG + alcohol} \overset{\text{catalyst}}{\rightleftharpoons} \text{MAG + RCOOR}'$$

Reaction 3 $$\text{MAG + alcohol} \overset{\text{catalyst}}{\rightleftharpoons} \text{GL + RCOOR}'$$

Overall reaction $$\text{TAG + alcohol} \overset{\text{catalyst}}{\rightleftharpoons} \text{GL + 3 RCOOR}'$$

Fig. 10. Chemistry of TAG transesterification

Few studies were concerned with detailed kinetic aspects of the transesterification of vegetable oils (Freedman et al., 1986; Noureddini & Zhu, 1997; Darnoko & Cheryan, 2000; Komers et al., 2002). Freedman et al. (1986) investigated the kinetics of acid- and base-catalyzed transesterification of TAG with methanol and 1-buthanol at 6:1 and 30:1 molar ratio alcohol:oil. The authors proposed pseudo first-order kinetics at high molar ratio alcohol:oil and second-order kinetics combined with a shunt-reaction at low alcohol:oil ratio. According to Noureddini & Zhu (1997), alkali-catalyzed methanolysis of oils can be described as follow: a) initially reaction is characterized by a mass transfer controlled regime (slow) that results from low miscibility of reactants; b) ester produced at beginning can act as mutual solvent and favor a kinetic controlled regime (fast) characterized by a sudden surge in products formation; c) in the final, an equilibrium regime (slow) is approached. Figure 11 shows typical distribution of reactants, intermediates and products during the course of transesterification, where a sigmoid behavior for ester production is exemplified.

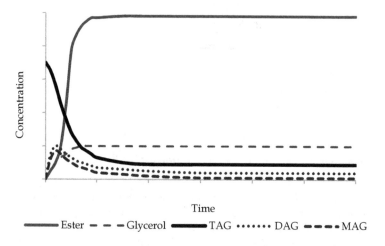

Fig. 11. Illustrative scheme of component concentration change during transesterification

Besides transesterification, reactions showed previously in this section can be involved during biodiesel preparation from lipid feedstocks depending on catalyst used. In fact, this reaction comprises a complex system. Komers and co-workers (2001b), in a fundamental research, were able to show that the reaction mixture of alkaline methanolysis of oils includes the following main components: TAG, DAG, MAG, methyl esters, methanol, soaps, KOH (in the form of OH-), CH_3OK (as CH_3O-), and water. Considering the system summarized in figure 12, the great issue is to establish appropriate conditions to minimize possible side reactions (hydrolysis and soaps formation) and, in consequence, drive the process toward ester production.

As is already well-know, transesterification may be influenced by several factors such as: feedstock composition; FFA content in raw materials, water concentration; alcohol to TAG molar ratio; catalyst type and concentration; type of alcohol; temperature; pressure; and mixing intensity. Researches have been intensively conducted to evaluate variables affecting ester yields and their respective interactions. Background about these parameters is detailed in several critical reviews (Schuchardt et al., 1998; Ma & Hanna, 1999; Fukuda et al., 2001; Van Gerpen & Knothe, 2005; Meher et al., 2006; Sharma et al., 2008; Vasudevan & Briggs, 2008; Demirbas, 2009b; Basha et al., 2009; Helwani et al., 2009; Vasudevan & Fu, 2010; Atadashi et al., 2010; Leung et al., 2010).

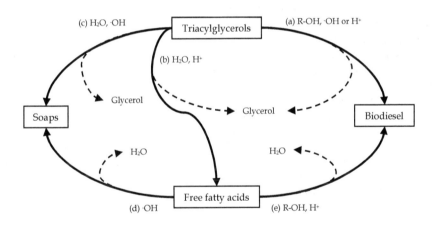

Fig. 12. Reactions involved in conventional biodiesel production: (a) alkali-catalyzed transesterification (expected route); (b) acid-catalyzed hydrolysis; (c) alkali-catalyzed hydrolysis; (d) acid-base neutralization; (e) acid-catalyzed esterification (expected route)

Homogeneous alkali-catalyzed transesterification is the most widely employed industrial process for biodiesel production (Helwani et al., 2009; Atadashi et al., 2010; Leung et al., 2010). This fact is because the base-catalyzed reation is faster than the acid one under mild conditions (Formo, 1954) resulting in a fuel-grade biodiesel. Alkaline catalysts are furthermore less corrosive than acidic compounds. Batch reactors are used for transesterification of refined vegetable oils with alcohol (molar ratio alcohol to oil 6:1) under anhydrous conditions. In summary, high esters conversion rates (>95%) are obtained in short times (after 1 h) at atmospheric pressure in temperatures ranged from 40 to 70 °C.

Metal hydroxides (NaOH and KOH) and methoxides ($NaOCH_3$ and $KOCH_3$) are generally applied as catalysts in concentrations ranging from 0.5 to 2% wt/wt of oil (Vicent et al., 2004; Dias et al., 2008). The most common acyl-receptor is methanol owing to its low cost. However, ethanol can be successfully used as well (Feuge & Gros, 1949; Wu et al., 1999; Encinar et al., 2002; Ghassan et al., 2004; Ferrari et al., 2005; Bouaid et al. 2007). Ethylic route is particularly interesting in countries with consolidated sugarcane industry like Brazil (Nass et al., 2007), allowing biodiesel production entirely based on biomass resources. Afterwards reaction achievement, spontaneous separation of biodiesel and rich-glycerol phases occurs by gravitational settling. In some cases, a centrifugation step may be used to speed up the separation of phases. Then biodiesel is isolated and purified by removal alcohol excess, water washing, drying, and vacuum distillation.

Rendered animal fats are attractive raw materials for biodiesel industry once they are immediately available and found in huge amounts at relative low-prices in regions with intensive livestock. The mentioned lipid sources are generated in meat-processing plants with different quality degrees. Often, inedible residual fats don't present specific requirements for direct application in conventional biodiesel approach mediated by alkalis. According to system showed in figure 12, feedstocks and reactants necessarily should meet suitable quality with respect to FFA and moisture. For that reason, refined vegetable oils are favored instead of lipid wastes.

The main technical restrictions with processing animal fat wastes are their relative high FFA (ranging from 5% to 30%) and water content. These two factors are key parameters for determining viability of transesterification process, because they may cause catalyst effectiveness and promote soaps formation. In fact, alkaline catalysts are consumed by neutralization with FFA in the reactional medium, leading to soaps and water formation. As a result of catalyst deactivation, ester yield is significantly reduced. In addition, post-treatment of the final mixture is more difficult by the occurrence of soaps, which prevents phase separation between esters and glycerol, promoting stable emulsion establishment in washing operations. Kusdiana & Saka (2004) were able to demonstrate this effect on TAG methanolysis using 1.5% NaOH (wt/wt) as illustrated in figure 13a. Restrictive limits of FFA ranging from <1% to <3%, as recently reported (Atadashi et al., 2010). According to reports involving fat residues, starting materials for basic-catalyzed transesterification should not exceed values beyond 0.5% FFA, which corresponds to an acid number of 1 mg KOH/g of oil (Ma et al., 1998; Canakci & Van Gerpen, 2001). For vegetable oils, a FFA value lower than 3% (6 mg KOH/g of oil) is recommended for good conversion efficiency (Dorado et al., 2002; Tamasevic & Siler-Marinkovic, 2003; Phan & Phan, 2008). In both cases, transesterification rate can be enhanced with bases if FFA is around 5%, although further quantity of catalyst must be added to compensate higher acidity and loss due soap formation (Van Gerpen, 2005). Particularly, this procedure involving excessive amount of catalyst is not recommended since it gives rise to the formation of gels that interfere in the reaction, hinder glycerol separation, and contribute to emulsification during water washing.

It's well-established that TAG transesterification with basic catalysis is also sensitive to water content. Water is one of the main causes for side reactions besides alcoholysis. The effective catalyzing agents in the alkaline catalyzed transesterification are alkoxide ions (RO^-). According to equilibrium study reported by Komers et al. (2001a), initial concentration of alkolate (RO^-) decreases with an increasing amount of water in methanol and KOH. This effect can also occur by water presence in oils and fats. Then, transesterification doesn't occur without catalyst generation and besides hydrolysis may take place as competitive reaction follow-on to soaps production. For the alkaline-catalyzed methanolysis of oils, ester conversion was slightly reduced when water concentration increased in reaction system, as showed in figures 13b and 14a (Canakci & Van Gerpen, 1999; Kusdiana & Saka, 2004). The effects of FFA and water content on alkali-transesterification of beef tallow were investigated by Ma and co-workers (1998). A significant interaction between two factors was clearly observed, characterizing synergistic negative effect on the reaction, according to data showed in figure 15b. With respect to the single effect, the apparent yield of beef tallow methyl esters (BTME) was the highest without addition of FFA and water. The apparent yield decreased with the increase of the water amount without addition of FFA. A similar behavior was noted without water addition when FFA level increased. Water generally can be removed from raw materials by drying, gravitational settling or with desiccant agents before processing transesterification.

The FFA content turn waste lipids unsuitable for conventional biodiesel route. Transesterification via acid catalysis is an alternative process claimed as more tolerant to high FFA levels (Lotero et al., 2005). The homogeneous acid-catalyzed transesterification is slower than alkaline process. Generally, this reaction is performed at high molar rations of alcohol:oil (50:1) at 80°C, and high catalyst concentrations (3% by weight of lipid feedstock). Besides, strong mineral acids (HCl and H_2SO_4) are corrosives, causing damages to reactors. As can be see in figure 13b, water is the major obstacle to this reaction, being more critical than in base catalysis. According to Canakci and Van Gerpen (1999), in order to achieve good ester conversion, the acid catalyst also requires water content lower than 0.5%, which is around the same for alkaline reaction. Only 0.1% of water in reaction medium is enough to result in some reduction of the methyl ester yield (Kusdiana & Saka, 2004). In acid-catalyzed transesterification mechanism, the key-step is the protonation of the carbonyl oxygen. This increases the electrophilicity of the carbonyl carbon, making it more prone to nucleophilic attack. When present in reactional medium, water can form clusters around protons with less acid strength than alcohol-only proton complexes. Therefore, the catalytic species (H^+) are deactivated by hydration, and don't allow TAG and their intermediates susceptible to alcohol attack (Helwani et al., 2009). On the other hand, acids are able to simultaneously catalyze both transesterification and esterification. Acid catalysts are effective at converting FFA to ester quickly. The integrated process is convenient to produce biodiesel from feedstocks having high FFA levels (Canakci & Van Gerpen, 2001; Zhang et al., 2003a, 2003b). The two-step approach includes an acid-catalyzed pre-treatment to esterification of FFA prior to alkali-catalyzed transesterification of TAG.

Nevertheless, as mentioned previously in this section, acid-catalyzed esterification is an equilibrium reaction, and hydrolysis occurs as inverse process. Water is produced in reactional medium when FFA react with alcohol to give esters. Canakci & Van Gerpen (1999), simulating FFA content in oil with palmitic acid, showed that water formed during acid-catalyzed esterification has similar negative effect on transesterification than when water was deliberatively added to reaction mixture. This fact is noted in figure 15a by coincident lines of acid-catalyzed transesterification (3% H_2SO_4, molar ratio 6:1, at 60°C) with water from reaction of palmitic acid and only with water addition. Then, water formed in the esterification limited FFA levels in the lipid source to 5%.

Even with all the above mentioned details regarding to raw materials' properties, several researches have stated that animal fat wastes are really important sources for biodiesel production. In Table 2, reactional conditions for biodiesel preparation from different animal fats are summarized.

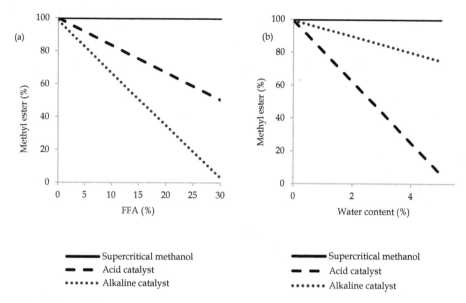

Fig. 13. Effects of FFA (a) and water (b) contents on the transesterification reaction of oils (adapted from Kusdiana & Saka, 2004).

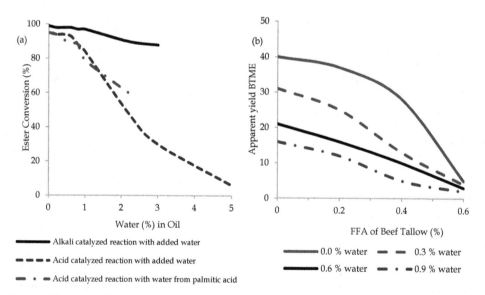

Fig. 14. (a) Water effect on transesterification of oils (adapted from Canakci & Van Gerpen, 2001); (b) FFA and water effects on alkali-catalyzed transesterification of beef tallow (adapted from Ma et al., 1998)

Feedstock	Catalyst (wt/wt of fat)	Alcohol	Molar ratio	T (°C)	Time (h)	Conv. (%)	Reference
Beef tallow	Step 1: NaOH 1% Step 2: NaOH 0.2%	MeOH	6:1 20%	70 60	0.5 1	-	Zheng & Hanna, 1996
Beef tallow	H_2SO_4 1%	MeOH	6:1	60	48	13.0	Alcantara et al., 2000
	NaOCH$_3$ 1%	MeOH	6:1	60	3	Quantitative	Alcantara et al., 2000
Beef tallow	KOH 2%	MeOH	-	65	1.5	>95	Moraes et al., 2008
Beef tallow	KOH 1.50%	MeOH	6:1	65	3	Quantitative	da Cunha et al., 2009
Beef tallow	Sulfonated polystyrene 20 mol%	MeOH	100:1	64	18	70.0	Soldi et al., 2009
Beef tallow: sunflower oil blends	NaOH 1%	MeOH	6:1	60	1	-	Taravus et al., 2009
Chicken tallow	H_2SO_4 25%	MeOH	30:1	50	24	Quantitative	Bhatti et al., 2008
	NaOH 1.5%	MeOH	-	30	1	88.1	
Feather meal fat	KOH 1%	MeOH	9:1	70	1/4	Quantitative	Kondamudi et al., 2009
Mutton tallow	H_2SO_4 25%	MeOH	30:1	60	24	93.2	Bhatti et al., 2008
	NaOH 1.5%	MeOH	-	30	1	78.3	
	KOH 1%	MeOH	6:1	65	3	79.7	
Duck tallow	NaOH 1%	MeOH	6:1	65	3	62.3	Chung et al., 2009
	NaOCH$_3$ 1%	MeOH	6:1	65	3	79.3	
Lard:soybean oil blends	NaOH 0.8%	MeOH	6:1	65	1	81.7-88.6	Dias et al., 2008
Lard:waste frying oil blends	NaOH 0.8%	MeOH	6:1	65	1	81.7-88.0	Dias et al., 2008
Lard	Step 1: H_2SO_4 2% Step 2: NaOH 1%	MeOH	6:1	65	Step 1: 5 Step 2: 1	66.2	Dias et al., 2009
Lard:soybean oil blend (25:75 wt/wt)	Step 1: H_2SO_4 2% Step 2: NaOH 1%	MeOH	6:1	65	Step 1: 5 Step 2: 1	64.4	Dias et al., 2009
Lard:soybean oil blend (25:75 wt/wt)	NaOH 1%	MeOH	6:1	65	1	77.8	Dias et al., 2009
Lard	Immobilized-lipase (Candida sp. 99-125)	MeOH	1:1(3x)	40	1/2	87.4	Lu et al., 2007
Lard	KOH 1.26%	MeOH	7.5:1	65	1/3	98.6	Jeong et al., 2009
Lard	KOH 0.9%	MeOH	6:1	60	1/3	89.2	Berrios et al., 2009
Leather	KOH 0.75%	MeOH	6:1	50	1/4	Quantitative	İsler et al., 2010
Poultry fat	Mg-Al hydrocalcite 10%	MeOH	30:1	120	8	93.0	Liu et al., 2007
Tallow	NaOH 0.5%	MeOH	6:1	60	3	-	Öner & Altun, 2009
Waste animal fat	H_2SO_4 2.25 M	EtOH	-	50	2	78.0	Ghassan et al., 2004
Waste animal fat	Step 1: H_2SO_4 0.08% Step 2: NaOH 0.01%	MeOH	-	62	2	89.0	Gürü et al., 2009

Table 2. Conditions of animal fats transesterification for biodiesel preparation

Figure 15 presents the raw materials employed for biodiesel production in Brazil from March 2010 until March 2011. As can be seen, soybean oil is the major feedstock. Additionally, beef tallow also plays an important role in this economic segment. The application of animal lipid sources in the Brazilian bioenergy sector is likely to increase because of accessibility to others profitable raw materials such as chicken and swine fat wastes. Recently, the simulation of investment in an industrial plant, made by Santos Filho et al. 2010, with processing capacity of 10,000 liters per day presented results that attest to the profitability of the enterprise.

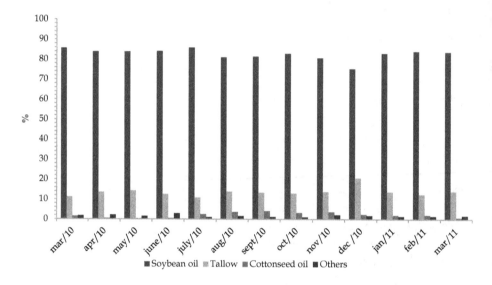

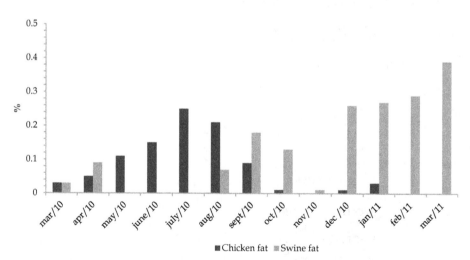

Fig. 15. Raw materials used for biodiesel production in Brazil
Source: (ANP, 2011)

The internal rate of return for an undertaking was 191%, the payback time was 1.51 years and the minimum price that enables the project was R$ 1,57, about US$ 1.00 (currency exchange August, 3rd, 2011), which is lower than the worst market since 2005. According to the authors, the results indicate that the use of acid fat from the slaughter of pigs and poultry for biodiesel production is technically and economically feasible, because there is high supply of raw material in different states of the country, facilitating logistics and providing a low cost transport of products. Increased demand for biofuels, especially for biodiesel, which year after year has been more requested for blending with diesel fuel, rising from 2% (2008) to 5% (2010) representing a consistent demand for production. Therefore, conversion of swine and poultry fats into biodiesel is advantageous for meat-processing industries that use this waste for burning and heat generation for boilers. Its use allows an increase in income and the chain will also be promoting an increase in the competitiveness of pork and poultry, turning a product with virtually no value into an income generator.

5. Conclusion

Worldwide vegetable oils are preferred as the main lipid starting materials for biodiesel production. However, animal fats have a great potential as feedstockes for biofuel segments, because they are not commodities, having a lower market value. Over the last years, meat production has increased significantly attaining 237.7 million tons in 2010, from which 42.7%, 33.4%, 23.9% corresponds respectively to pork, poultry and beef. Then a larger amount of residues from animal processing-plants has been generated in countries with intensive livestock production. Within agroindustrial residues, lipid sources may be used to solve inappropriate environmental disposal, besides contributing to energy supply. Brazilian government demands increasing addition of biodiesel into fossil diesel, taking place in 2010 a novel regulatory mark which raised the level up to 5%. Therefore, it has been encouraged the search for other renewable raw materials for application in the biofuel industry, such as non-edible oils and waste animal fats. Brazil is one of the main meat producers account to 9.1 beef, 3.2 pork, and 12.3 poultry million tons, dominating the world market together with the USA. In Brazil there is a broad range of residual lipid sources from slaughterhouse and rendering establishments ready available for application in biodiesel synthesis, including tallow, lard, poultry fat, mixed animal fat (mammal and poultry fat), and floating material from wastewater treatment plants. In a couple of years, researches focusing on fat residues should be accomplished mainly in order to improve feedstocks standardization process, because FFA and water content are decisive factors determining economic viability and biodiesel quality. Besides, researches in the field of prominent process such as heterogeneous catalysis (Di Serio et al., 2008; Cordeiro et al., 2011), enzyme-based process (Shimada et al., 2002), and supercritical fluids (Demirbas, 2006) should be carried out using animal fat wastes turning these raw materials more and more attractive.

6. Acknowledgements

Authors would like to acknowledge EMBRAPA (Empresa Brasileira de Pesquisa Agropecuária) and FAPESC (Fundação de Amparo a Pesquisa e Inovação do Estado de Santa Catarina) for the financial support.

7. References

Alcantara, R., Amores, J., Canoira, L., Fidalgo, E., Franco, M. J., & Navarro, A. (2000). Catalytic production of biodiesel from soy-bean oil, used frying oil and tallow. *Biomass and Bioenergy*, Vol.18, No.6, (June 2000), pp. 515-527, ISSN 0961-9534

ANP, 2011. Boletim mensal do biodiesel. Agência Nacional do Petróleo, Gás Natural e Biocombustível. Superintendência de Refino e Processamento de Gás Natural – SRP. Ministério de Minas e Energia. April 2011. Available from: <http://www.udop.com.br/download/estatistica/anp__biodiesel/2011/0411_bol etim_mensal.pdf>

Atadashi, I. M., Aroua, M. K., & Aziz, A. A. (2010). High quality biodiesel and its diesel engine application: A review. *Renewable and Sustainable Energy Reviews*, Vol.14, No.7, (September 2010), pp.1999-2008, ISSN 1364-0321

Balat, M. & Balat, H. (2010). Progress in biodiesel processing. *Applied Energy*, Vol.87, No.6, (June 2010), pp.1815-1835, ISSN 0306-2619

Basha, S. A.; Gopal, K. R., & Jebaraj, S. (2009). A review on biodiesel production, combustion, emissions and performance. *Renewable and Sustainable Energy Reviews*. Vol.13, No.6-7, (August /September 2009), pp. 1628-1634, ISSN 1364-0321

Berrios, M., Gutiérrez, M. C., Martín, M. A., & Martin, A. (2009). Application of the factorial design of experiments to biodiesel production from lard. *Fuel processing and Technology*, Vol.90, No.12, (December 2009), pp. 1447-1451, ISSN 0378-3820

Bhatti, H. N., Hanif, M. A., Qasim, M., & Rehman, A. U. (2008). Biodiesel production from waste tallow. *Fuel*, Vol.87, N.13-14, (October 2008), pp.2961–2966, ISSN 0016-2361

Bouaid, A., Martinez, M., & Aracil, J. (2007). A comparative study of the production of ethyl esters from vegetable oils as a biodiesel fuel optimization by factorial design. *Chemical Engineering Journal* , Vol.134, No.1-3 (November 2007) pp. 93–99, ISSN 1385-8947

Canakci, M. & Gerpen, J. V. (1999). Biodiesel production via acid catalysis. *Transactions of the ASABE*, Vol.42, No.5, pp. 1203-1210, ISSN 0001-2351

Canakci, M. & Gerpen, J. V. (2001). Biodiesel production from oils and fats with high free fatty acids. *Transactions of the ASAE*, Vol.44, No.6, (November/December 2001), pp. 1429-1436, ISSN 0001-2351

Canakci, M. (2007). The potential of restaurant waste lipids as biodiesel feedstocks. *Bioresource Technology*, Vol.98, No.1, (January 2007), pp. 183-190, ISSN 0960-8524

Canakci, M. & Sanli, H. (2008). Biodiesel production from various feedstocks and their effects on the fuel properties. *Journal of Industrial Microbiology and Biotechnology*, Vol.35, No.5, (May 2008), pp. 431-441, ISSN 1476-5535

Carey, F. A., Sundberg, R. J. (1983). *Advanced Organic Chemistry* (2nd edition), Plenum Press, ISBN 0-306-41088-5, New York

Carlsson, A. S. (2009). Plant oils as feedstock alternatives to petroleum – A short survey of potential oil crop platforms. *Biochimie*, Vol.91, No.6, (June 2009), pp. 665–670, ISSN 0300-9084

Chung, K.-H. & Kim, J. (2009). Biodiesel production by transesterification of duck tallow with methanol on alkali catalysts. *Biomass and Bioenergy*, Vol.33, No.1, (January 2009), pp. 155-158, ISSN 0961-9534

Cordeiro, C. S., da Silva, F. R., Wypych, F. & Ramos, L. P. (2011). Heterogeneous catalysts for biodiesel production. *Química Nova*, Vol.34, No.3, (March, 2011), pp. 477-486, ISSN 0100-4042

da Cunha, M. E., Krause L. C., Moraes, M. S. A., Faccini, C. S., Jacques, R. A., Almeida, S. R., Rodrigues, M. R. A., & Caramão, E. B. (2009). Beef tallow biodiesel produced in a pilot scale. *Fuel Processing Technology*, Vol.90, No.4, (April 2009), pp. 570-575, ISSN 0378-3820

Darnoko, D. & Cheryan, M. (2000). Kinetics of palm oil transesterification in a batch reactor. *Journal of the American Oil Chemists' Society*, Vol.77, No.12, (December 2000), pp. 1263-1267, ISSN 0003-021X

Demirbas, A. (2006). Biodiesel production via non-catalytic SCF method and biodiesel fuel characteristics. *Energy Conversion and Management*, Vol.47, No.15-16, (September 2006), pp. 2271-2282, ISSN 0196-8904

Demirbas, A. (2008). Comparison of transesterification methods for production of biodiesel from vegetable oils and fats. *Energy Conversion and Management*, Vol.49, N.1, (January 2008), pp. 125–130, ISSN 0196-8904

Demirbas, A. (2009a). Biofuels. In: Demirbas A, editor. Biofuels: Securing the planet's future energy needs. London (England): Springer; 336 p, e-ISBN 978-1-84882-011-1

Demirbas, A. (2009b). Progress and recent trends in biodiesel fuels. *Energy Conversion and Management*, Vol.50, No.1, (January 2009), pp. 14–34, ISSN 0196-8904

Dias, J. M., Alvim-Ferraz, M. C. M., & Almeida, M. F. (2008). Comparison of the performance of different homogeneous alkali catalysts during transesterification of waste and virgin oils and evaluation of biodiesel quality. *Fuel*, Vol.87, No.17-18, (December 2008), pp. 3572–3578, ISSN 0016-2361

Dias, J. M., Alvim-Ferraz, M. C., & Almeida, M. F. (2009). Production of biodiesel from acid waste lard. *Bioresource Technology*, Vol.100, No.24, (August 2009), pp. 6355-6361, ISSN 0960-8524

Di Serio, M., Tesser, R., Pengmei, L., & Santacesaria, Elio. (2008). Heterogeneous catalyst for biodiesel production. *Energy & Fuels*, Vol.22, No.1, (December 2007), pp. 207-217, ISSN 1520-5029

Dorado, M. P., Ballesteros, E., de Almeida, J. A., Schellert, C., Löhrlein, H. P., & Krause, R. (2002). An alkaline-catalyzed transesterification process for high free fatty acid waste oils. *Transactions of the ASABE*, Vol.45, No.3, pp. 525-529, ISSN 0001-2351

Encinar, J. M., González, J. F., Rodríguez, J. J., & Tejedor, A. (2002). Biodiesel fuels from vegetable oils: transesterification of *Cynara cardunculus* L. oils with ethanol. *Energy & Fuels*, Vol.16, No.2, (January 2002), pp. 443-450, ISSN 1520-5029

European Parliament (October 10 2002), 2011 Regulation (EC) 1774/2002 of the European Parliament and of the Council of 3 October 2002 (laying down health rules concerning animal by-products not intended for human consumption), Available from: <http://eur-lex.europa.eu/LexUriServ/ LexUriServ.do?uri=OJ:L:2002:273:0001:0095:EN:PDF>

FAO, 2010. Faostat. March 2010. Available from <http://www.fao.org/corp/statistics/en/>

Fazal, M.A., Haseeb, A. S. M. A., & Masjuki, H. H. (2011). Biodiesel feasibility study: An evaluation of material compatibility; performance; emission and engine durability.

Renewable & Sustainable Energy Reviews, Vol.15, No.2, (February 2011), pp. 1314-1324, ISSN 1364-0321

Ferrari, R. A., Oliveira, V. S., & Scabio, A. (2005). Biodiesel de soja — Taxa de conversão em ésteres etílicos, caracterização físico-química e consumo em gerador de energia. *Química Nova*, Vol.28, No.1, (January/February 2005), pp 19-23, ISSN 1678-7064

Feuge, R. O. & Gros, A. T. (1949). *Journal of the American Oil Chemists' Society*, Vol.26, No.3, (March 1949), pp. 97-102, ISSN 0003-021X

Formo, M. W. (1954). Ester reactions of fatty materials. *Journal of the American Oil Chemists' Society*, Vol.31, No.11, (November 1954), pp. 548-559, ISSN 0003-021X

Freedman, B., Pryde, E. H., & Mounts, T. L. (1984). Variables affecting the yields of fatty esters from transesterified vegetable oils. *Journal of the American Oil Chemists' Society*, Vol.61, No.10, (October 1984), pp. 1638-1643, ISSN 0003-021X

Freedman, B., Butterfield, R. O., & Pryde, E. H. (1986). Transesterification Kinetics of Soybean Oil. *Journal of the American Oil Chemists' Society*, Vol.63, No.10, (October 1986), pp. 1375-1380, ISSN 0003-021X

Fukuda, H., Kondo, A., & Noda, H. (2001). Biodiesel fuel production by transesterification of oils. *Journal of Bioscience and Bioengineering*, Vol.92, No.5, pp. 405-416, ISSN 1389-1723

Gürü, M., Artukoglu, B. D., Keskin, A., & Koca, A. (2009). Biodiesel production from waste animal fat and improvement of its characteristics by synthesized nickel and magnesium additive. *Energy Conversion and Management* Vol.50, No.3, (March 2009), pp. 498-502, ISSN 0196-8904

Helwani, Z., Othman, M. R., Aziz, N., Fernando, W. J. N., & Kim, J. (2009). Technologies for production of biodiesel focusing on green catalytic techniques: A review. Fuel Processing Technology, Vol.90, No.12, (December 2009), pp. 1502-1514, ISSN 0378-3820

Hoffmann, R. (2000). Elasticidade-renda das despesas com alimentos em regiões metropolitanas do Brasil em 1995-96. *Informações Econômicas*, Vol.30, No.2, (Fevereiro 2000), pp. 17-24, ISSN 1678-832X

Homem de Mello, F. (1990). O crescimento agrícola brasileiro dos anos 80 e as perspectivas para os anos 90. *Revista de Economia Política*, Vol. 10, No 3, (Julho-Setembro 1990), pp. 22-30, ISSN 0101-3157

Isler, A., Sundu, S., Tuter, M., Karaosmanoglu, F. (2010). Transesterification reaction of the fat originated from solid waste of the leather industry. Waste Management, Vol.30, No.12, (December 2010), pp. 2631-2635, ISSN 0956-053X

Janaun, J. & Ellis, N. (2010). Perspectives on biodiesel as a sustainable fuel. *Renewable and Sustainable Energy Reviews* Vol.14, No.4, (May 2010), pp.1312-1320, ISSN 1364-0321

Jeong, G.-T., Yang, H.-S., & Park, D.-H. (2009). Optimization of transesterification of animal fat ester using response surface methodology. *Bioresource Technology*, Vol.100, No.1, (January 2009), pp. 25-30, ISSN 0960-8524

Kerihuel, A, Kumar, M. S., Bellettre, J., & Tazerout, M. (2006). Ethanol animal fat emulsions as a diesel engine fuel – Part 1: Formulations and influential Parameters. *Fuel*, Vol.85, No.17-18, (December 2006), pp. 2371-2684, ISSN 0016-2361

Komers, K., Machek, J., & Stloukal, R. (2001a). Biodiesel from rapeseed oil, methanol and KOH 2. Composition of solution of KOH in methanol as reaction partner of oil.

European Journal of Lipid Science and Technology, Vol.103, No.6, (June 2001), pp. 359-362, ISSN 1438-9312

Komers, K., Stloukal, R., Machek, J., & Skopal, F. (2001b). Biodiesel from rapeseed oil, methanol and KOH 3. Analysis of composition of actual reaction mixture. *European Journal of Lipid Science and Technology*, Vol.103, No.6, (June 2001), pp. 363-371, ISSN 1438-9312

Komers, K., Skopal, F., Stloukal, R., & Machek, J. (2002). Kinetics and mechanism of the KOH – catalyzed methanolysis of rapeseed oil for biodiesel production. *European Journal of Lipid Science and Technology*, Vol.104, No.11, (November 2002), pp. 728-737, ISSN 1438-9312

Kondamudi, N., Strull, J., Misra, M., & Mohapatra, S. K. (2009). A green process for producing biodiesel from feather meal. *Journal of Agricultural and Food Chemistry*, Vol.57, No.14, (June 2009), pp. 6163-6166, ISSN 0021-8561

Kusdiana, D. & Saka, S. (2004). Effects of water on biodiesel fuel production by supercritical methanol treatment. *Bioresource Technology*, Vol.91, No.3, (February 2004), pp. 289-295, ISSN 0960-8524

Lam, M. K., Lee, K. T., & Mohamed, A. R. (2010). Homogeneous, heterogeneous and enzymatic catalysis for transesterification of high free fatty acid oil (waste cooking oil) to biodiesel: A review. *Biotechnology Advances* Vol.28, No.4, (July-August 2010), pp.500-518, ISSN 0734-9750

Lapuerta, M., Armas, O., & Rodríguez-Fernández, J. (2008). Effect of biodiesel fuels on diesel engine emissions. *Progress in Energy and Combustion Science*, Vol.34, No.2, (April 2008), pp. 198-223, ISSN 0360-1285

Leung, D. Y C., Wu, X., & Leung, M. K. H. (2010). A review on biodiesel production using catalyzed transesterification. *Applied Energy*, Vol.87, N.4, (April 2010), pp. 1083-1095, ISSN 0306-2619

Liu, Y., Lotero, E., Goodwin Jr, J. G., & Mo, X. (2007). Transesterification of poultry fat with methanol using Mg-Al hydrotalcite derived catalysts. *Applied Catalysis A: General*, Vol.331, pp. 138-148, ISSN 0926-860X

Lotero, E., Liu, Y., Lopez, D. E., Suwannakarn, K., Bruce, D. A., & Goodwin Jr., J. G. (2005). Synthesis of biodiesel via acid catalysis. *Industrial & Engineering Chemistry Research*, Vol.44, No.14, (January 2005), pp.5353-5363, ISSN 1520-5045

Lu, J., Nie, K., Xie, F., Wang, F., & Tan, T. (2007). Enzymatic synthesis of fatty acid methyl esters from lard with immobilized Candida sp. 99-125. *Process Biochemistry*, Vol.42, No.9, (September 2007), pp. 1367-1370, ISSN 0032-9592

Ma, F., Clements, L. D., & Hanna, M. A. (1998). The effects of catalyst, free fatty acids, and water on transesterification of beef tallow, *Transactions of the ASAE*, Vol.41, No.5, pp. 1261-1264, ISSN 0001-2351

Ma, F. & Hanna, M. A. (1999). Biodiesel production: a review. *Bioresource Technology*, Vol.70, No.1, (October 1999), pp. 1-15, ISSN 0960-8524

Martins, R., Nachiluk, K., & Bueno, C. R. F. & de Freitas, S. F. (2011). O biodiesel de sebo bovino no Brasil. *Informações Econômicas*, Vol.41, No.5, (Maio 2011), pp. 56-70, ISSN 1678-832X

Meher, L. C., Sagar, D. V., & Naik, S. N. Technical aspects of bidiesel production by transesterification – a review. (2006). *Renewable & Sustainable Energy Reviews*, Vol.10, No.3, (June 2006), pp. 248-268, ISSN 1364-0321

Monteiro, M. R., Ambrozin, A. R. P., Lião, L. M., & Ferreira, A. G. (2008). Critical reviews on analytical methods for biodiesel characterization. *Talanta*, Vol.77, No.2, (December 2008), pp. 593-605, ISSN 0039-9140

Moraes, M. S. A., Krause, L. C., da Cunha, M. E., Faccini, C. S., de Menezes, E. W., Veses, R. C., Rodrigues, M. R. A., & Caramão, E. B. (2008). Tallow Biodiesel: Properties Evaluation and Consumption Tests in a Diesel Engine. *Energy & Fuels*, Vol.22, No.3, (March 2008), pp. 1949-1954, ISSN 1520-5029

Morrison, R. & Boyd, R. (1996). *Química Orgânica* (13th edition), Fundação Calouste Gulbenkian, ISBN 972-31-0513-6, Lisboa

Nass, L. L., Pereira, P. A. A., & Ellis, D. (2007.) Biofuels in Brazil: An Overview. *Crop Science*, Vol.47, No.6 (November/December 2007), pp. 2228-2237, ISSN 1435-0653

Noureddini, H. & Zhu, D. (1997). Kinetics of transesterification of soybean oil. *Journal of the American Oil Chemists' Society*, Vol.74, No.11, (December 1997), pp. 1457-1463, ISSN 0003-021X

Öner, C. & Altun, S. (2009). Biodiesel production from inedible animal tallow and an experimental investigation of its use as alternative fuel in a direct injection diesel engine. *Applied Energy*, Vol. 86, No. 10, (October 2009), pp. 2114-2120, ISSN 0306-2619

Pacheco, J. W. (2006). *Guia técnico ambiental de graxarias – Série P+L*, CETESB, São Paulo.

Phan, A. N. & Phan, T. M. (2008). Biodiesel production from waste cooking oils. *Fuel*, Vol.87, No.17-18 (December 2008), pp. 3490-3496, ISSN 0016-2361

Pinto-Payeras, J. A. Estimação do sistema quase ideal de demanda para uma cesta ampliada de produtos empregando dados da pof de 2002-2003. Economia Aplicada, v. 13, n. 2, 2009, pp. 231-255

Rodrigues, R. C. & Ayub, M. A. Z. (2011). Effects of the combined use of *Thermomyces lanuginosus* and *Rhizomucor miehei* lipases for the transesterification and hydrolysis of soybean oil. *Process Biochemistry*, Vol.46, No.3, (March 2011), pp. 682-688, ISSN 1359-5113

Rostagno, H. S., Albino, L. F. T., Donzele, J. L., Gomes, P. C., Oliveira, R. F., Lopes, D.C., Ferreira, A. S., Barreto, S. L. T., & Euclides, R. F. (2011). *Brazilian Tables for poultry and swine: Food Composition and Nutritional Requirements*. Viçosa Federal University (UFV), CDD 22. ed. 636.085, 3rd ed, 252 p., Viçosa, MG, Brazil

Santos Filho, J. I. (2010). Viabilidade econômica da utilização de gordura ácida de frangos e suínos na produção de biodiesel, *Procedings of 4° Congresso da Rede Brasileira de Tecnologia de Biodiesel*, Belo Horizont (MG), October 2010

Schlindwein, M. M. Influência do custo de oportunidade do tempo da mulher sobre o padrão de consumo alimentar das famílias brasileiras. 2006. 118 p. Tese (Doutorado em Economia Aplicada) – Escola Superior de Agricultura "Luiz de Queiroz", Universidade de São Paulo, Piracicaba

Schuchardt, U., Sercheli, R., & Vargas, R. M. (1998). Transesterification of vegetable oils: a Review. *Journal of the Brazilian Chemical Society*, Vol.9, No.3, (May/June 1998), pp. 199-210, ISSN 0103-5053

Sharma, Y.C., Singh, B., & Upadhyay. (2008). Advancements in development and characterization of biodiesel: A review. *Fuel*, Vol.87, No.12 (September 2008), pp. 2355-2373, ISSN 0016-2361

Shimada, Y., Watanabe, Y., Sugihara, A., & Tominaga, Y. (2002). Enzymatic alcoholysis for biodiesel fuel production and application of the reaction to oil processing. *Journal of Molecular Catalysis B: Enzymatic*, Vol.17, No.3-5, (June 2002), PP. 133-142, ISSN 1381-1177

Shrestha, D. & Gerpen, J. V. (2010). Biodiesel from Oilseed Crops, In: *Industrial Crops and Uses*, B.P. Singh, (Ed.), 140-156, ISBN-13 9781845936167, CABI, Wallingford, U.K. 512 p.

Soldi, R. A., Oliveira, A. R. S., Ramos, L. P., & César-Oliveira, M. A. F. (2009). Soybean oil and beef tallow alcoholysis by acid heterogeneous catalysis. *Applied Catalysis A: General*, Vol.361, No.1-2 (June 2009), pp. 42–48, ISSN 0926-860X

Taravus, S., Temur, H. & Yartasi, A. (2009). Alkali-Catalyzed Biodiesel production from Mixtures of Sunflower Oil and Beef Tallow, *Energy & Fuels*, Vol.23, No.8, (July 2009), pp. 4112–4115, ISSN 1520-5029

Tashtoush, G. M., Al-Widyan, M. I., Al-Jarrah, M. M. (2004). Experimental study on evaluation and optimization of conversion of waste animal fat into biodiesel. *Energy Conversion and Management* Vol.45, No.17, (October 2004), pp. 2697-2711, ISSN 0196-8904

Teixeira, R., Almeida, E., Oliveira Jr., L., & Fernandes, H. (2010). Sep 28. Produtividade e logística na produção de biodiesel. *Ensaios FEE*, Porto Alegre, Vol.31, No.1, (August 2010), pp. 7-30. Available at:
http://revistas.fee.tche.br/index.php/ensaios/article/view/2213/2731

Tickell, J. (2006). *Biodiesel America: How to achieve energy security, Free America from middle-east oil dependence and make money growing fuel*. ISBN 9780970722744, Yorkshire Press, USA

Tomasevic, A. V. & Siler-Marinkovic, S. S. (2003). Methanolysis of used frying oil. *Fuel Processing Technology*, Vol.81, No.1, (April 2003), pp. 1-6, ISSN 0378-3820

UBABEF. Brazilian Poultry Union. Meat production in Brazil. May 27 2011, Available from: <http://www.brazilianchicken.com.br/english/the-poultry-industry/the-world-of-chicken.php>

USDA. United States Department of Agriculture. Livestock and poultry: World markets and trade. October 2010. May 23 2011, Available from: <http://www.fas.usda.gov/psdonline/circulars/livestock_poultry.pdf>

USDA. Foreign Agricultural Service, March 18 2010, Available from: <http://www.fas.usda.gov/psdonline/>

van Gerpen, J. & Knothe, G. (2005). Basics of the transesterification reaction. In: Knothe G, Van Gerpen J. H. and Krahl, J., editors. The Biodiesel Handbook. Illinois: AOCS Press; 302 p, ISBN 0-000000-00-00

Van Gerpen, J. (2005). Biodiesel processing and production. *Fuel Processing Technology*, Vol. 86, No.10, (June 2005), pp. 1097-1107, ISSN 0378-3820

Vasudevan, P. T. & Briggs, M. (2008). Biodiesel production—current state of the art and challenges. *Journal of Industrial Microbiology and Biotechnology*, Vol.1, No.1, (March 2010), pp. 47–63, ISSN 1367-5435

Vasudevan, P. T. & Fu, B. (2010). Environmentally Sustainable Biofuels: Advances in Biodiesel Research. *Waste Biomass Valorization*, Vol.35, No.5, (May 2008), pp. 421-430, ISSN 1877-2641

Vicente, G., Martínez, M., & Aracil, J. (2004). Integrated biodiesel production: a comparison of different homogeneous catalysts systems. *Biresource Technology*, Vol.92, No.3, (May 2004), pp. 297-305, ISSN 0960-8524

Wang, L. (2009). *Biodiesel production from waste oils and fats*, CRC Press, ISBN 978-1-4200-6339-4

Wu, W. H., Foglia, T. A., Marmer, W. N., & Phillips, J. G. (1999). Optimizing production of ethyl esters of grease using 95% ethanol by response surface methodology. *Journal of the American Oil Chemists' Society*, Vol.76, No.4, (April 1999), pp. 517-521, ISSN 0003-021X

Xu, X. (2003). Engineering of enzymatic reactions and reactors for lipid modification and synthesis. *European Journal of Lipid Science and Technology*, Vol.105, No.6, (June 2003), pp. 289-304, ISSN 1438-7697

Zhang, Y. Dubé, M. A., McLean, D. D., & Kates, M. (2003a). Biodiesel production from waste cooking oil: 1. Process design and technological assessment. *Bioresource Technology*, Vol.89, No.1, (August 2003), pp. 1-16, ISSN 0960-8524

Zhang, Y. Dubé, M. A., McLean, D. D., & Kates, M. (2003b). Biodiesel production from waste cooking oil: 2. Economic assessment and sensitivity analysis. *Bioresource Technology*, Vol.90, No.3, (December 2003), pp. 229-240, ISSN 0960-8524

Zheng, D. & Hanna, M. A. (1996). Preparation and properties of methyl esters of beef tallow. *Bioresource Technology*, Vol.57, No.2, (August 1996), pp. 137-142, ISSN 0960-8524

3

Biodiesel Production from Waste Cooking Oil

Carlos A. Guerrero F., Andrés Guerrero-Romero and Fabio E. Sierra
National University of Colombia,
Colombia

1. Introduction

Biodiesel refers to all kinds of alternative fuels derived from vegetable oils or animal fats. The prefix bio refers to renewable and biological nature, in contrast to the traditional diesel derived from petroleum; while the diesel fuel refers to its use on diesel engines. Biodiesel is produced from the triglycerides conversion in the oils such as those obtained from palm oil, soybean, rapeseed, sunflower and castor oil, in methyl or ethyl esters by transesterification way. In this process the three chains of fatty acids of each triglyceride molecule reacts with an alcohol in the presence of a catalyst to obtain ethyl or methyl esters.

The ASTM (American Society for Testing and Materials Standard) describes the biodiesel as esters monoalkyl of fatty acids of long chain that are produced from vegetable oil, animal fat or waste cooking oils in a chemical reaction known as transesterification.

Biodiesel has the same properties of diesel used as fuel for cars, trucks, etc. This may be mixed in any proportion with the diesel from the oil refined. It is not necessary to make any modifications to the engines in order to use this fuel.

"The use of pure biodiesel can be designated as B100 or blended with fuel diesel, designated as BXX, where XX represents the percentage of biodiesel in the blend. The most common ratio is B20 which represents a 20% biodiesel and 80% diesel"(Arbeláez & Rivera, 2007 pp 4). Colombia in South America, is taking advantage of the opportunities that biofuels will open to the agriculture. With more than a million liters a day, Colombia is the second largest producer of ethanol in Latin America, after Brazil. This has decongested the domestic market of sugar at more than 500 thousand tons. The result is strong revenue for the 300,000 people who derive their livelihood from the production of panela (from sugar cane).

In Colombia the biodiesel is produced from the palm oil and methanol, "being the last imported to meet the demand in the biodiesel production". In the past two years, the biodiesel production from Palm was between 300000 liters/day to 965000 liters per day, distributed in four plants located in the Atlantic coast and in the country center.

In the biodiesel production is technically possible to use methanol and ethanol alcohol (Cujia & Bula, 2010. pp 106).

The palm oil is one of oilseeds trade more productive on the planet; it is removed between six and ten times more oil than the other as soy, rapeseed and sunflower. Colombia has more than 300,000 hectares planted in Palm oil, generating permanent and stable employment for more than 90,000 people.

The biodiesel advantages are that it is a renewable and biodegradable biofuel; it produces less harmful emissions to the environment than those that produce fossil fuels. Specifically the Palm biodiesel pure or mixed with diesel fuel reduces the emissions of CO_2, nitrogen oxides (NOx) and particulate material. Table 1, shows the world production of vegetable oils.

OILS	MILLION TONS
Palm oil (fruit)	43.20
Soy oil	38.11
Rapeseed oil	19.38
Sunflower oil	11.45
Cotton oil	4.94
Palm oil (seed)	5.10
Peanut oil	4.93
Coconut oil	3.62
Olive oil	2.97

Table 1. World production of vegetable oils, 2008/2009. (Source: "Oilseeds: World markets and trade". FAS-USDA, October 2008)

The estimated consumption of diesel in the world at the end of the year 2005 was 960 billion liters. On the other hand, the production of biodiesel during the same year was 4.2 billion liters (Figure 1). For example, assuming that 2% of diesel was replaced with biodiesel, it would mean an increase of 15 billion liters in the biodiesel global production. This amount of biodiesel has other impacts, including overproduction of glycerin, the use of more land, etc.

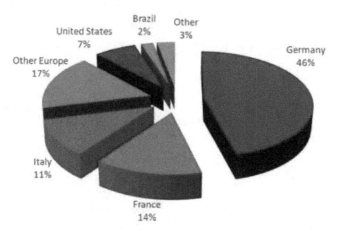

Fig. 1. World production of biodiesel (Source: National Federation of Oil Palm Growers (FEDEPALMA)).

ASTM has specified different fuel tests needed to ensure their proper functioning.
Table 2, lists the specifications established for biodiesel and the corresponding test method.

FEATURES	UNIT	LIMITS		TEST METHOD
		Minimum	Maximum	
Ester content	%(m/m)	96.5	-	EN 14103
Density a 15°C	Kg/m²	860	900	EN ISO 3675 EN ISO 12185
Viscosity a 40°C	Mm²/g	3.50	5.00	EN ISO 3104
Flash point	°C	120	-	Pr EN ISO 3679
Sulfur content	mg/kg	-	10.0	PrEN ISO 20846 pr EN ISO 20884
Carbon residue (in 10% of distilled residue)	% (m/m)	-	0.30	EN ISO 10370
Cetane index		51.0		EN ISO 5165
Sulphated ash content	% (m/m)	-	0.02	ISO 3987
Water content	mg/kg	-	500	EN ISO 12937
Total contamination	mg/kg	-	24	EN 12662
Cooper band corrosion (3 h at 50°C)	Classification	Class 1		EN ISO 2160
Oxidation stability 110°C	Hours	6.0		EN 14122
Acid index	mg KOH/g		0.50	EN 14111
Iodine index	g de iodine/100 g		140	EN 14103
Methyl ester of linoleic acid	%(m/m)		12.0	EN 14103
Methyl esters of methylpoli-unsaturated (> = 4 double bonds)	%(m/m)		1	
Methanol content	%(m/m)		0.20	EN 14110
Monoglycerides content	%(m/m)		0.80	EN 14105
Diglycerides content	%(m/m)		0.20	EN 14105
Triglycerides content	%(m/m)		0.20	EN 14105
Free glycerin	%(m/m)		0.20	EN 14105 EN 14105
Total glycerin	%(m/m)		0.25	EN 14105
Metals of group 1 (Na+K)	mg/kg		5.0	EN 14108 EN 14109
Metals of group 2 (Ca+Mg)	mg/kg		5.0	PrEN 14538
Phosphorus content	mg/kg		10.0	EN 14107

Table 2. ASTM Features

1.1 Environmental problems for disposing used cooking oil

Used cooking oil causes severe environmental problems, "a liter of oil poured into a water course can pollute up to 1000 tanks of 500 liters". It's feasible to demonstrate the contamination with the dumping of these oils to the main water sources.

The oil which reaches the water sources increases its organic pollution load, to form layers on the water surface to prevent the oxygen exchange and alters the ecosystem. The dumping of the oil also causes problems in the pipes drain obstructing them and creating odors and increasing the cost of wastewater treatment. For this reason, has been necessary to create a way to recover this oil and reuse it. Also due to the wear and tear resulting in sewer pipes may cause overflows of the system, "generating diseases that can cause mild stomach cramps to diseases potentially fatal, such as cholera, infectious hepatitis and gastroenteritis, due to the sewage contains water which can transport bacteria, viruses, parasites, intestinal worms and molds" (Peisch. Consulted: http://www.seagrantpr.org/catalog/files/fact_sheets/54-aguas-usadas-de-PR.PDF). The dangerous odors generate impact negatively on health, "is formed hydrogen sulfide (H_2S), which can cause irritation of the respiratory tract, skin infections, headaches and eye irritation" (Peisch. Consulted: http://www.seagrantpr.org/ catalog/files/fact_sheets/54-aguas-usadas-de-PR.PDF).

2. Types of cooking oil

Among the alternatives as a vegetal raw material to extract the oil are: oil palm, soybean, sesame, cotton, corn, canola, sunflower and olives.

2.1 Palm oil

Palm oil is retrieved from the mesocarp of the Palm fruit, this oil is regarded as the second most widely produced only surpassed by the soybean oil. The oil palm is a tropical plant characteristic of warmer climates that grows below 500 meters above sea level. "Its origin is located in the Guinea Gulf in West Africa." "Hence its scientific name, Elaeis guineensis Jacq and its popular name: African oil palm" (FEDEPALMA. Consulted: http://www.fedepalma.org/palma.htm).

Colombia is the largest producer of palm oil in Latin America and the fourth in the world. "The extracted oil from the palm contains a relationship 1:1 between saturated and unsaturated fatty acids, is also a major source of natural antioxidants as tocopherols, tocotrienols and carotenes"(FEDEPALMA. Consulted: http://www.fedepalma.org/ palma.htm). It has been proven that Palm oil is natural source vitamin E, in the form of tocopherols and tocotrienols. The tocotrienol act as protectors against cells aging, arthrosclerosis, cancer and some neurodegenerative diseases such as Alzheimer's disease. Unrefined palm oil is the richest in beta-carotene natural source; its consumption has proved to be very useful for preventing and treating the deficiency of vitamin A in risk populations.

2.1.1 Characteristics of plant

The oil palm presents fruit by thousands, spherical, ovoid or elongates, to form compact clusters of between 10 and 40 kilograms of weight. Inside, they kept a single seed, almonds or palmist, to protect with the fart, a woody endocarp, surrounded in turn by a fleshy pulp. Both, pulp and almond oil generously provide. The productive life of the oil palm can be most of fifty years, but from the twentieth or twenty-five the stem reaches a height that hinders the work of harvest and marks the beginning of the renewal in commercial plantations. 25 to 28 °C on average monthly temperatures are favorable, if the minimum average temperature is below 21 °C. Temperatures of 15 °C stop the growth of the seedlings from greenhouse and decrease the performance of adult palms. Between 1,800 and 2,200 mm precipitation is optimal, if it is well distributed in every month. Like the coconut palm, the palm oil is favored by deep, loose and well drained soils. A superficial phreatic level limits the development and nutrition of roots. In general, the physical characteristics good, texture and structure, are preferable to the level of fertility, as it can be corrected with mineral fertilization. The palm oil resists low acidity levels, up to pH 4. Too alkaline soils are harmful. Although you can plant with success on land of hills with slopes above of 20 °, are preferred levels or slightly wavy, with no more than 15 ° gradients.

2.1.2 Pests
The major pest of palm oil and its damage are:
- Acaro: They are located on the underside of the leaves, mainly in vivarium palms. The damages are identified by the discoloration of the leaves, which reduces the photosynthetic area. We can fight it with Tedión.
- Arriera ant: it is common in tropical areas. This animal can cause serious defoliations in palms of all ages. We can fight it with bait poisoned as Mirex, applied to the nest mouths.
- Estrategus: Is a beetle of 50 to 60 mm long, black, with two horns. This animal drills in the ground, at the foot of the Palm, a gallery of even 80 cm; penetrates the tissues of the trunk base and destroys it. It is controlled with 200 g of heptachlor powdered 5%, slightly buried around the Palm.
- Rats: This animal can cause damage at the trunk base of young palms. Controlled with baits of coumarine, which must be changed regularly.
- Yellow beetle or alurnus: attacks the young leaves of the plant heart as well as on the coconut tree. It is controlled with sprayings of Thiodan 35 EC, solution of 800 cc in 200 liters of water. Apply 2 to 4 liters in palm.
- Beetles or black palm weevil: In Palm oil causes the same damage to the coconut palm.
- Lace bug: is 2.5 mm long. It is an insect of transparent grey color. It is located in the underside of the leaves. Their stings favor infections by various fungi, which may cause draining of the leaves.

2.2 Rapeseed or canola oil
Rapeseed is a "specie oilseed in the cruciferous family. Many of the species of this family have been cultivated since long time ago that their roots, stems, flowers and seeds are edible" (Iriarte, Consulted: http://www.inta.gov.ar/ediciones/idia/oleaginosa /colza01.pdf). Ideally grows in climates that go from temperate to slightly cold and wet (minimum of 0 °C and maximum of 40 °C). When the seeds of rapeseed are crushed we can obtain oil and a kind of pulp or prized residue from always to feed livestock, since that gives a 34% protein and 15% crude fiber. The biodegradable properties of rapeseed or canola oil make it ideal to be used on the basis of paints, herbicides, lubricants, food packaging, etc.

2.2.1 Characteristics of plant
Oilseed rape (Brassica napus) is a crucifer of deep and pivoting root. The stem has a size of 1.5 m approximately. The lower leaves are petiolate but the superiors entire and lanceolate. The flowers are small, yellow, and are grouped in terminal racemes. The fruits have a number of grains by pod around 20-25, depending on the variety. The rapeseed composition is showed in the table 3:

COMPOSITION	%
Proteins	21,08
Fat	48,55
Fiber	6,42
Ashes	4,54
Nitrogen-free extracts	19,41
TOTAL	100,00

Table 3. Rapeseed composition.

The seeds are spherical of 2 to 2.5 mm in diameter and when are mature have a reddish or black brown color. Rapeseed has a proportion (39%) of oil where there are a large number of fatty acids of long-chain, which quantitatively the most important is the erucic acid. The cultivation of rapeseed has ability to grow in temperate climates to temperate cold with good humidity. It adapts to different soil types, the ideals are the franc soils of good fertility and permeable which is a very sensitive crop to the superficial flooding.

2.2.2 Pests

- Rape stem weevil (Ceuthorrhynchus napi): the grub of this insect deforms the stem of the rape, which is curved and often indenting in a certain length.
- Terminal bud Weevil (Ceuthorrhynchus picitarsis): adults do not cause damage, but the larvae destroy the terminal bud and force the plant to produce side shoots. The treatments are made with endosulfan and Fosalón.
- The siliques weevil (Ceuthorrhynchus assimilis): adults bite the young siliques and the larvae gnaw seeds causing a significant decrease in the harvest. Endosulfan and Fosalón are used in treatments.
- Cecydomia (Dasyneura brassiceae): The larvae of this insect destroy the siliques totally. The endosulfan and fosalon control this plague.
- Meligetos of the cruciferous (Meligethes sp): adults are in charge of gnawing the buttons of the rapeseed; these attacks are more important younger are the buttons. When begin the flowering the damage decrease.
- Flea of rapeseed (Psyllodes chrysocephala): adults appear in autumn rape fields, generally shortly after birth gnawing the leaves and can destroy large number of plants. Karate to doses of 40-80 cc/hL is recommended for the treatment.
- Flea of the cabbage (Phylotreta sp): adult insects wintering in the soil in September and appear in April. Karate works very well against these insects.

2.3 Sunflower oil

The oil extracted from sunflower seeds is considered to be of high quality for a low percentage of saturated fatty acids and a high percentage of unsaturated fatty acids. It also contains essential fatty acids and a considerable amount of tocopherols that gives it stability. The acidic composition of the sunflower depends on the genotype and the environment. There are currently three groups of genotypes: traditional, oleic medium and oleic high.

2.3.1 Characteristics of plant

The sunflower belongs at the family "Asteraceae, whose scientific name is Helianthus annuus. It is an annual plant with a vigorous development in all its organs. Within this species there are many types or subspecies grown as ornamental plants, oilseeds and forage plants" (INFOAGRO, Consulted: http://www.infoagro.com/herbaceos/oleaginosas /girasol.htm). Average sunflower cycle includes between 100 and 150 days according to genotypes, dates of planting, latitude and availability of water and nutrients. The "temperature is the most important factor in the control of the seeds germination being the optimal near to 26 °C with maximum temperatures of 40 °C and minimum from 3 to 6 °C. The threshold for soil temperature (0 to 5 cm) from which normally starts sowing is between 8 and 10 °C" (Diaz-Zorita et al, Consulted: http://www.asagir.org.ar /Publicaciones/cuadernillo_web.pdf). The availability of water acts on the soaking of seeds, on the subsequent growth of the seedling. The water excess decreases the amount of air in the soil.

2.3.2 Pests

Pests of early-onset (e.g. cutting caterpillars, leafcutter ants, velvety larvae, worm wire, tenebrionido of the sunflower, underground grille, weevils, black beetle, slugs, etc.) produce damage in seeds and seedlings. Slugs cause great damage to the leaves. The control is convenient with treatments of seeds or specific toxic baits.

3. Biodiesel production process

The biodiesel production is given by the transesterification reaction which consists of three consecutive and reversible reactions. First, the triglyceride is converted in diacylglycerol, and running at monoglyceride and glycerin. In each reaction one mole of methyl ester is released as shown in Figure 2.

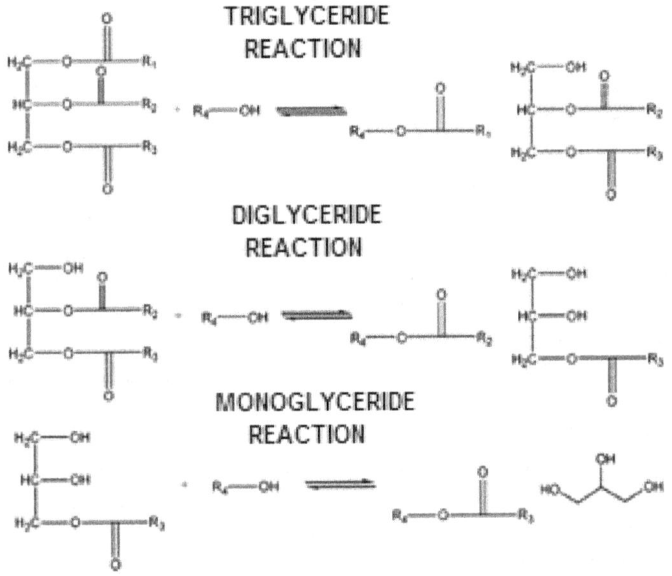

Fig. 2. Stages of the transesterification reaction (Arbeláez & Rivera, 2007. pp 13)

Figures 3 and 4 show the secondary reactions that may occur: the saponification reaction and the neutralization reaction of free fatty acids.

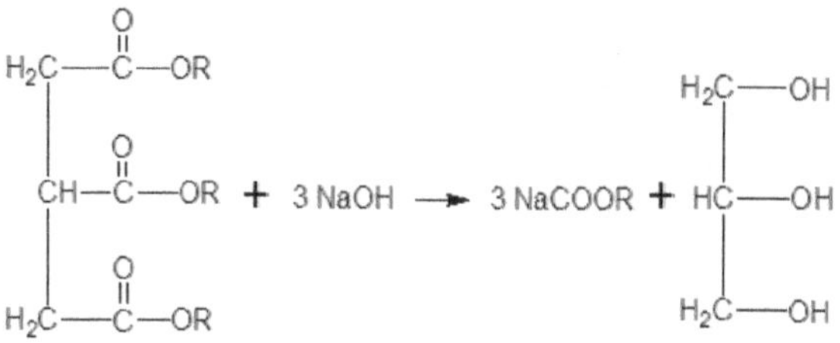

$$
\begin{array}{c}
\mathrm{H_2C-\overset{\overset{\displaystyle O}{\|}}{C}-OR} \\[2mm]
\mathrm{CH-\overset{\overset{\displaystyle O}{\|}}{C}-OR} \quad + \; 3\,\mathrm{NaOH} \; \longrightarrow \; 3\,\mathrm{NaCOOR} \; + \\[2mm]
\mathrm{H_2C-\overset{\overset{\displaystyle O}{\|}}{C}-OR}
\end{array}
\qquad
\begin{array}{c}
\mathrm{H_2C-OH} \\[2mm]
\mathrm{HC-OH} \\[2mm]
\mathrm{H_2C-OH}
\end{array}
$$

Fig. 3. Saponification reaction (Arbeláez & Rivera, 2007. pp 13)

$$\mathbf{RCOOH + NaOH \rightarrow R-COONa + H_2O}$$

Fatty acid   *Sodium carboxylate*

Fig. 4. Neutralization reactions of free fatty acids (Arbeláez & Rivera, 2007. pp 13)

3.1 Raw materials

The biodiesel production comes mostly from oils extracted oilseed plants especially sunflower, soy, rapeseed and animal fats. However, any material that contains triglycerides can be used for the biodiesel production. "In addition to the oil or fat is needed an alcohol and catalyst to convert oils and fats in alkyl esters". (Arbeláez & Rivera, 2007. pp7)

3.1.1 Alcohol

"Primary and secondary alcohols with string of 1-8 carbons are used for the biodiesel production, among the alcohols that can be used in this process are: methanol, ethanol" (Cujia & Bula, 2010. pp 106), propanol y butanol. "When are used alcohols such as ethanol is more complicated the recovery of pure alcohol in the process because the azeotrope that forms with water"(Cheng et al. 2008. pp 4) and the performance of ethyl esters is less compared to the methyl esters due methanol has a lower molecular weight (32.04 g/mole) compared to ethanol (46.07 g/mole)."On the other hand if you use methanol, not would contribute to environmental issues and sustainability, biodiesel would not be completely bio, by having a fossil component provided by the alcohol, because methanol is made from natural gas, which is fossil" "(Cheng et al. 2008. pp 4).To use methanol or ethanol is needed "a mechanical agitation to encourage the transfer of mass"(Arbeláez & Rivera, 2007. pp 10)."In the course of the reaction form emulsions, using methanol is easy and quickly dissolved, forming a glycerol-rich bottom layer and a higher layer in methyl esters, while using ethanol these emulsions are more stable making the process of separation and purification of ethyl esters more difficult"(Arbeláez & Rivera, 2007. pp 10).

Is preferred to use methanol in the biodiesel production because of their low viscosity (0.59 m * Pa * s at 20 °C), because using alcohols such as ethanol with high viscosity (1,074 m * Pa * s at 20 °C), the biodiesel viscosity increases and as a result a "fuel of high viscosity not will be pulverized properly by injection systems that have diesel engines. Also increase the opacity of fumes which limits their application in automotive engines"(Benjumea et al. 2007. pp 149).

In the reaction performance is feasible to reach "higher conversions with methanol, ethanol using the process is more complex, expensive, requires a higher consumption of energy and time"(EREN. 2003.pp 38). "We found that it requires less reaction time when using methanol rather than ethanol, either in acid or alkaline catalysis, reaching high yields"(Giron et al. 2009.pp 18).

With the above, the methanol is selected to be used in the biodiesel production due to its lower cost, better performance and less time and energy during the reaction.

3.1.2 Catalysts

Homogeneous, heterogeneous or enzyme catalysts are used in the biodiesel production. Homogeneous catalysts are soluble in the middle of reaction, i.e. they are in a single phase either liquid or gaseous. "One of the advantages of homogeneous catalysis is the high speed of reaction, and moderate temperature and pressure conditions" (EREN. 2003.pp 4). The catalysts can be acids or alkalis, the acid catalysts are effective but require a time interval extremely long and temperatures exceeding 100 °C for its action. "Getting conversions of 99% with a concentration of 1% sulfuric acid in relation to the amount of oil, it takes about 50 hours" "(EREN. 2003.pp 13). We can use this catalytic process when the oils have a high degree of acidity and "harm the action of alkali catalysts with acidity greater than 10 %"(EREN. 2003.pp 39).We can use sulfuric acid (H_2SO_4), phosphoric acid (H_3PO_4), among others. When is used "acid catalysts with alcohol excess is that the recovery of glycerin is more difficult as the quantities of alcohol are quite large compared to other type of catalyst" (Arbeláez & Rivera, 2007. pp13).

"Using HCl are achieved yields of 61% and with H_2SO_4 we can obtain 80%"(Liu et al. 2006a pp 186), but these "catalysts are more corrosive than alkali catalysts"(Errazu et al. 2005 pp 1305). In comparison with the acidic catalysts, the basic catalysts accelerate the reaction rate, the disadvantage of basic catalysts is that produces soaps due to the high amounts of free fatty acids and water by which we must add the appropriate amount of base to neutralize fatty acids free. The most commonly used are sodium hydroxide (NaOH), potassium hydroxide (KOH) and inappropriate for industrial application (CH_3ONa) sodium methoxide since this is more expensive and "requires total absence of water" (EREN. 2003 pp 40). "The catalysts are dissolved in the reaction mixture alcohol-oil what does that not can be recovered at the end of the transesterification reaction" (Arbeláez & Rivera, 2007. pp13). "By using KOH as a catalyst we can produce potassium fertilizers such as potassium chloride, potassium sulphate and potassium nitrate if the product with phosphoric acid is neutralized" (Arbeláez & Rivera, 2007. pp14).

"The maximum yield found with NaOH is 85% at a sodium hydroxide concentration of 1,0%. Adding an excess in the amount of the catalyst, it gives rise to the formation of an emulsion which increases viscosity and leads to the gel formation" "(Cheng et al. 2008. pp 2210)."With regard to the use of catalyst as ($NaOCH_3$) sodium methoxide and ($KOCH_3$) potassium methoxide we can observe high efficiency compared with other alkali catalysts"(Cheng et al. 2008. pp 2210). The temperature of the transesterification reaction "should not exceed the boiling point of alcohol, because it vaporizes and forms bubbles which limit the reaction in the interfaces alcohol/oil/biodiesel"(Giron et al. 2009.pp 18)

"To be used as catalyst NaOH with methanol, has been found that the optimum temperature to achieve high yields was 60 °C, while using KOH to this same temperature not achieved such high yields and higher catalyst concentrations should be used to using NaOH" (Liu et al. 2006b pp 110). "In an alkali catalyzed process is reached high purity and yields in short periods of time ranging between 30 - 60 minutes" (Liu et al. 2006a pp 186).

Heterogeneous catalysts are found in two phases and a contact area, "the use of these catalysts simplifies and makes more economical the purification process due the easy separation of the products and reactants. The disadvantage is the difficulty to temperature control for very exothermic reactions, limitations on mass transfer of reactants and products, as well as high mechanical resistance to the catalyst" (Arbeláez & Rivera, 2007. pp12). Among the most common catalysts are the metal oxides (MgO, CaO), acids of Lewis (SnCl$_2$), etc. For example, by using zinc oxide are obtained yields of 50.7%, when using Al$_2$O$_3$ is obtained 57.5% and using CaO yield of 65%"(Rojas & Torres. 2009 pp 15). "These catalysts have limitations on transfer of mass of reactants and products" (Arbeláez & Rivera, 2007. pp12), but they have the advantage that they are not corrosive to the reactor"(Guan et al. 2009 pp 520).The easy separation of the products generates a "simplification of the manufacturing process since the catalyst can be separate from the products of reaction with a simple filtration process"(Lles et al. 2008 pp 63). "Don't generate byproduct of soap by reaction with free fatty acids (AGL)". (Bournay et al. 2005. pp 191) "Using CaO is achieved a yield of 65% and by using MgO a yield of 64%"(Bournay et al. 2005. pp 192). To achieve high yields the reaction must be carried out "to a higher temperature increasing energy costs" (Bournay et al. 2005. pp 191).Reported high reaction times, because the "speed of transesterification reaction with these catalysts is lower in comparison with homogeneous catalysts, due to the mass transfer resistance" (Guan et al. 2009 pp 522).

Finally, the lipases being effective for the transesterification reaction can be used between the enzyme catalysts. "This type of catalysis has the advantage of allowing the use of alcohol with high content of water (more than 3%), low temperatures, which is an energy-saving and high degrees of acidity in oils" (EREN. 2003. pp 41).

3.1.3 Waste cooking oil

The waste cooking oil is generated from the fried food, which need large amounts of oil because it requires the full immersion of food at temperatures greater than 180 °C. Accordingly to the high temperatures are generated changes in its chemical and physical composition, as well as in its organoleptic properties which affect both the food and oil quality.

Reuse of domestic oil has a high risk to the health of consumers as depending on the type of food subjected to frying, "this absorbs between 5% and 20% of the used oil, which can increase significantly the amount of hazardous compounds that provide degraded oil to food" (EREN. 2003. pp 31)."In an alkali catalyzed process is reached high purity and high yields in short periods of time ranging between 30 - 60 minutes"(Liu et al. 2006 pp 186).

Used cooking oil is normally black, a strong odor and does not have large amount of solids because its collection is passed through a fine mesh. In Figure 5, we can see a sample of used oil from the hotel sector.

Fig. 5. Sample of waste cooking oil

3.1.3.1 Domestic waste oil treatment

Wastes containing these types of oils are products of decomposition that impair the oil quality causing reduction in productivity in the transesterification reaction and may also generate undesirable by-products which hurt the final product. For these reasons, it is important to refine the waste domestic oil for the biodiesel production. "This type of refinement has a right effect on the yield of the reaction from 67% to 87% after bleaching". (EREN. 2003. pp 36). For the treatment of adequacy of waste domestic oil, the operations that can be applied are filtration, de-acidification or neutralization and whitening. The processes of degumming and deodorization aren't needed because the oils have already been treated prior to use and although during degradation odors occur, the removal is not essential for the biodiesel production.

- *Filtration.* The operation is for removing solids, inorganic material, and other contaminants in the oil. It can be carried out at temperatures higher than 60 °C, where substances carbonaceous produced from burnt organic material, pieces of paper, waste food and other solids are removed or occur at low temperatures which depend on the physical condition of the oil. In addition, we can delete solid fats or products of low melting points from the frying process.
- *Desacidification* It is the process by which free oils fatty acids are removed, various methods are used:
  a. Neutralization with alkaline solution: in this process the acids are removed in the form of soaps.
  b. Esterification with glycerin: seeks to regenerate the triglyceride.
  c. Extraction by solvents: where it is used ethanol in proportions 1.3 times the amount of oil.
  d. The distillation of fatty acids, this method requires a high energy cost.
  e. Removal of fatty acids with ion-exchange: a resin of strongly basic character for the removal of free fatty acids and the color of the oil is used.

  Method that provides greater account of productivity in the removal of free fatty acids is the neutralization by caustic soda, since it not only are obtained high relations, but also helps in the bleaching of the oil, because made soaps help dragging the color generators. There are basically two procedures:
- Neutralization with dilute alkali: are used concentrations of 0.75 to 2 N.
- Neutralization with concentrated alkali, where the concentration of caustic soda vary between 2 and 5 N.

In each of the procedures mentioned above neutralization is carried out hot, with oil at a temperature between 50 - 60 °C and addition of caustic soda between 70-80 °C.

3.1.3.2 Chemical characteristics of the used oil

Chemical characterization of the used oil, is presented in table 4

FEATURES	OIL COLLECTED BY THE HOTEL SECTOR
Acidity (%)	0.56
Moisture	0.25
Viscosity at 37°C (centistokes)	44.78
Iodine index (CgI_2/g)	108.22
Peroxide index (meq. Oxygen active/Kg of sample)	16.61
Unsaponifiable material (%)	1.70
Saponification index (mg KOH/g)	195.87
Ash (%)	0.030
Refractive index 25°C	1.4700
Density 15°C (g/mL)	0.9216

Table 4. Characterization of cooking oil collected by the hotel sector (Source: Avalquímico Ltda.2010)

4. Biodiesel production from used cooking oils

According to table 5, the catalyst with higher industrial scaling, economic cost, high yields and short reaction time, is the alternative of basic catalysis, using sodium hydroxide. Although soap can be formed using sodium hydroxide in the transesterification reaction, this occurs if the content of free fatty acids is greater than 1% and the type of oil collected from the hotel sector has a percentage of acidity of 0.54%, so it is not problem to use this type of catalyst for the biodiesel production.

SELECTION PARAMETERS	Comparison of alternatives		
	Alkaline	Heterogeneous	Acid
Catalyst	NaOH	CaO	H_2SO_4
Alcohol	Methanol	Methanol	Methanol
Scaling	High	Low	Low
Catalyst separation	Low	High	Low
By-products formation	Glycerin Soap, pasty	Glycerin	Glycerin
Environmental impact	High	Low	High
Cost	Economic	High	Moderate
Availability	Medium	Low	Medium
Catalyst concentration (%w/w)	Low	Medium	High
Molar ratio alcohol/oil	Medium	Low	High
Temperature (°C)	Low	High	Medium
Yield	High	Low	Medium
Reaction time (hours)	Low	Low	High
Safety level	High	Low	High

Table 5. Comparison of alternatives for the biodiesel production.

To select the best alternative for the biodiesel production are defined three ranges for the operation conditions to be used in the transesterification reaction (table 6).

SELECTION PARAMETERS	RANGES		
	Low	Middle	High
Catalyst concentration (% w/w)	0.2% - 1%	>1% - 3%	>3% - 15%
Molar ratio alcohol/oil	3:1 - 6:1	>6:1 - 12:1	>12:1 - 80:1
Temperature (°C)	50 - 60	>60 - 100	>100 - 200
Yield (%)	20 – 70	>70 - 90	> 90 – 100
Reaction time (hours)	0.16 – 1	>1 - 2	> 2 – 40

Table 6. Ranges established for the operation conditions of the transesterification reaction (Rojas & Torres. 2009. pp 18).

4.1 Experimental design

It is a key to make a design from which the most appropriate values for each of the design factors can be established. The selected factors were: molar ratio alcohol/oil, percentage of catalyst, temperature and washing agent, where the first three are design variables and the latter is a design condition. Before starting the design is important defines the ranges and levels for these factors, for this reason, we search the experimental phase in scientific articles related to the project (table 7).

DESIGN FACTORS	RANGE
Molar ratio alcohol/oil	6:1- 15:1
Catalyst Percentage (% wt.)	0,4-1
Temperature (°C)	40-70
Washing agent	Water (40°C) - Acetic acid

Table 7. Ranges for the design factors

Based on the ranges set out in table 7, we provide that the factorial design appropriate for the process is the factorial design 2^k, which can be solved by the technique of Yate contrasts which establish two levels for each of the design factors, these levels are high (+) and low (-) (see table 8)

Design Factor	High level (+)	Low level (-)
Molar ratio alcohol/oil	9:1	6:1
Catalyst Percentage(% wt)	0.7	0.5
Temperature(°C)	60	50
Washing agent	Acetic acid	Water (40°C)

Table 8. Levels for each design factor

For the molar ratio alcohol/oil is found a ratio of 6: 1 that is optimum for achieving high conversions, some articles display that lower ratio not is possible to reach a complete transesterification reaction. There are also good results with ratios ranging between 9:1 and 12:1, while if we use higher than 15:1 molar ratio there are difficulties in the separation of glycerin and methyl esters.

For the catalyst concentration the values vary in a range of 0.4 - 2% being the concentration 1% better but the reaction yield is not very high.

For the design factorial mentioned we can set the number of trials, having clear that 2 is the number of levels and k is the number of factors, i.e. which has a total of 2^4 treatments and is carried out a duplicated for each one, as should be taken into account the time limit and the project costs, which in total we have 16 experiments. Then is defined the signs matrix (table 9), where it is necessary to enumerate the trials and then is assigned a combination of treatments that aims to relate the design factors.

Consecutively is assigned the level for each combination, positive for the design factor that is being evaluated in the trial and negative for whose are not related. For this type of design the first trial has low levels and the final test has the higher levels.

Trial	Combination of Treatments	Design Factors			
		A	B	C	D
1	1	-	-	-	-
2	A	+	-	-	-
3	B	-	+	-	-
4	C	-	-	+	-
5	D	-	-	-	+
6	AB	+	+	-	-
7	AC	+	-	+	-
8	AD	+	-	-	+
9	BC	-	+	+	-
10	BD	-	+	-	+
11	CD	-	-	+	+
12	ABC	+	+	+	-
13	ABD	+	+	-	+
14	ACD	+	-	+	+
15	BCD	-	+	+	+
16	ABCD	+	+	+	+

Table 9. Matrix of signs

Performance is evaluated according to the treatments combination, considering that a duplicated is made by treatment. With data from the response variable, which is the yield, we carry out a statistical analysis by means of the ANOVA table or analysis of variance for data, which gives the more appropriate conditions for each factors of design and allows establishing that trials were the best.

For each main effect and interaction effect we have associated a single degree of freedom, so this is calculated using the following expression:

$$GL_i = N^0{}_{niveles} - 1 \qquad (1)$$

Where:

i: is any combination of treatment

$N^o_{levels} = 2$, because we have a high level and one low

To determine that interactions are significant an f for each source of variation is calculated and compared to $f_{0,05}$ (1,16) = 4,49, with this we can determine in which region (probable region RP or critical region RC) is each treatment and thus be able to establish that treatment is accepted or rejected with the help of figure 6.

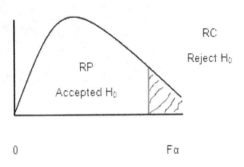

Fig. 6. Function f

4.2 Experimental development at the laboratory level

Table 10, shows the quantities determined at the laboratory level for each compound that is involved in the biodiesel production.

AMOUNT OF REAGENTS	
Waste cooking oil (mL)	150
Amount of methanol (6:1) (mL)	40.63
Amount of methanol (9:1) (mL)	60.95
Amount of NaOH grams (0.5%p/p)	0.823
Amount of NaOH grams (0.7%p/p)	1.105
Amount of HCl grams (0.5% p/p)	1.466
Amount of HCl grams (0.7% p/p)	2.053

Table 10. Amount of reagents

First is the filtration of used oil, then mixing alcohol/catalyst to add it to the reactor which contains the oil at the temperature of the transesterification reaction, then is the separation of biodiesel and glycerin, washes the biodiesel and finally is the distillation of the biodiesel. "The dissolution is agitated at low rpm because that at high revolutions sodium hydroxide can be oxidized" (Arbeláez & Rivera, 2007. pp45), should also be covered because amount of methanol due to its volatility can be lost. For the transesterification reaction are used reactors of four mouths with capacity of 500 mL and 1000 mL, magnetic stirrers, plates of agitation, spiral capacitors, mercury thermometers, thermostat bath and temperature controller. Figure 7, shows the setup for the biodiesel production.

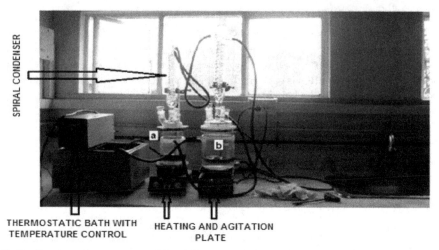

Fig. 7. Biodiesel production assembly: a) Reactor of 500 mL b) Reactor of 1L.

To carry out the transesterification reaction is loaded oil at the reactors and heated up to reaction temperature, while it reaches the temperature is made the mixture of the catalyst with alcohol, then it is added to the reactor. At the end of the reaction time is added HCl to 37 per cent in order to neutralize the reaction.

Completed reaction, the product is poured into the separation funnels and let a minimum time of 8 hours, to ensure good separation of the phases (Figure 8). Separation times were not equal for all runs varied between 10-24 hours.

Fig. 8. Biodiesel-Glycerin separation

Once separate the glycerin from biodiesel, it is carried out the washing on the funnels. Separate the glycerin, the biodiesel must be washed because that may contain residues of catalyst, methanol, soaps and glycerides without reacting. Two types of washings are established according to the experimental design.

In one of the washing, water at 40 °C is used for three washes. Other tests of washing have been conducted with acetic acid 10% wt, where is used the same amount of used cooking oil; two washes are carried out with acid solution and the third is done with deionized water.

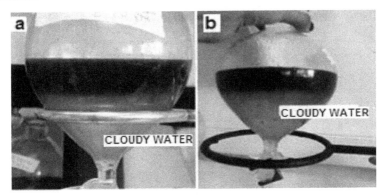

Fig. 9. First biodiesel wash: (a) Water at 40 °C (b) acetic acid solution

The distillation is carried out at 40 °C, temperature which is below the boiling point of methanol. The vacuum pump is used in order to minimize the time of distillation and vacuum trap is used to prevent waste of alcohol and water to the pump.

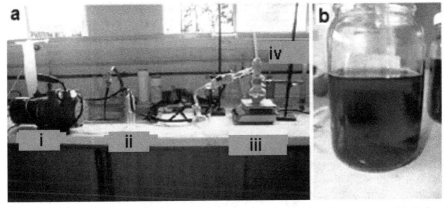

Fig. 10. Biodiesel distillation (a) Mounting: i. vacuum pump ii. Vacuum trap iii. Hot plate iv. Thermometer (b) Distilled biodiesel

4.3 Results and analysis

We show the results and analysis of tests conducted at the laboratory level, as its density and the analysis of variance. According to the literature retrieved biodiesel is "liquid, transparent and reddish color without any content of solids or gels" (Arbeláez & Rivera, 2007. pp37).

Taking into account the results of the biodiesel appearance, it can be concluded that the catalyst percentage influences in the biodiesel stability because that similar conditions were equal in appearance. The majority of samples that had contained solids, gels and their appearance was opaque, were the samples where used 0.5% NaOH.

Opaque samples as the 3, 4 and 7 are because have much water content, this might be for the washes with water at 40 °C and its water content was increased. Opacity is an indicator that the methyl esters have presence of water.

The table 11 shows the densities for each test sample.

TEST	DUPLICATE 1	DUPLICATE 2
	Biodiesel density (g/ml)	Biodiesel density (g/ml)
1	0,902	0,910
2	0,902	0,877
3	0,867	0,887
4	0,893	0,881
5	0,897	0,903
6	0,885	0,887
7	0,883	0,906
8	0,906	0,860
9	0,871	0,887
10	0,885	0,893
11	0,914	0,906
12	0,897	0,891
13	0,900	0,895
14	0,885	0,875
15	0,883	0,862
16	0,873	0,891

Table 11. Density for each sample

The biodiesel density according to standard ASTM D-1298 must be in a range of (0.86 - 0.90 g/ml), the density does not guarantee that retrieved biodiesel is of good quality, but is taken to compare samples with each other, due there is evidence that satisfy with the provided density but its appearance is not adequate or are samples which do not comply with the permitted density value but its appearance is appropriated. As for example, the duplicate 2 from sample 1 exceeds the density range allowed for biodiesel, but the appearance fulfill with the stipulated features. For the sample 11, both the duplicate 1 and the 2 exceed the density level and its appearance is a little opaque.

Samples 11 and 12 are in the density range allowed but the appearance does not meet any features because have high solids content and are highly opaque.

Some samples contain solids and do not fulfill with the physical characteristics of biodiesel, have a density within the level agreed by the ASTM D-1298 standard for a biodiesel of good quality, which must be due to that the solids are smaller proportion than the liquid phase that is rich in methyl esters.

It can be concluded that the density and appearance analysis are not reliable parameters to determine the biodiesel quality. For this reason, based on the order of the table 12, we can determine the variable response (table 13), which is expressed as the ratio between the mass of the biodiesel produced and the mass of oil used for the production (productivity per cent).

TEST	COMBINATION OF TREATMENTS	DESIGN FACTORS			
		A	B	C	D
1	1	6:1	0,5	50	Water (40°C)
2	A	9:1	0,5	50	Water (40°C)
3	B	6:1	0,7	50	Water (40°C)
4	AB	9:1	0,7	50	Water (40°C)
5	C	6:1	0,5	60	Water (40°C)
6	AC	9:1	0,5	60	Water (40°C)
7	BC	6:1	0,7	60	Water (40°C)
8	ABC	9:1	0,7	60	Water (40°C)
9	D	6:1	0,5	50	Acetic Acid (T amb)
10	AD	9:1	0,5	50	Acetic Acid (T amb)
11	BD	6:1	0,7	50	Acetic Acid (T amb)
12	ABD	9:1	0,7	50	Acetic Acid (T amb)
13	CD	6:1	0,5	60	Acetic Acid (T amb)
14	ACD	9:1	0,5	60	Acetic Acid (T amb)
15	BCD	6:1	0,7	60	Acetic Acid (T amb)
16	ABCD	9:1	0,7	60	Acetic Acid (T amb)

Table 12. Rearranged experimental matrix

TEST	COMBINATION OF TREATMENTS	PRODUCTIVITY (%)	
		DUPLICATE 1	DUPLICATE 2
1	1	70	44,9
2	A	79,5	74,3
3	B	85,6	84,5
4	AB	93,7	93,9
5	C	69,4	68,4
6	AC	62,3	88,2
7	BC	64,5	62,6
8	ABC	88,7	85,5
9	D	62,1	63,1
10	AD	89,5	60,7
11	BD	76,2	80,7
12	ABD	88,9	88,5
13	CD	61,3	57,5
14	ACD	81,2	80,3
15	BCD	74,8	71,8
16	ABCD	86,2	81,8

Table 13. Reaction productivity

We carry out the ANOVA statistical analysis (table 14) and the test of hypothesis (table 15).

COMBINATION OF TREATMENTS	EFECTS	SQUARES SUM	FREEDOM DEGREES	MEAN SQUARE	f CALCULATED
A	14,1125	1593,30	1	1593,30	22,78
B	12,2	1190,72	1	1190,72	17,02
C	-3,225	83,205	1	83,21	1,19
D	-0,7125	4,06125	1	4,06	0,06
AB	-0,8	5,12	1	5,12	0,07
AC	1,375	15,125	1	15,12	0,22
AD	-0,4125	1,36125	1	1,36	0,02
BC	-6,2875	316,26125	1	316,26	4,52
BD	-0,55	2,42	1	2,42	0,03
CD	1,375	15,125	1	15,13	0,22
ABC	2,437500	47,531250	1	47,53	0,68
ABD	-2,425000	47,045000	1	47,05	0,67
ACD	0,950000	7,220000	1	7,22	0,10
BCD	3,212500	82,561250	1	82,56	1,1803
ABCD	-4,5375	164,71125	1	164,71	2,35
ERROR		1119,21	16	69,950625	
TOTAL		4694,97875	31		

Table 14. ANOVA TABLE

COMBINATION OF TREATMENTS	EFECTS	f CALCULATED	f alfa	DECISION
A	14,1125	22,77751271	4,49	NOT ACCEPTED
B	12,2	17,02229251	4,49	NOT ACCEPTED
C	-3,225	1,189481867	4,49	ACCEPTED
D	-0,7125	0,058058809	4,49	ACCEPTED
AB	-0,8	0,073194485	4,49	ACCEPTED
AC	1,375	0,216223944	4,49	ACCEPTED
AD	-0,4125	0,019460155	4,49	ACCEPTED
BC	-6,2875	4,521206923	4,49	NOT ACCEPTED
BD	-0,55	0,034595831	4,49	ACCEPTED
CD	1,375	0,216223944	4,49	ACCEPTED
ABC	2,437500	0,679497145	4,49	ACCEPTED
ABD	-2,425000	0,672545814	4,49	ACCEPTED
ACD	0,950000	0,103215661	4,49	ACCEPTED
BCD	3,212500	1,180278947	4,49	ACCEPTED
ABCD	-4,5375	2,354678747	4,49	ACCEPTED

Table 15. Hypothesis test

The factor A, the molar ratio alcohol/oil, has a significant effect on the reaction productivity, since the effect is positive by increasing the molar ratio increases productivity, for this reason the highest 9:1 ratio is selected.

The factor B, the catalyst concentration has a significant effect on the reaction productivity so increasing the catalyst concentration increases the reaction productivity for this reason we select 0.7% w/w. The factor C, the reaction temperature has no significant effect, but is recommended to work with the lowest level, i.e. at a temperature 50 °C to prevent a further loss of methanol due to its volatility. The factor D, the washing agent has no significant effect because the effect is negative, we select the low level (-) water at 40 °C.

The combination of treatments AB, AC, AD, CD, ABC, ABD, ACD, BCD, ABCD, do not has significant effects on the reaction productivity. The treatment combination BC has significant effects on the reaction productivity, so it is advisable to work with levels higher for each factor, catalyst concentration of 0.7 %p/p and reaction temperature of 50 °C.

Table 16, shows the made characterization of the sample obtained to a molar ratio alcohol/oil 9:1, catalyst concentration 0.7% w/w, reaction temperature of 50 °C and water at 40 °C as washing agent that is the biodiesel with a best properties.

Properties	Unit	Results	Standard (ASTM D-6751)
Density at 15.6 °C	kg/m^3	889.9	860- 900
Grades API		27	Minimum 32
Cinematic Viscosity at 40°C	Cst	5.21	1.9 – 6
Cetane index		48	Minimum 47
Caloric value	J/g	40,873.00	37,216.00
Temperature 90% distilled	°C	332	Maximum 360

Table 16. Properties of the biodiesel sample to the best conditions

API gravity for the biodiesel retrieved is in a range of 32-34 degrees API, the analyzed sample showed a value below the reported ranges, which indicates that the biodiesel retrieved from this sample has a high density, which as we see in the analysis is of 889.9 kg/m^3, taking into account that the API gravity is inversely proportional to the density. As for the other properties analyzed, these can be found within the values reported by the literature, which guarantees the quality of biodiesel.

5. Conclusions

We can conclude that using acetic acid or water as a washing agent does not affect the reaction productivity, similar to the reaction temperature has no effect on the variable response within the levels used in the research. The unique variables that affect the biodiesel production are the catalyst concentration and the molar ratio alcohol/oil.

According with the above, the best conditions of operation are:

- Molar ratio alcohol/aceite: 9:1
- Catalyst concentration of: 0.7% w/w
- Reaction temperature: 50 °C
- Washing agent: water at 40 °C

6. Acknowledgment

This research was carried out in the laboratory of the Natural Resources Energetic Exploitation research group of the Chemistry Department's National University of

Colombia and sponsored for the research group CDM & EG of the Mechanical Department of Engineering Faculty of the National University of Colombia.

7. References

Arbeláez, M., Rivera, M. (2007). *Diseño conceptual de un proceso para la obtención de biodiesel a partir de algunos aceites vegetales colombianos.* Universidad Eafit. Medellín.

Benjumea P, Benavides A Y Pashova V. (2007). *El biodiesel de aceite de higuerilla como combustible alternativo para motores diesel,* Universidad Nacional De Colombia Sede Medellín.

Bournay L, Casanave D, Chodorge J, Delfort D Y Hillion G. (2005), *new heterogeneous process for biodiesel production: a way to improve the quality and the value of the crude glycerin produced by biodiesel plants,* France.

Chen G, Meng X Y Wang Y. (2008). *Biodiesel production from waste cooking oil via alkali catalyst and its engine test,* China Architecture Design & Research Group, Beijing, China.

Cujia G., Bula A. (2010). *Potencial obtención de gas de síntesis para la producción de metanol a partir de la gasificación de residuos de palma africana.* Asociación Interciencia Venezuela.

Diaz-Zorita, M., Et Al., El cultivo De Girasol, In: *ASAGIR,* 2003, <Http://Www.Asagir.Org.Ar/Publicaciones/Cuadernillo_Web.Pdf>

Ente Regional De La Energía De Castilla Y León (EREN) (2003). *Plan piloto de recogida de aceites de cocina usados para la producción de biodiesel.* España.

Errazu A, Marchetti J Y Miguel V.(2005). *Possible methods for biodiesel production,* Planta Piloto De Ingeniería Química, Argentina.

FEDEPALMA(n.d.), Biodiesel, In: *FEDEPALMA.* 20 of june of 2011, Available from: <Http://Www.Fedepalma.Org/>

Girón E, Rojas A Y Torres Harlen. (2009).*variables de operación en el proceso de transesterificación de aceites vegetales: una revisión. In: Catálisis Química,* Vol 29 No 3.

Guan G, Kusakabe K Y Yamasaki S.(2009). *Tri-potassium phosphate as a solid catalyst for biodiesel production from waste cooking oil,* Department Of Living Environmental Science, Fukuoka Women's University,

INFOAGRO, El Cultivo Del Girasol (1ª Parte) 20 of june of 2011, In: INFOAGRO, Available from: <Http://Www.Infoagro.Com/Herbaceos/Oleaginosas/Girasol.Htm>

Iriarte, L. El Cultivo De Colza En Argentina, In: *ANTI,* 20 of june of 2011, Available from: <Http://Www.Inta.Gov.Ar/Ediciones/Idia/Oleaginosa/Colza01.Pdf>

Liu P, Ou S, Wang, Y. Zhang Z. (2006a). *Preparation of biodiesel from waste cooking oil via two-step catalyzed process,* Department Of Food Science And Engineering.

Liu P, Ou S, Tang S, Wang Y Y Xue F.(2006b) *comparison of two different processes to synthesize biodiesel by waste cooking oil,* Department Of Food Science And Engineering, Jinan University.

Peisch, S. (n.d.). Aguas Usadas En Puerto Rico: Amenaza A La Salud Publica Y Al Ambiente, In: *Datos Marinos #54.* 20 of june of 2011, Available from: <Http://Www.Seagrantpr.Org/Catalog/Files/Fact_Sheets/54-Aguas-Usadas-De-Pr.Pdf>

Rojas A Y Torres H. (2009). *Variables de operación en el proceso de transesterificación de aceites vegetales: una revisión In: Catálisis Química.*

Lles M, Pires E, Royo C. (2008). *Estudio de catalizadores heterogéneos en la síntesis de biodiesel a partir de aceite de girasol.*

4

Microbial Biodiesel Production - Oil Feedstocks Produced from Microbial Cell Cultivations

Jianguo Zhang and Bo Hu
University of Minnesota,
USA

1. Introduction

Crude oil price has increased to over $100 per barrel, which is causing serious negative impact on the global and national economy. In 2005, the United States produced 8.3 million bbl/day, but consumed 20.8 million bbl/day, the balance of which was imported from other countries. For these fossil fuels, the U.S. used about 138 billion gallons of gasoline in 2006, accounting for about 44 percent of the world's gasoline consumption (EarthTrends, 2008). The annual U.S. usage of jet fuel was 21 billion gallons in 2006 (Energy Information Administration, 2008). The U.S. annual consumption of diesel fuel in 2006 was about 50 billion gallons (Energy Information Administration, 2008). Massive consumption of fossil fuels has already caused serious concern over global warming caused by greenhouse gases.

Biofuel offers an alternative to fossil fuels. It provides several benefits, such as alleviation from foreign oil dependence, carbon neutral process without greenhouse emission, and profits to local farmers. Bioethanol production from starch and lignocellulosic materials is a kind of an alternative to fossil fuels. It can be blended with gasoline in varying quantities up to pure ethanol (E100). The first generation of ethanol biofuel has been massively commercialized and dominated by the U.S. and Brazil. Fuel ethanol in the U.S. is primarily produced from corn, while Brazilian ethanol is produced mainly from sugarcane. These raw materials are in direct competition with human diet or the land to produce food, which triggers the controversy of food versus fuel. The second generation of ethanol is proposed to be produced from lignocellulosic biomass, which can be obtained from agricultural residue or other woody and herbal biomass from marginal land. Intense scientific research has been carried out over the past decade, focusing on this route in order to decrease the overall process cost and this process is gradually focusing on commercialization.

2. Biodiesel and current feedstocks

2.1 Biodiesel production

Another approach for alternative biofuels is biodiesel. The most common type of biodiesel is the methyl esters of fatty acid (FAME), obtained by transesterification of lipid with methanol or ethanol. It can be used in pure form (B100) or may be blended with fossil diesel at any rate. The commonly used biodiesel is B99 because 1% of fossil fuel is applied to inhibit mold growth, which shortens its shelf life. Biodiesel is around 5-8% less efficient than conventional fossil diesel; yet, biodiesel has potential as a total or partial (in the cold regions) replacement to the fossil diesel, compatible to the current diesel engine. There are numerous environmental benefits to replace fossil diesel to biodiesel since combustion of biodiesel emits far less pollutants comparing to fossil diesel (except the NO_x emission) and the entire process is close to carbon neutral considering that the plant oil used to produce biodiesel is synthesized in the agriculture from CO_2 in the air. The production and utilization of biodiesel are significant in many aspects, for instance, increasing oilseed crop market, providing domestic job opportunities to the rural community, and decreasing the dependence on imported oil; therefore biodiesel has been commercialized around the globe. Another type of biodiesel is the hydrocarbon generated with direct decarboxylation of fatty acid or lipid. Although it shows superior features than current FAME biodiesel in many aspects, the chemical decarboxylation process to produce diesel still needs further development compared to the transesterification process for biodiesel production.

2.2 Agricultural feedstocks: Vegetable oil, animal fat and recycled grease

The biodiesel industry suffered due to limited raw materials such as soybean oil or vegetable oil. Although there are numerous potential renewable and carbon neutral feedstocks for the production of biodiesel, none seems capable of displacing fossil diesel. Table 1 shows a comparison of the oil yields of some sources of biodiesel, the land area for their cultivation, and the percentage of existing United States (U.S.) cropping area to meet half of the transport fuel needs. Using soybean as feedstock for the production of biodiesel requires 326% of the U.S. cropping area to meet the 50% of all U.S. transport fuel needs. Based on this rough estimation, none of the terrestrial crops is able to completely substitute crude oil.

Crop	Oil yield [Liters per hectare]	Land area needed [Mha][a]	Percent of existing US cropping area[a]
Corn	172	1540	846
Soybean	446	594	326
Canola	1190	223	122
Jatropha	1892	140	77
Coconut	2689	99	54
Oil palm	5950	45	24
Microalgae[a]	136,900	2	1.1
Microalgae[c]	58,700	4.5	2.5

[a] For meeting 50% of all transport fuel needs of the U.S.
[b] 70% oil (by wt) in biomass.
[c] 30% oil (by wt) in biomass.

Table 1. Comparison of some sources of biodiesel (Chisti 2007)

Various raw materials are proposed to produce biodiesel, including waste cooking oil and various oil-accumulating plants, which could help solve the shortage for raw materials but are not sufficiently available. New approaches need to be developed to use an alternative biomass as the substrate for biofuel production, for example, biomass leftover from agricultural harvest or the biomass produced in non-traditional agricultural land as the source of sugars for biofuel and bioproducts. This potential has been gaining much attention recently as it is been defined as "second generation of bioenergy", compared to current technologies that primarily are limited to food materials. Oil accumulation with microalgae and other similar technologies started to show some potential to produce biofuel products with massive scales (Benemann, 1996; Chisti, 2007; Jarvis et al., 1994).

3. Microorganisms for microbial lipid production

3.1 Microalgae
Algae oil is being seriously considered because of the large oil yields it shows compared with other oilseeds. The Office of Fuels Development, a division of the Department of Energy, funded a program from 1978 through 1996 under the National Renewable Energy Laboratory known as the "Aquatic Species Program". The focus of this program was to investigate high-oil algaes that could be grown specifically for the purpose of widescale biodiesel production. NREL's research showed that one quad (7.5 billion gallons) of biodiesel could be produced from 200,000 hectares of desert land (200,000 hectares is equivalent to 780 square miles, roughly 500,000 acres). It would be preferable to spread the algae production around the country, to lessen the cost and energy used in transporting the feedstocks. Algae farms could also be constructed to use waste streams (either human waste or animal waste from animal farms) as a food source, which would spread algae production around the country. Nutrients can also be extracted from the algae for the production of a fertilizer high in nitrogen and phosphorous. By using waste streams (agricultural, farm animal waste, and human sewage) as the nutrient source, these farms essentially also provide a means of recycling nutrients from fertilizer to food to waste and back to fertilizer.

Microalgae are generally autotrophic eukaryotic cells, and contain one or more types of chlorophyll plus additional pigments known as carotenoids and biloproteins (also called phycobilins). Carotenoids are yellow, orange, or red water-insoluble linear hydrocarbons; biloproteins are blue or red water-soluble pigment-protein complexes. The color of the different groups of algae depends on the ratio of these pigments. Almost all the microalgae cells contain chlorophyll, but the green color can be masked by the carotenoids, giving them a brown or red color. Microalgae cells are mostly unicellular, but some species are colonial or filamentous. They can grow autotrophically and/or heterotrophically, with a wide range of tolerance to different temperature, salinity, pH and nutrient availabilities. Cyanobacteria, although sometimes also referred as algae, specifically as blue-green algae, are by definition prokaryotic bacteria, instead of microalgae. The eukaryotes microalgae we are referring to include green algae, diatoms, yellow–green algae, golden algae, red algae, brown algae, dinoflagellates and others; and only limited species have the capability to accumulate high content of lipids in their cell biomass. Microbial species that can accumulate over 20% of lipids in their cell biomass are considered oleaginous species.

One of the characteristics of algae that separate them from other oilseeds is the quantity of lipids and fatty acids algae have as membrane components; some algae have been found to contain over 80% of lipids, which is of great interest for a sustainable feedstock for biodiesel production. Different microalgae species are listed in Table 2 as examples of lipid accumulation research. For example, *Chlorella vulgaris* is a commercially important green microalgae because it has the potential to serve as a food and energy source due to its high photosynthetic efficiency, which can, in theory, reach 8%. It can grow with both autotrophic and heterotrophic modes, and its mixotrophic growth rate is the sum of its autotrophic growth rate and heterotrophic growth rate, separately. *Chlorella protothecoides* is another single-cell green microalgae, which has high potential for the energy and food production. Heterotrophic growth of *C. protothecoides* supplied with acetate, glucose, or other organic compounds as carbon source, results in high biomass and high content of lipid in cells.

Microalgae	Cultures			Substrates	Growth Rate	Lipid Content
	AC	MC	HC			
Chlorella protothecoides	X		X	glucose, acetate / CO_2	3.74g/L - 144h	55.2%
Chlorella vulgaris	X	X	X	glucose, acetate, lactate / CO_2	0.098/h	
Crypthecodinium cohnii			X	glucose/ CO_2	40g/L - 60-90h	15-30%
Scenedesmus obliquus	X		X	glucose/ CO_2	double in 14h after adaptation	14-22%
Chlamydomonas reinhardtii	X	x	X	acetate/ CO_2	exponential during the first 20 Hr	21%
Micractinium pusillum		x	X	CO_2	0.94g/L - 24h	
Euglena Gracilis			X	CO2		14-20%
Schizochytrium sp				glycerol/ CO_2	130 -140h (28°C)	55%
Spirulina platensis	X		X	glucose/ CO_2	0.008 /h	
Botryococcus Braunii			X	CO_2	low growth rate	20 - 86%
Dunaliella salina	X	x		CO_2		~70%

AC: autotrophic cultures; MC: mixotrophic cultures; HC: heterotrophic cultures

Table 2. Some examples of microalgae cultivation for oil accumulation

Lipid accumulation occurs within the microalgae cells and it varies from strain and growth conditions. There are many nutritional and environmental factors controlling the cell growth and lipid contents, such as organic and inorganic carbon sources, nitrogen source, and other essential macro- and micro-nutrients like magnesium and copper, temperature, pH level, salinity, agitation speed (dissolved oxygen). Many microalgae species, for example, *C. protothecoides*, accumulated a higher content of lipids in cells and achieved higher growth rate when the culture was under heterotrophic mode. Like yeast and fungi, heterotrophic algae can accumulate biomass and lipids using organic carbon as its source instead of carbon dioxide and sunlight. Compared with autotrophic algae, the heterotrophic growth process has the advantages of no light limitation, a high degree of process control, higher productivity, and low costs for biomass harvesting (Barclay, Meager et al. 1994). Table 2 shows the oil production of autotrophically and heterotrophically cultured microalgae. Most heterotrophically cultured algae have greater than ten times the biomass concentration, while the lipid productivity is also significantly higher than the theoretical data for autotrophic cultivation. For certain algae strains, it was suggested the heterotrophically cultured cells exhibited better capability for biomass and lipid production. Miao and Wu (2006) reported that the oil content of heterotrophically cultured *C. prototthecoides* was approximately four times greater than that in the corresponding autotrophic culture. Liu et al. (2010) demonstrated that the heterotrophically cultured cells of *Chlorella zofingiensis* showed 411% and 900% increases in dry cell weight and lipid yield, respectively, compared to autotrophically cultured cells. Moreover, biodiesel produced from heterotrophically cultured algae oils had similar properties to diesel fuel in terms of density, viscosity, heating value, and H/C ratio (Xu, Miao et al. 2006). In addition to lipid production, high value byproducts can be obtained from heterotrophically cultured microalgae, including polyunsaturated fatty acids and carotenoids (Chen and Chen 2006).

3.2 Yeast and fungi

Besides microalgae, many yeast and fungi species (e.g., *Mucor circillenous* or *Mortierella isabellina*) also can accumulate a high content of lipids (Xia 2011; Heredia-Arroyo, Wei et al. 2011). Many oleaginous yeasts were studied for lipid accumulation on different substrates, such as industrial glycerol (Meesters, Huijberts et al. 1996; Papanikolaou and Aggelis 2002), sewage sludge (Angerbauer, Siebenhofer et al. 2008), whey permeate (Ykema, Verbree et al. 1988; Akhtar, Gray et al. 1998), sugar cane molasses (Alvarez, Rodriguez et al. 1992), and rice straw hydrolysate (Huang, Zong et al. 2009). The use of non-starch biomass is critical so that lignocelluloses can be used for organic carbon supply without concern of using food crops for fuel sources. Recent studies detailed conversion of hemicellulose hydrolysate into lipids by oleaginous yeast strains and their tolerance degrees to lignocellulose degradation compounds (Chen, Li et al. 2009; Hu, Zhao et al. 2009; Huang, Zong et al. 2009). However, these strains were unable to efficiently produce lipids in the presence of inhibitors in the hydrolysate, necessitating detoxification treatment prior to fermentation, which increases the cost of the process. Thus, using strains capable of growing in the non-detoxified hydrolysate is necessary for viable microbial lipid production in an industrial context. In addition, previous reports indicate that temperature is an key factor in regulating the fatty acid composition in fungi (Kendrick and Ratledge 1992; Weinstein, Montiel et al. 2000).

Similarly, some oleaginous filamentous fungi can also produce lipids by utilizing glycerol, acetic acid, soluble starch, wheat straw, and wheat bran. Dey et al screening two endophytic oleaginous fungi *Colletotrichum sp.* and *Alternaria sp.* with lipid content 30% and 58% respectively (Dey, Banerjee et al. 2011). Fifteen eukaryotic microorganism were tested for waste glycerol assimilation to produce lipid. Fungi accumulated lipid inside their mycelia (lipid content ranging between 18.1 and 42.6%) (Chatzifragkou, Makri et al. 2011). Such capabilities provide potential to utilize sugars in the pretreated lignocellulosic materials hydrolysate. Moreover, because the fatty acid profile of microbial oils is quite similar to that of conventional vegetable oils, oleaginous filamentous fungi are suggested as a favorable feedstock for a sustainable biodiesel industry (Peng and Chen 2008; Zhao, Hu et al. 2010).

4. Feedstocks for microbial lipid production

4.1 Light and carbon dioxide

Microalgae cells can generally utilize sunlight, carbon dioxide and nutrients from waste water for their cell growth (Brennan and Owende 2010). Lipid accumulation with microalgae cultivation is relatively efficient due to its high production efficiency and less demand of agricultural land. Most microalgal ponds have a solar energy conversion efficiency of 1-4% under normal operating conditions and higher efficiencies can be achieved with closed photo-bioreactor systems. There is a considerable margin for improvement, which is being targeted through accelerated breeding programs and genetic modification. Autotrophic microalgae cells also absorb CO_2 as their carbon source to support their cell growth, which makes the microalgae an attractive option for the biological CO_2 fixation. Atmospheric CO_2 accumulation, derived mainly from fossil fuel combustion, is proved as the leading cause of global warming. Current mitigation methods such as physicochemical adsorption, injection into deep oceans and geological formations are not economically feasible due to the high cost of implementing these methods and possible CO_2 leakage. After CO_2 is fixed via microalgae assimilation, the cell biomass can be utilized to generate methane gas, biochar, or the oil can be extracted to generate biodiesel. Microalgae can fix not only atmospheric CO_2, but also the CO_2 from industrial exhaust gases and from the carbonate salt, which are chemically fixed. Atmospheric CO_2 levels (0.0387%) are not sufficient to support the high microalgae growth rates and productivities needed for commercial biofuel production, adding CO_2 into autotrophic microalgae culture is an effective method to accelerate the microalgae growth rate. Utilization of CO_2, for example, flue gas from electrical plant, by means of microalgae alleviates the impact of CO_2 on the environment (greenhouse effect) and renders algal biomass production less expensive. It was also reported that the CO_2 content was reduced from 44-48% to 2.5-11% when the microalgae cultivation was integrated with anaerobic digestion to remove impurities of biogas produced from anaerobic digestion. Microalgae also assimilated other impurities such as ammonia and hydrogen sulfide; and the gas leaving the algae pond had 88-97% by volume of methane [23]. Although there are conflicting results from different references about the toxicity of ammonia, hydrogen sulfide and other impurities on the growth of microalgae cells, integration of microalgae cultivation together with this industrial processes can be more sustainable and economically feasible than the individual microalgae cultivation process to generate oil for biodiesel production.

4.2 Wastewater

Culturing microalgae with nutrients from wastewater, such as nitrogen and phosphate, can decrease the cost of the raw materials and also provide some environmental benefits. Agricultural effluent and municipal wastewater, even after treated with anaerobic digestion (AD), cannot be disposed directly because of their high nutrient level (Levine, Costanza-Robinson et al. 2011). In contrast, these wastewater can be considered as a cost-effective candidate of raw materials for biodiesel production (Siddiquee and Rohani 2011). Cultivation of microalgae, yeast, or fungi can be integrated with AD system to reduce the remaining COD, phosphate, and ammonia. Microalgae were also studied to absorb metal ion, waste pharmaceutical chemicals and dye into their cell biomass in order to remove the pollutants from wastewater once their cell biomass were stabilized and harvested. A typical example is the recently developed high-rate algae pond (HRAP) in the tertiary wastewater treatment facility (El Hamouri 2009). HRAP functions behind a two-step upflow anaerobic reactor (pre-treatment) and was followed by one maturation pond (MP) for polishing. The HRAP was revealed to have no activity for removing the COD from the wastewater; however, it removed 85% of total N and 63% of total P. Nitrogen removal was discovered due to the assimilation of microalgae for their growth, and denitrification did not play any role in removing the nitrogen in this process. Phosphorus removal in this process was attributed to chemical precipitation and biological assimilation (around 50% each). In removing ammonia, the HRAP is superior to the traditional bacterial nitrification-denitrification process, which requires the assimilation of extra-organic carbon as a carbon source. Phosphorus removal by microalgae is largely thought to be due to its uptake for normal growth, as an essential element required for making cellular constituents such as phospholipids, nucleotides, and nucleic acids. Under certain conditions microalgae can be triggered to uptake more phosphorus than is necessary for survival, in the form of polyphosphate (Powell, Shilton et al. 2011). Phosphorus removal by luxury uptake (amount of P uptake more than growth required) was confirmed to occur in the microalgae growing in the wastewater treatment facility; further research is needed in this prosperous field about the detailed mechanism and applications.

4.3 Lignocellulosic biomass

Several research studies revealed that organic carbon sources, if the algae have the capability to grow on heterotrophic mode, can significantly increase the cell growth rate and dramatically enhance the lipid content of the biomass. For example, *Chlorella protothecoides* can only accumulate 18-25% lipid in the autotrophic culture, while the lipid content can reach to 55.2% dry cell biomass if cultured with addition of organic carbons (Xiong, Li et al. 2008). However, the heterotrophic cultivation of microalgae mostly request pure mono-sugars, which are usually costly and limited. Alternative materials with relative abundance and zero or negative valued organic materials, such as lignocellulosic biomass from agriculture, are the only choice for this route. Agricultural feedstocks contribute a large part of renewable resource for biodiesel production, which are the target of biodiesel resource of cost reduction of biodiesel and non-human food resource discovery. Most of these substrates are locally available and thus are expected to support mainly small production facilities. Lignocellulosic materials are mainly composed of cellulose, hemicellulose, and lignin, which make up approximately 90% of the dry weight of most plant materials (Kumar, Barrett et al. 2009). Cellulose and hemicellulose can be converted to fermentable sugars for microbial lipid production. However, the direct enzymatic hydrolysis of cellulose and hemicellulose to sugars is impeded by the cell wall physico-chemical and structural composition. Thus, biomass pretreatment prior to enzymatic hydrolysis is essential to enhance enzymatic digestibility. Distinctly different from autotrophic microalgae cultures, this process is more similar to cellulosic ethanol production, where hydrolysis and fermentation are needed for conversion.

5. Conversion processes for microbial lipid production

5.1 Autotrophic microalgae cultivation and lipid accumulation

Three cultivation processes were designed to culture microalgae and other oleaginous cells, including open-pond system, photobioreactor, and fermentation. The open-pond system is typically a closed loop with a pump to create microalgae flow in the channel. The channel is 0.2-0.5 meter deep, and the pump keeps the microalgae cell well mixed for continuous growth. This type of open-pond system has been used for several years because it is easy to operate and inexpensive to maintain. The open-pond system also can be upgraded to large-scale microalgae production. On the other hand, a high cell density of microalgae cannot be reached because of limited capability to assimilate the sun light and low carbon dioxide concentration of air. Increasing the CO_2 concentration by using flue gas instead of air can increase the microagale cell concentrations, but the final cell density is still limited to the mutual shading effects where light cannot penetrate through dense microalgae cell broth. Another problem is biological contamination during the long period of cultivation. The bacteria contamination or other non-oleaginous microalgae invasion can occur in stressed cultural conditions, where lipid accumulation usually is stimulated, such as nitrogen depletion or other nutrient imbalance. There is now extensive evidence that open-pond systems can operate for more than six months without significant contamination using a wide range of microalgae. Prolific strains of *Chlorella*, for example, are often dominant because they outgrow their competitors (and indeed can often be contaminants themselves in *Arthrospira* cultures or other microalgal strains). Extreme halophiles, such as *Dunaliella salina*, are also dominant in their optimal environments because they do not encounter much competition at high salinities. However, in the context of the wider microalgal industry, contamination issues are still of significant interest.

To enhance the productivity of microalgae, closed photobioreactor systems (tubular flat plate, Orcolumn are designed to increase the surface of microalgae broth exposed to sunlight. Closed photobioreactors are more costly than open-pond systems, but they have potential for higher productivity of cell biomass with less chance of contamination. The flat plate photobioreactors can receive greater sunlight for microalgae growth although there is potential for cell mixing. The microalgae cell density could reach up to 80g/L dry cell weight, significantly higher than the cell density of a pond system, which ranges within several g/L (Hu, 1998). Another design for the photoreactor is the tubular photoreactor, made with a diameter less than 0.1 m to maximize the sunlight harvest by microalgae. The tubular reactor can also expose the microalgae cells to sunlight from all the directions (Miron, Gomez et al. 1999; Ugwu, Ogbonna et al. 2002). There are a few reports about scale-up test of tubular photoreactor, such as the one in Hawaii with a size of 25M^3 (Olaizola 2000), and 700 M^3 in Germany (Pulz 2001). However, the tubular photobioreactors cannot scale-up indefinitely because of oxygen accumulation, carbon dioxide limitation, and pH changes (Eriksen 2008). The third type of photoreactor is the column photoreactor, the most controllable type among three because it most closely resembles the traditional bioreactor. The column containing microalgae is vertical, and the air is bubbled from the bottom. Sunlight is provided horizontally (Eriksen 2008).

In addition to individual open-pond system and closed photobioreactor, the hybrid system of microalgae cultivation is currently under intense investigation because it combines the open-pond system and closed photobioreactor to increase the cell productivity and reduce the cost. The first stage is autotrophy to avoid biological contamination, and the second stage is heterotrophy, which provides the stress condition for lipid accumulation.

5.2 Oleaginous microbial fermentation and lipid accumulation
Besides the pond system and photobioreactor, heterotrophic cell cultivation, including heterotrophic microalgae, oleaginous yeast and fungi, is usually limited to industrial fermentation tanks. These cells can be cultured in a dark fermentor with optional sunlight, and can usually reach a very high cell density (up to 200g/L). Due to its excellent controllability of all operation parameters, most of the industrial microalgae cultivations for nutraceutical production (e.g., polyunsaturated fatty acid) have been switched to the heterotrophic fermentations where sugar is provided to produce high-valued products, such as Docosahexaenoic acid (DHA).

Factor	Raceway	Photobioreactor	Fermenter
Cell density in culture	Low	Medium	High
Limiting factor for growth	Light	Light	Oxygen
Culture volume necessary to harvest a unit weight of cells	High	Medium	Low
Surface area-to-volume ratio	High	Very high	Not applicable
Control over parameters	Low	Medium	Very high
Commercial availability	Readily available	Usually custom built	Readily available
Construction costs per unit volume produced	Medium	High	Low
Operating costs	Medium	High	Low
Technology base	Readily available	Under development	Readily available
Risk of contamination	High	Medium	Low
Evaporative water losses	High	High[33]	Low
Weather dependence	High	Medium	Low
Maintenance	Easy to maintain	Difficult to maintain	Requires specialized maintenance
Susceptibility to overheating	Low	High	N/A
Susceptibility to excessive O_2 levels	Low	High[34]	N/A
Ease of cleaning	Very easy	Difficult	Difficult (must be sterilized)
Ease of Scale-up	High	Variable[35]	High
Land requirement	High	Variable	Low
Applicability to different species	Low	High	Low

Table 3. Comparison of different algae cultivation systems (Alabi. 2009)

Table 3 compares three cultivation methods and shows that the process cost of fermentation can be high due to its requirement of raw materials and oxygen, and sterilization of culture media during the cell growth. It is readily available both in the lab and in the industry, but is only suitable to produce high-valued products, of which biofuel products are not. The key barriers to apply this technology to biofuel production is the cost and availability of raw materials. Considering the competition with human diet, sugars cannot serve as the raw material for biofuel production; and alternative materials such as lignocellulosic materials should be used for the heterotrophic oil production. If the oleaginous cells are capable of generating the hydrolytic enzymes for lignocelluloses degradation, it will be the big plus for biodiesel production via oleaginous fermentation the overall system. Otherwise, external hydrolytic enzymes have to be used to release the monosugar, followed by lipid accumulation via olgeaginous microorganisms. Separated hydrolysis and fermentation (SHF) is a common working model to have these two steps separated. Two bioreactors will be necessary because the hydrolytic degradation of lignocellulose is preferred at 50°C, while the oleaginous microorganisms grow at much lower temperature (28°C to 30°C for most of the fungus). Simultaneous saccharification and fermentation is another working model currently under intense investigation, in which two steps are integrated into one.

Different fermentation processes are applied to obtain high productivity of lipids and high conversion ratio of substrate for the fermentation, such as batch cultivation, fed-batch cultivation, and continous cultivation. Fed-batch cultivatioin is a modified batch model that can reach high cell density and it has many applications in the fermentative lipid accumulation process. For example, *Rhorosporidum toruloides* reach much higher cell density with 48% lipid compared to its batch cultivation (Li, Zhao et al. 2007). The high productivity of fed-batch cultivation was conformed by *Phodotorula glutinis* (Xue, Miao et al. 2008), *C. curvatus* (Meesters, Huijberts et al. 1996), and *L. starkeyi* (Yamauchi, Mori et al. 1983). Continuous cultivation has advantages of easy maintenance and time-saving, although it is difficult to control the contamination. It has limited applicatioins in the fermentative lipid accumulation.

Besides the commonly used submerged cultivations, solid state fermentation, as a compact process for lipid production, showed many advantages, such as low requirements to the raw materails; low capital cost; low energy expenditure; less expensive downstream processing; less water usage and low water output; potential higher volumetric productivity; less fermentation space; easy operation and maintenance. The research for *Aspergillus oryzae* growing on rice bran and wheat bran through solid state fermentation resulted to the lipid content of cell biomass at about 10-11% (Da Silveira, Oliveira et al. 2010). The lipid yield reached 62.87 mg/gds in solid state fermentation on the 6th day after Plackett-Burman design (PBD) by *A. oryzae* A-4 (Lin, Cheng et al. 2010) . Currently, the solid state fermentation research is still in its infancy and many barriers are hindering this process from commercilization. The lipid yield is relatively low compared to submerge cultivation. Modern biotechnological approaches, such as heterogenous expression of hydrolytic enzymes and UV radiation, are available to enhance the hydrolytic enzymes production (Li, Yang et al. 2010; Awan, Tabbasam et al. 2011). Semi-solid state fermentation is used to avoid high sugar concentration on the surface of lignocellulose. An oleaginous fungus *M. isabellina* was cultured at semi-solid state fermentation with the results of 11g oil per 100g sweet sorghum (Economou, Makri et al. 2010) .

6. Cell harvest and lipid extraction

6.1 Cell harvest methods

The algae cell harvest from pond water and the subsequent water reuse have been one of the major obstacles for the algae-to-fuel approach. Microalgae cell harvest is technically challenging, especially considering the low cell densities (typically in the range of 0.3-5 g/L) of autotrophic microalgae due to limited light penetration, the small size of the oleaginous algal cells (typically in the range of 2-40 um), and their similar density to water (Li, Horsman et al. 2008). Oleaginous microalgae cells are usually suspended in the water and are hard to settle by natural gravity force due to their negative charges. The recovery of microalgae biomass generally requires one or more solid–liquid separation steps, and usually accounts for 20–30% of the total costs of production, according to one source (Uduman, Qi et al. 2010).

How to harvest microalgae cells is dependent on the characteristics of the microalgae, such as size and density(Olaizola 2003). All of the available harvest approaches, which include flocculation, flotation, centrifugal sedimentation, and filtration, have limitations for effective, cost-efficient production of biofuel (Shelef, Sukenik et al. 1984). For instance, flotation methods, based on the trapping of algae cells using dispersed micro-air bubbles, is very limited in its technical and economic viability. Most conventional and economical separation methods such as filtration and gravitational sedimentation are widely applied in wastewater treatment facilities to harvest relatively large (>70 μm) microalgae such as *Coelastrum* and *Spirulina*. However, they cannot be used to harvest algae species approaching bacterial dimensions (<30 μm) like *Scenedesmus*, *Dunaliella*, and *Chlorella* (Brennan and Owende 2010), to which most oleaginous microalgae species belong. Centrifugation is a method widely used to recover microalgae biomass, especially small-sized algae cells; however, its application is restricted to algae cultures for high-value metabolites due to intensive energy needs and high equipment maintenance requirements. While flocculation is used to harvest small-sized microalgae cells, it is a preparatory step to aggregate the microalgae cells and increase the particle size so that other harvesting methods such as filtration, centrifugation, or gravity sedimentation can be applied (Molina Grima, Belarbi et al. 2003). Several flocculants have been developed to facilitate the aggregation of microalgae cells, including multivalent metal salts like ferric chloride (FeCl$_3$), aluminium sulphate (Al$_2$(SO$_4$)$_3$), and ferric sulphate (Fe$_2$(SO$_4$)$_3$), and organic polymers such as Chitosan (Li, Horsman et al. 2008). Chemical flocculation can be reliably used to remove small algae cells from pond water by forming large-sized (1–5 mm) flocs (Sharma, Dhuldhoya et al. 2006). However, the chemical reactions are highly sensitive to pH and the high doses of flocculants required produce large amounts of sludge and may leave a residue in the treated effluent. In summary, most technologies including chemical and mechanical methods greatly increase operational costs for algal production and are only economically feasible for production of high-value products (Park, Craggs et al. 2011).

Besides traditional methods mentioned above, there are several new technology developments in this field. DOE-ARPA-E recently funded a research project for Algae Venture Systems (AVS) to develop a Harvesting, Dewatering, and Drying (AVS-HDD) technology by using the principles of liquid adhesion and capillary action to extract water from dilute microalgae solutions. Attached algal culture systems have been developed for growing microalgae on the surface of polystyrene foam (Wilkie and Mulbry 2002) (Johnson and Wen 2010) to simplify the cell harvest. New bioflocculants, which are more environmentally friendly, are also proposed to address the cost and environmental concerns for current flocculation method (Uduman, Qi et al. 2010). All these methods are innovative and will decrease the harvest cost to some extent if developed successfully, but heavy investments on equipment and chemical supplies are still needed.

Dr. Bo Hu's research group at University of Minnesota developed an innovative approach to enhance natural algae aggregation and to encourage simple gravity settling or filtration by co-culturing filamentous fungal cells at the end of the microalgae cultures. Instead of suspended culture, this approach uses pelletized or granulized culture where cells form pellets in culture medium. In submerged cultures, many filamentous microorganisms tend to aggregate and grow as pellets/granules. They are spherical or ellipsoidal masses of hyphae with variable internal structure, ranging from loosely packed hyphae, forming "fluffy" pellets, to tightly packed, compact, dense granules (Hu and Chen 2007; Hu and Chen 2008; Hu, Zhou et al. 2009; Chunjie Xia 2011). Besides merits from the cell immobilization, there are several other advantages, especially for the micro-oil production: a). easy to harvest cells, and b). easy to re-use pond water (Johnson and Wen 2010; Xia 2011). As the first research group to introduce pelletized liquid fermentation (PLF) into biofuel production, this research group at University of Minnesota found key operational conditions that induce the fungal pelletization. They discovered that changing conditions during cell cultivation can force fungal cells to aggregate and form pellets. This method avoids traditional approaches that use $CaCO_3$ powder to induce the fungal pelletization (Liao, Liu et al. 2007; Liu, Liao et al. 2008), which are costly and cause solid waste disposal issues. Self aggregated pelletization/granulation dramatically improves mass transfer and cell cultivation performance and facilitates cell harvest and separation. A simple filtration can be used to separate the cell biomass from the fermentation broth. This approach brings tremendous advantages to decrease the harvest cost of biofuel production, especially when the raw materials only contain very diluted sugar, (which are the cases for many agricultural waste). This would appear to be the most promising option to achieve both a high-quality treated effluent in terms of total suspended solids and economically recovering algal biomass for biofuel use (Uduman, Qi et al. 2010). It will also be more environmentally sound than current procedures which may need chemical addition.

6.2 Lipid extraction methods
Oil extraction also contributes a large part of the cost in the process to generate microbial biodiesel. Several oil extraction technologies are currently available to process the microbial biomass in order to meet the requirement of being low cost, easy and safe to operate, and environmentally friendly.

6.2.1 Mechanical methods
Mechanical methods include pressing, bead milling, and homogenization. Pressing is a technology to harvest lipids out of cells by high pressure. Bead milling works in a container to destruct the cell wall by high speed small beads. Homogenization provides a sudden pressure change when cells go through an orifice. The mechanical technologies are often used in combination with solvent methods to separate the lipid from the cell biomass. The mechanical methods are energy intensive and better operated at the high cell density condition; in addition, pretreatments are necessary to obtain high recovery ratio (Greenwell, Laurens et al. 2010).

6.2.2 Solvent extraction methods
Solvent extraction is a commonly used method for soybean processing, and it is also used to extract lipids from microbial cells. Organic solvents should be insoluble in water, be easy to obtain, have a low boiling point, and be reusable. Current industrial solvents for microlipids accumulation include hexane, chloroform, acetone, benzene, and cyclohexane, can dissolve lipid without residual cell. The extraction process is significantly affected by operation condition, such as temperature and pressure. Accelerated solvent extraction (ASE) is named when the operation temperature is higher than that of solvent boiling point, which can be used for oil extraction from dry biomass (Cooney, Young et al. 2009). Mixture chloroform and methanol (Bligh and Hyer method) is the most common organic solvent to extract oil from biomass. This organic mixture can extract oil not only from dry biomass but also from wet biomass. However, the efficiency is different at certain condition (Zhu, Zhou et al. 2002). The efficiency of oil extraction was not working well at wet *Mortierella alpina* biomass. The process generated large amounts of wastewater and solvent often contaminated the final products. Simultaneous extraction and tranesterification is more efficient (15-20%) than the separate process (Belarbi, Molina et al. 2000); however, the important point of the simultaneous process is to balance the reaction time for the best components of product (Lewis, Nichols et al. 2000).

6.2.3 Supercritical fluid extraction

Supercritical fluid extraction takes advantage that some chemicals behave as both a liquid and a gas, and have increasing solvating power when they are raised above their critical temperature and pressure points. Carbon dioxide is the most commonly used supercritical fluid, sometimes modified by co-solvents such as ethanol or methanol. Critical temperature and critical pressure of carbon dioxide is at 31°C and 74 bar, respectively (Cooney, Young et al. 2009). Supercritical fluids produce highly purified extracts without using toxic solvent; and the process is fast and safe for thermally sensitive products. Supercritical CO_2 extraction efficiency is affected by four main factors: pressure, temperature, CO_2 flow rate, and extraction time. Ethanol (10 -15%), co-solvent, lead to similar results of Bligh and Hyer method at extracting oil from *Arthrospira maxima* and *Spirulina platensis* (Mendes, Reis et al. 2006; Sajilata, Singhal et al. 2008). The limitation of supercritical fluid extraction is high capital cost and high cost for maintainence.

6.2.4 Other methods

Besides the methods mentioned above, numerous technologies are being tested at different labs to harvest lipids from cells. Genetic engineering has been applied to improve the porosity of the cell membranes in order to increase the release of lipids directly from the cells (Greenwell, Laurens et al. 2010). Enzymes treatment and pulsed electric field technology are other effective methods to break the cell wall and membrane and enhance the mass transfer across the cell membrane for oil extraction (Shah, Sharma et al. 2004; Guderjan, Elez-Martinez et al. 2007). Microwave technology is a portential pretreatment method, which heats the cell components in order to increase the release of oil. Oil yield increased from 4.8% to 17.7% from microalgae *Crypthecodinium chnii* when microwave was applied (Cravotto, Boffa et al. 2008). Microwave technology is featured for its time-savings, but its disadvantages include the oxidative damage to products and its intense energy need. Sonnication is a timely and efficient method, free of toxic materials. Cavitation occurs when high voltage is applied into cell lipids. Vapour bubbles form with negative pressure and cause a violent collapse when compressed under positive pressure while growing; then the cell contents are released (Wei, Gao et al. 2008). The sonnication is, however, difficult to scale-up.

7. Techno-economic analysis and life cycle assessment

A complete techno-econoic analysis for the microbial biodiesel production is difficult, especially considering that most of the technologies are still in the early research stage. Initial investment into microalgal biofuels has mostly failed and several early start-up companies have closed. Different versions of economic analysis for microalgae biofuel production have been published recently, and Table 4 lists an analysis conducted by Seed Science Ltd, sporsored by the British Columbia Innovation Council in Canada.

Table 4 shows that although photobioreactor has a higher cell concentration and utilizes CO_2, its cost to produce lipid is the highest of all methods. Heterotrophic fermentation, however, appears to be the most economically feasible route to produce microbial biodiesel. Techno-economic analyses may vary from different research group, but their conclusions are similar. The biomass and oil generated from heterotrophic fermentation are more close to current fossil fuel cost. Heterotrophic fermentation relies less on local climate conditions and can be carried out in close fermentors, which may facilitate their commercialization.

More effective, cost-efficient, and environmentally sound fermentation means to produce lipids are urgently needed, as well as adaptation of the fermentation cells to utilize lignocellulosic biomass. It is also widely indicated that currently microalgal biofuel systems are dependent on the production of coproducts (e.g., biochar, pigments, and nutriceuticals) for profitability. Considering the large scale of biofuel production, the market of the valuable byproducts will be the primary concern.

	Raceway		Photobioreactor		Fermentor	
Initial invertment ($/L)	52		111		2	
Production cost						
Labor cost	$4.03	26.69%	$2.96	11.90%	$0.29	10.88%
Other production cost	$3.71	24.59%	$6.37	25.59%	$2.07	78.45%
Capital cost	$7.35	48.71%	$15.56	62.50%	$0.28	10.66%
Total cost	$15.09		$24.89		$2.63	
Credit from sale of algae cake*	$0.65		0.29		$0.05	
Net total cost	$14.44		$24.60		$2.58	
Lipid content	15%		25%		50%	
Cost per kg of algae	$2.66		$7.32		$1.54	

*Assumes that the algae cake is sold to an ethanol producer for its carbohydrate content

Table 4. Cost comparison among different microalagae cultivation methods (Alabi. 2009)

8. Conclusions

Although microalgal biofuel systems theoretically have the potential to address both the food versus fuel challenges, to date no microbial biofuel system has achieved economic viability. Microbial lipid productivity must increase tremendously and the overall cost must significantly decrease before this approach can be commercially available.

9. References

Akhtar, P., Gray, J.I.; et al. (1998). "Synthesis of lipids by certain yeast strains grown on whey permeate." *Journal of Food Lipids* 5(4): 283-297.

Alabi., A. O. (2009). "Microalgae technologies and processes for biofuels/bioenergy production in British Columbia."

Alvarez, R. M., B. Rodriguez, et al. (1992). "Lipid accumulation in Rhodotorula glutinis on sugar cane molasses in single-stage continuous culture." *World Journal of Microbiology & Biotechnology* 8(2): 214-215.

Angerbauer, C., M. Siebenhofer, et al. (2008). "Conversion of sewage sludge into lipids by Lipomyces starkeyi for biodiesel production." *Bioresource Technology* 99(8): 3051-3056.

Awan, M. S., N. Tabbasam, et al. (2011). "Gamma radiation induced mutagenesis in Aspergillus niger to enhance its microbial fermentation activity for industrial enzyme production." *Molecular Biology Reports* 38(2): 1367-1374.

Barclay, W., K. Meager, et al. (1994). "Heterotrophic production of long chain omega-3 fatty acids utilizing algae and algae-like microorganisms." *J. Appl. Phycol.* 6(2): 123-129.

Belarbi, E. H., E. Molina, et al. (2000). "A process for high yield and scaleable recovery of high purity eicosapentaenoic acid esters from microalgae and fish oil." *Enzyme and Microbial Technology* 26(7): 516-529.

Brennan, L. and P. Owende (2010). "Biofuels from microalgae--A review of technologies for production, processing, and extractions of biofuels and co-products." *Renewable and Sustainable Energy Reviews* 14(2): 557-577.

Chatzifragkou, A., A. Makri, et al. (2011). "Biotechnological conversions of biodiesel derived waste glycerol by yeast and fungal species." *Energy* 36(2): 1097-1108.

Chen, G.-Q. and F. Chen (2006). "Growing Phototrophic Cells without Light." *Biotechnology Letters* 28(9): 607-616.

Chen, X., Z. Li, et al. (2009). "Screening of oleaginous yeast strains tolerant to lignocellulose degradation compounds." *Applied Biochemistry and Biotechnology* 159(3): 591-604.

Chisti, Y. (2007). "Biodiesel from microalgae." *Biotechnol Adv* 25(3): 294-306.

Cooney, M., G. Young, et al. (2009). "Extraction of Bio-oils from Microalgae." *Separation and Purification Reviews* 38(4): 291-325.

Cravotto, G., L. Boffa, et al. (2008). "Improved extraction of vegetable oils under high-intensity ultrasound and/or microwaves." *Ultrasonics Sonochemistry* 15(5): 898-902.

Da Silveira, C. M., M. D. Oliveira, et al. (2010). "Lipid content and fatty acid profile of defatted rice bran and wheat bran submitted to solid state fermentation by *Aspergillus oryzae*." *Boletim Do Centro De Pesquisa De Processamento De Alimentos* 28(1): 133-140.

Dey, P., J. Banerjee, et al. (2011). "Comparative lipid profiling of two endophytic fungal isolates - Colletotrichum sp. and Alternaria sp. having potential utilities as biodiesel feedstock." *Bioresource technology* 102(10): 5815-5823.

Economou, C. N., A. Makri, et al. (2010). "Semi-solid state fermentation of sweet sorghum for the biotechnological production of single cell oil." *Bioresource technology* 101(4): 1385-1388.

El Hamouri, B. (2009). "Rethinking natural, extensive systems for tertiary treatment purposes: The high-rate algae pond as an example." *Desalination and Water Treatment* 4(1-3): 128-134.

Eriksen, N. T. (2008). "The technology of microalgal culturing." *Biotechnology Letters* 30(9): 1525-1536.

Greenwell, H. C., L. M. L. Laurens, et al. (2010). "Placing microalgae on the biofuels priority list: a review of the technological challenges." *Journal of the Royal Society Interface* 7(46): 703-726.

Guderjan, M., P. Elez-Martinez, et al. (2007). "Application of pulsed electric fields at oil yield and content of functional food ingredients at the production of rapeseed oil." *Innovative Food Science & Emerging Technologies* 8(1): 55-62.

Heredia-Arroyo, T., W. Wei, et al. (2011). "Mixotrophic cultivation of Chlorella vulgaris and its potential application for the oil accumulation from non-sugar materials." *Biomass & Bioenergy* 35(5): 2245-2253.

Hu, B. and S. L. Chen (2007). "Pretreatment of methanogenic granules for immobilized hydrogen fermentation." *International Journal of Hydrogen Energy* 32(15): 3266-3273.

Hu, B. and S. L. Chen (2008). "Biological hydrogen production using chloroform-treated methanogenic granules." *Applied Biochemistry and Biotechnology* 148(1-3): 83-95.

Hu, B., X. Zhou, et al. (2009). "Changes in Microbial Community Composition Following Treatment of Methanogenic Granules with Chloroform." *Environmental Progress & Sustainable Energy* 28(1): 60-71.

Hu, C., X. Zhao, et al. (2009). "Effects of biomass hydrolysis by-products on oleaginous yeast Rhodosporidium toruloides." *Bioresource Technology* 100(20): 4843-4847.

Hu, Q., Kurano, N., Iwasaki, I., et al. (1998). Ultrahigh cell density culture of a marine green alga, Chlorococcum littorale in a flat plate photobioreactor. *Applied Microbiology Biotechnology* 49(6):655-662.

Huang, C., M.-h. Zong, et al. (2009). "Microbial oil production from rice straw hydrolysate by Trichosporon fermentans." *Bioresource Technology* 100(19): 4535-4538.

Johnson, M. B. and Z. Y. Wen (2010). "Development of an attached microalgal growth system for biofuel production." *Appl. Microbiol. Biotechnol.* 85(3): 525-534.

Kendrick, A. and C. Ratledge (1992). "Lipid formation in the oleaginous mould Entomophthora exitalis grown in continuous culture: effects of growth rate, temperature and dissolved oxygen tension on polyunsaturated fatty acids." *Applied Microbiology and Biotechnology* 37(1): 18-22.

Kumar, P., D. M. Barrett, et al. (2009). "Methods for Pretreatment of Lignocellulosic Biomass for Efficient Hydrolysis and Biofuel Production." *Industrial & Engineering Chemistry Research* 48(8): 3713-3729.

Levine, R. B., M. S. Costanza-Robinson, et al. (2011). "Neochloris oleoabundans grown on anaerobically digested dairy manure for concomitant nutrient removal and biodiesel feedstock production." *Biomass & Bioenergy* 35(1): 40-49.

Lewis, T., P. D. Nichols, et al. (2000). "Evaluation of extraction methods for recovery of fatty acids from lipid-producing microheterotrophs." *Journal of Microbiological Methods* 43(2): 107-116.

Li, X. H., H. J. Yang, et al. (2010). "Enhanced cellulase production of the Trichoderma viride mutated by microwave and ultraviolet." *Microbiological Research* 165(3): 190-198.

Li, Y. H., Z. B. Zhao, et al. (2007). "High-density cultivation of oleaginous yeast Rhodosporidium toruloides Y4 in fed-batch culture." *Enzyme and Microbial Technology* 41(3): 312-317.

Li, Y., M. Horsman, et al. (2008). "Biofuels from microalgae." *Biotechnol Prog* 24(4): 815-820.

Liao, W., Y. Liu, et al. (2007). "Studying pellet formation of a filamentous fungus Rhizopus oryzae to enhance organic acid production." *Applied Biochemistry and Biotechnology* 137: 689-701.

Lin, H., W. Cheng, et al. (2010). "Direct microbial conversion of wheat straw into lipid by a cellulolytic fungus of Aspergillus oryzae A-4 in solid-state fermentation." *Bioresource technology* 101(19): 7556-7562.

Liu, J., J. Huang, et al. (2010). "Differential lipid and fatty acid profiles of photoautotrophic and heterotrophic Chlorella zofingiensis: Assessment of algal oils for biodiesel production." *Bioresour. Technol.* 102(1): 106-110.

Liu, Y., W. Liao, et al. (2008). "Co-production of lactic acid and chitin using a pelletized filamentous fungus Rhizopus oryzae cultured on cull potatoes and glucose." *Journal of Applied Microbiology* 105(5): 1521-1528.

Meesters, P. A. E. P., G. N. M. Huijberts, et al. (1996). "High cell density cultivation of the lipid accumulation yeast Cryptococcus curvatus using glycerol as a carbon source." *Appl. Microbiol. Biotechnol.* 45(5): 575-579.

Meesters, P. A. E. P., G. N. M. Huijberts, et al. (1996). "High cell density cultivation of the lipid accumulation yeast Cryptococcus curvatus using glycerol as a carbon source." *Applied Microbiology and Biotechnology* 45(5): 575-579.

Mendes, R. L., A. D. Reis, et al. (2006). "Supercritical CO2 extraction of gamma-linolenic acid and other lipids from Arthrospira (Spirulina)maxima: Comparison with organic solvent extraction." *Food Chemistry* 99(1): 57-63.

Miao, X. and Q. Wu (2006). "Biodiesel production from heterotrophic microalgal oil." *Bioresour. Technol.* 97(6): 841-846.

Miron, A. S., A. C. Gomez, et al. (1999). "Comparative evaluation of compact photobioreactors for large-scale monoculture of microalgae." *Journal of Biotechnology* 70(1-3): 249-270.

Molina Grima, E., E. H. Belarbi, et al. (2003). "Recovery of microalgal biomass and metabolites: process options and economics." *Biotechnology Advances* 20(7-8): 491-515.

Olaizola, M. (2000). "Commercial production of astaxanthin from Haematococcus pluvialis using 25,000-liter outdoor photobioreactors." *Journal of Applied Phycology* 12(3-5): 499-506.

Olaizola, M. (2003). "Commercial development of microalgal biotechnology: from the test tube to the marketplace." *Biomolecular Engineering* 20(4-6): 459-466.

Papanikolaou, S. and G. Aggelis (2002). "Lipid production by Yarrowia lipolytica growing on industrial glycerol in a single-stage continuous culture." *Bioresource Technology* 82(1): 43-49.

Park, J. B. K., R. J. Craggs, et al. (2011). "Wastewater treatment high rate algal ponds for biofuel production." *Bioresource technology* 102(1): 35-42.

Peng, X. and H. Chen (2008). "Single cell oil production in solid-state fermentation by Microsphaeropsis sp. from steam-exploded wheat straw mixed with wheat bran." *Bioresource Technology* 99(9): 3885-3889.

Powell, N., A. Shilton, et al. (2011). "Luxury uptake of phosphorus by microalgae in full-scale waste stabilisation ponds." *Water Science and Technology* 63(4): 704-709.

Pulz, O. (2001). "Photobioreactors: production systems for phototrophic microorganisms." *Appl. Microbiol. Biotechnol.* 57(3): 287-293.

Sajilata, M. G., R. S. Singhal, et al. (2008). "Supercritical CO2 extraction of gamma-linolenic acid (GLA) from Spirulina platensis ARM 740 using response surface methodology." *Journal of Food Engineering* 84(2): 321-326.

Samson, R. and A. Leduy (1985). "Multistage continuous cultivation of blue-green-alga spirulina-maxima in the flat tank photobioreactors with recycle." *Canadian Journal of Chemical Engineering* 63(1): 105-112.

Shah, S., A. Sharma, et al. (2004). "Extraction of oil from Jatropha curcas L. seed kernels by enzyme assisted three phase partitioning." *Industrial Crops and Products* 20(3): 275-279.

Sharma, B. R., N. C. Dhuldhoya, et al. (2006). "Flocculants—an Ecofriendly Approach." *Journal of Polymers and the Environment* 14(2): 195-202.

Shelef, G., A. Sukenik, et al. (1984). Microalgae harvesting and processing: a literature review: Medium: ED; Size: Pages: 70.

Siddiquee, M. N. and S. Rohani (2011). "Lipid extraction and biodiesel production from municipal sewage sludges: A review." *Renewable & Sustainable Energy Reviews* 15(2): 1067-1072.

Uduman, N., Y. Qi, et al. (2010). "Dewatering of microalgal cultures: A major bottleneck to algae-based fuels." *Journal of Renewable and Sustainable Energy* 2(1).

Ugwu, C. U., J. C. Ogbonna, et al. (2002). "Improvement of mass transfer characteristics and productivities of inclined tubular photobioreactors by installation of internal static mixers." *Appl. Microbiol. Biotechnol.* 58(5): 600-607.

Wei, F., G. Z. Gao, et al. (2008). "Quantitative determination of oil content in small quantity of oilseed rape by ultrasound-assisted extraction combined with gas chromatography." *Ultrasonics Sonochemistry* 15(6): 938-942.

Weinstein, R. N., P. O. Montiel, et al. (2000). "Influence of growth temperature on lipid and soluble carbohydrate synthesis by fungi isolated from fellfield soil in the maritime Antarctic." *Mycologia* 92(2): 222-229.

Wilkie, A. C. and W. W. Mulbry (2002). "Recovery of dairy manure nutrients by benthic freshwater algae." *Bioresource Technology* 84(1): 81-91.

Xia, C. J., J. G. Zhang, et al. (2011). "A new cultivation method for microbial oil production: cell pelletization and lipid accumulation by Mucor circinelloides." *Biotechnology for Biofuels* 4.

Xiong, W., X. F. Li, et al. (2008). "High-density fermentation of microalga Chlorella protothecoides in bioreactor for microbio-diesel production." *Applied Microbiology and Biotechnology* 78(1): 29-36.

Xu, H., X. Miao, et al. (2006). "High quality biodiesel production from a microalga Chlorella protothecoides by heterotrophic growth in fermenters." *J. Biotechnol.* 126(4): 499-507.

Xue, F. Y., J. X. Miao, et al. (2008). "Studies on lipid production by Rhodotorula glutinis fermentation using monosodium glutamate wastewater as culture medium." *Bioresource technology* 99(13): 5923-5927.

Yamauchi, H., H. Mori, et al. (1983). "Mass-production of lipids by lipomyces-starkeyi in microcomputer-aided fed-batch culture." *Journal of Fermentation Technology* 61(3): 275-280.

Ykema, A., E. C. Verbree, et al. (1988). "Optimization of lipid production in the oleaginous yeast Apiotrichum curvatum in whey permeate." *Applied Microbiology and Biotechnology* 29(2-3): 211-218.

Zhao, X., C. Hu, et al. (2010). "Lipid production by Rhodosporidium toruloides Y4 using different substrate feeding strategies." *Journal of Industrial Microbiology and Biotechnology*: 1-6.

Zhu, M., P. P. Zhou, et al. (2002). "Extraction of lipids from Mortierella alpina and enrichment of arachidonic acid from the fungal lipids." *Bioresource technology* 84(1): 93-95.

5

Algal Biomass and Biodiesel Production

Emad A. Shalaby

Biochemistry Dept., Facult. Of Agriculture, Cairo University
Egypt

1. Introduction

Biodiesel has become more attractive recently because of its environmental benefits and the fact that it is made from renewable resources. The cost of biodiesel, however, is the main hurdle to commercialization of the product. The used cooking oil and algae are used as raw material, adaption of continuous transesterification process and recovery of high quality glycerol from biodiesel by-product (glycerol) are primary options to be considered to lower the cost of biodiesel. There are four primary ways to make biodiesel, direct use and blending, microemulsions, thermal cracking (pyrolysis) and transesterification. The most commonly used method is transesterification of vegetable oils and animal fats. The transesterification reaction is affected by molar ratio of glycerides to alcohol, catalysts, reaction temperature, reaction time and free fatty acids and water content of oils or fats. In the present chapter we will focus on how algae have high potentials in biodiesel production compared with other sources.

2. Algae as biological material

Microalgae are prokaryotic or eukaryotic photosynthetic microorganisms that can grow rapidly and live in harsh conditions due to their unicellular or simple multicellular structure. Examples of prokaryotic microorganisms are Cyanobacteria (Cyanophyceae) and eukaryotic microalgae are for example green algae (Chlorophyta) and diatoms (Bacillariophyta) [Richmond, 2004]. A more in depth description of microalgae is presented by Richmond [Richmond, 2004]. Microalgae are present in all existing earth ecosystems, not just aquatic but also terrestrial, representing a big variety of species living in a wide range of environmental conditions. It is estimated that more than 50,000 species exist, but only a limited number, of around 30,000, have been studied and analyzed [Richmond, 2004]. Algae are aquatic plants that lack the leaves, stem, roots, vascular systems, and sexual organs of the higher plants. They range in size from microscopic phytoplankton to gain kelp 200 feet long. They live in temperatures ranging from hot spring to arctic snows, and they come in various colors mostly green, brown and red. There are about 25,000 species of algae compared to 250,000 species of land plants. Algae make up in quantity what they lack in diversity for the biomass of algae is immensely greater than that of terrestrial plants (Lowenstein, 1986). Phytoplankton comprises organisms such as diatome, dinoflagellates and macrophytes include: green, red and brown algae. As photosynthetic organisms, these groups play a key role in productivity of ocean and constitute the basis of marine food chain. On the other hand, the use of macroalgae as a potential source of high value chemicals and in therapeutic purpose has a long history.

Recently, macroalgae have been used as a noval food with potential nutritional benefits and in industry and medicine for various purposes.

Furthermore, macroalgae have shown to provide a rich source of natural bioactive compounds with antiviral, antifungal, antibacterial, antioxidant, anti-inflammatory, hypercholesterolemia, and hypolipidemic and antineoplasteic properties. Thus, there is a growing interest in the area of research on the positive effect of macroalgae on human health and other benefits. In Egypt, the macroalgae self grown on the craggy surface near to the seashore of the Mediterranean and Red Seas. Macroalgae have not used as healthy food, while in Japan and China the macroalgae are tradionally used in folk medicine and as a healthy food in addition to, biofuel production (Lee-Saung *et al.*, 2003). The present study was conducted to evaluate the potentialities of micro and macroalgae species for biodiesel production and study the effect of biotic and a biotic stress on biodiesel percentage and the difference between biodiesel production from vegetable sources and algae.

Algae were promising organisms for providing both novel biologically active substances and essential compounds for human nutrition (Mayer and Hamann, 2004). Therefore, an increasing supply for algal extracts, fractions or pure compounds for the economical sector was needed (Dos Santos *et al.*, 2005). In this regard, both secondary and primary metabolisms were studied as a prelude to future rational economic exploitation as show in Fig. 1.

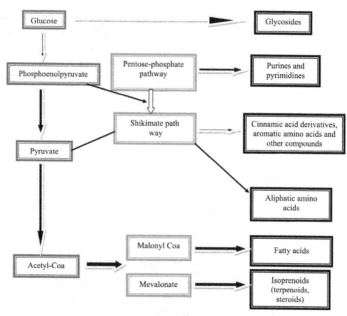

Fig. 1. Secondary and primary metabolites produced from algal cell

3. Diesel production problems

The transportation and energy sectors are the major anthropogenic sources, responsible in European Union (EU) for more than 20% and 60% of greenhouse gas (GHG) emissions, respectively [European Environmental Agency, 2004]. Agriculture is the third largest anthropogenic source, representing about 9% of GHG emissions, where the most important gases are nitrous oxide (N_2O) and methane (CH_4) [European Environmental Agency, 2007]. It is expected that with the development of new growing economies, such as India and China, the global consumption of energy will raise and lead to more environmental damage [International Energy Agency, 2007].

GHG contributes not only to global warming (GW) but also to other impacts on the environment and human life. Oceans absorb approximately one-third of the CO_2 emitted each year by human activities and as its levels increase in the atmosphere, the amount dissolved in oceans will also increase turning the water pH gradually to more acidic. This pH decrease may cause the quick loss of coral reefs and of marine ecosystem biodiversity with huge implications in ocean life and consequently in earth life [Ormerod *et al.*, 2002].

As GW is a problem affecting different aspects of human life and the global environment, not only a single but a host of solutions is needed to address it. One side of the problem concerns the reduction of crude oil reserves and difficulties in their extraction and processing, leading to an increase of its cost [Laherrere, 2005]. This situation is particularly acute in the transportation sector, where currently there are no relevant alternatives to fossil fuels. To find clean and renewable energy sources ranks as one of the most challenging problems facing mankind in the medium to long term. The associated issues are intimately connected with economic development and prosperity, quality of life, global stability, and require from all stakeholders tough decisions and long term strategies. For example, many countries and regions around the world established targets for CO_2 reduction in order to meet the sustainability goals agreed under the Kyoto Protocol. Presently many options are being studied and implemented in practice, with different degrees of success, and in different phases of study and implementation. Examples include solar energy, either thermal or photovoltaic, hydroelectric, geothermal, wind, biofuels, and carbon sequestration, among others [Dewulf *et al.*, 2006]. Each one has its own advantages and problems and, depending on the area of application.

4. Biodiesel instead of diesel

One important goal is to take measures for transportation emissions reduction, such as the gradual replacement of fossil fuels by renewable energy sources, where biofuels are seen as real contributors to reach those goals, particularly in the short term. Biofuels production is expected to offer new opportunities to diversify income and fuel supply sources, to promote employment in rural areas, to develop long term replacement of fossil fuels, and to reduce GHG emissions, boosting the decarbonisation of transportation fuels and increasing the security of energy supply. The most common biofuels are biodiesel and bio-ethanol, which can replace diesel and gasoline, respectively, in today cars with little or none modifications of vehicle engines. They are mainly produced from biomass or renewable energy sources and contribute to lower combustion emissions than fossil fuels per equivalent power output. They can be produced using existing technologies and be distributed through the available distribution system. For this reason biofuels are currently pursued as a fuel alternative that can be easily applied until other options harder to implement, such as hydrogen, are available.

Although biofuels are still more expensive than fossil fuels their production is increasing in countries around the world. Encouraged by policy measures and biofuels targets for transport, its global production is estimated to be over 35 billion liters [COM, 2006]. The main alternative to diesel fuel in EU is biodiesel, representing 82% of total biofuels production and is still growing in Europe, Brazil, and United States, based on political and economic objectives. Biodiesel is produced from vegetable oils (edible or non-edible) or animal fats. Since vegetable oils may also be used for human consumption, it can lead to an increase in price of food-grade oils, causing the cost of biodiesel to increase and preventing its usage, even if it has advantages comparing with diesel fuel.

The potential market for biodiesel far surpasses the availability of plant oils not designated for other markets. For example, to fulfill a 10% target in EU from domestic production, the actual feedstocks supply is not enough to meet the current demand and the land requirements for biofuels production, would be more than the potential available arable land for bio-energy crops [Scarlat et al., 2008]. The extensive plantation and pressure for land use change and increase of cultivated fields may lead to land competition and biodiversity loss, due to the cutting of existing forests and the utilization of ecological importance areas [Renewable Fuel Agency, 200]. Biodiesel may also be disadvantageous when replacing crops used for human consumption or if its feedstocks are cultivated in forests and other critical habitats with associated biological diversity. The negative impacts of global warming, now accepted as a serious problem by many people, have clearly been observed for past decade and seem to intensify every year. The release of the carbon oxides and related inorganic oxides are more than the amount that could be absorbed by the natural sinks in the world since 88% of the world energy demand is provided by carbon based non-renewable fuels (Baruch, 2008). It is vital to develop solutions to prevent and/or reduce the emission of greenhouse gases, such as carbon dioxide, to the atmosphere. Carbon dioxide neutral fuels like biodiesel could replace fossil fuels.

Biodiesel, an alternative diesel fuel, is made from renewable biological sources such as vegetable oils and animal fats. It is biodegradable and nontoxic, has low emission profiles and so is environmentally beneficial (Krawczyk, 1996). One hundred years ago, Rudolf Diesel tested vegetable oil as fuel for his engine (Shay, 1993). With the advent of cheap petroleum, appropriate crude oil fractions were refined to serve as fuel and diesel fuels and diesel engines evolved together. In the 1930s and 1940s vegetable oils were used as diesel fuels from time to time, but usually only in emergency situations. Recently, because of increases in crude oil prices, limited resources of fossil oil and environmental concerns there has been a renewed focus on vegetable oils and animal fats to make biodiesel fuels. Continued and increasing use of petroleum will intensify local air pollution and magnify the global warming problems caused by CO2 (Shay, 1993). In a particular case, such as the emission of pollutants in the closed environments of underground mines, biodiesel fuel has the potential to reduce the level of pollutants and the level of potential or probable carcinogens (Krawczyk, 1996). Edible vegetable oils such as canola, soybean, and corn have been used for biodiesel production and found to be a diesel substitute [Lang et al., 2002]. However, a major obstacle in the commercialization of biodiesel production from edible vegetable oil is its high production cost, which is due to the higher cost of edible oil. Waste cooking oil, which is much less expensive than edible vegetable oil, is a promising alternative to edible vegetable oil [Canakci et al., 2003]. Waste cooking oil and fats set forth significant disposal problems in many parts of the world. This environmental problem could be solved by proper utilization and management of waste cooking oil as a fuel. Many developed countries have set policies that penalize the disposal of waste cooking oil the waste drainage [Kulkarni et al., 2006]. The Energy Information Administration in the United States estimated that around 100 million gallons of waste cooking oil is produced per day in

USA, where the average per capita waste cooking oil was reported to be 9 pounds [Radich *et al.*, 2006]. The estimated amount of waste cooking oil collected in Europe is about 700,000–100,000 tons/year [Supple *et al.*, 2002]

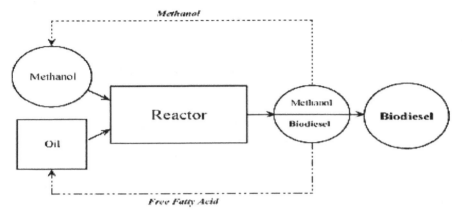

Fig. 2. Biodiesel production process

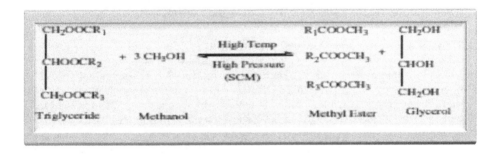

Fig. 3. Transesterification of triglycerides

Biodiesel is made from biomass oils, mostly from vegetable oils. Biodiesel appears to be an attractive energy resource for several reasons. First, biodiesel is a renewable resource of energy that could be sustainably supplied. It is understood that the petroleum reserves are to be depleted in less than 50 years at the present rate of consumption [Sheehan *et al.*, 1998]. Second, biodiesel appears to have several favorable environmental properties resulting in no net increased release of carbon dioxide and very low sulfur content [Antolin *et al.*, 2002]. The release of sulfur content and carbon monoxide would be cut down by 30% and 10%, respectively, by using biodiesel as energy source. Using biodiesel as energy source, the gas generated during combustion could be reduced, and the decrease in carbon monoxide is owing to the relatively high oxygen content in biodiesel. Moreover, biodiesel contains no aromatic compounds and other chemical substances which are harmful to the environment. Recent investigation has indicated that the use of biodiesel can decrease 90% of air toxicity and 95% of cancers compared to common diesel source. Third, biodiesel appears to have significant economic potential because as a non-renewable fuel that fossil fuel prices will increase inescapability further in the future. Finally, biodiesel is better than diesel fuel in terms of flash point and biodegradability [Ma *et al.*, 1999].

5. Algae as potentials for biodiesel production

5.1 Separation of biodiesel from algae
5.1.1 Extraction of oil

Extraction of oil was carried out using two extraction solvent systems to compare the oil content in each case and select the most suitable solvent system for the highest biodiesel yield (Afify *et al.*, 2010).

5.1.1.1 Chloroform/methanol (2:1, v/v) method

A known weight of each ground dried algal species (10 g dry weight) was mixed separately with the extraction solvent mixture; chloroform/methanol (100 ml, 2:1, v/v) for 20 min. using shaker, followed by the addition of mixture of chloroform/water (50 ml, 1:1, v/v) for 10 min. filter and the algal residue was extracted three times by 100 ml chloroform followed by filtration (Fig.1) according to Bligh and Dayer (1959).

5.1.1.2 Hexane/ether (1:1, v/v) method

A known weight of each ground dried algal species (10 g dry weight) was mixed with the extraction solvent mixture, hexane/ether (100 ml, 1:1, v/v), kept to settle for 24 hrs, followed by filtration (Fig. 1) according to Hossain and Salleh (2008).

5.1.2 Transesterification and biodiesel production

The extracted oil was evaporated under vaccum to release the solvent mixture solutions using rotary evaporator at 40- 45 °C. Then, the oil produced from each algal species was mixed with a mixture of catalyst (0.25g NaOH) and 24 ml methanol, a process called transesterification (Fig. 2, 3,4, 5 and Table 2), with stirring properly for 20 min. The Mixture was kept for 3hrs in electric shaker at 3000 rpm. (National Biodiesel Board, 2002). After shaking the solution was kept for 16 hrs to settle the biodiesel and the sediment layers clearly. The biodiesel layer was separated from sedimentation by flask separator carefully. Quantity of sediments (glycerin, pigments, etc) was measured. Biodiesel (Fig. 6) was washed by 5% water many times until it becomes clear then Biodiesel was dried by using dryer and finally kept under the running fan for 12 h. the produced biodiesel was measured (using measuring cylinder), pH was recorded and stored for analysis.

6. Biodiesel from algae

Sustainable production of renewable energy is being hotly debated globally since it is increasingly understood that first generation biofuels, primarily produced from food crops and mostly oil seeds are limited in their ability to achieve targets for biofuel production, climate changemitigation and economic growth. These concerns have increased the interest in developing second generation biofuels produced from non-food feedstocks such as microalgae, which potentially offer greatest opportunities in the longer term. This paper reviews the current status of microalgae use for biodiesel production, including their cultivation, harvesting, and processing. The microalgae species most used for biodiesel production are presented and their main advantages described in comparison with other available biodiesel feedstocks. The various aspects associated with the design of microalgae production units are described, giving an overview of the current state of development of algae cultivation systems (photo-bioreactors and open ponds). Other potential applications and products from microalgae are also presented such as for biological sequestration of CO_2, wastewater treatment, in human health, as food additive, and for aquaculture (Mata et al., 2010).

Biodiesel seem to be a viable choice but its most significant drawback is the cost of crop oils, such as canola oil, that accounts for 80% of total operating cost, used to produce biodiesel (Demirbas, 2007). Besides, the availability of the oil crop for the biodiesel production is limited (Chisti, 2008). Therefore, it is necessary to find new feedstock suitable for biodiesel production, which does not drain on the edible vegetable oil supply. One alternative to oil crops is the algae because they contain lipids suitable for esterification/ transesterification. Among many types of algae, microalgae seem to be promising (Table 1) because:

1. They have high growth rates; e.g., doubling in 24 h (Rittmann, 2008).
2. Their lipid content could be adjusted through changing growth medium composition (Naik et al., 2006).
3. They could be harvested more than once in a year (Schenk et al., 2008).
4. Salty or waste water could be used (Schenk et al., 2008).
5. Atmospheric carbon dioxide is the carbon source for growth of microalgae (Schenk et al., 2008).
6. Biodiesel from algal lipid is non-toxic and highly biodegradable (Schenk et al., 2008).
7. Microalgae produce 15–300 times more oil for biodiesel production than traditional crops on an area basis (Chisti, 2007).

Strain	Protein	Carbohydrates	Lipid	Nucleic acid
Scenedesmus obliquus	50–56	10–17	12–14	3–6
Scenedesmus quadricauda	47	-	1.9	-
Scenedesmus dimorphus	8–18	21–52	16–40	-
Chlamydomonas rheinhardii	48	17	21	-
Chlorella vulgaris	51–58	12–17	14–22	4–5
Chlorella pyrenoidosa	57	26	2	-
Spirogyra sp.	6–20	33–64	11–21	-
Dunaliella bioculata	49	4	8	-
Dunaliella salina	57	32	6	-
Euglena gracilis	39–61	14–18	14–20	-
Prymnesium parvum	28–45	25–33	22–39	1–2
Tetraselmis maculata	52	15	3	-
Porphyridium cruentum	28–39	40–57	9–14	-
Spirulina platensis	46–63	8–14	4–9	2–5
Spirulina maxima	60–71	13–16	6–7	3–4.5
Synechoccus sp.	63	15	11	5
Anabaena cylindrica	43–56	25–30	4–7	-

Note: Algal-oil is very high in unsaturated fatty acids. Some UFA's found in different algal-species include: Arachidonic acid (AA), Eicospentaenoic acid (EPA), Docasahexaenoic acid (DHA), Gamma-linolenic acid (GLA) Linoleic acid (LA).

Table 1. Biochemical composition of algae expressed on a dry matter basis (Becker, 1994)

Algae are made up of eukaryotic cells. These are cellswith nuclei and organelles. All algae have plastids, the bodies with chlorophyll that carry out photosynthesis. But the various strains of algae have different combinations of chlorophyll molecules. Some have only Chlorophyll A, some A and B, while other strains, A and C [Benemann et al., 1978]. Algae biomass contains three main components: proteins, carbohydrates, and natural oil. The chemical compositions of various microalgae are shown in Table 1. While the percentages vary with the type of algae, there are algae types that are comprised of up to 40% of their overall mass by fatty acids [Becker, 1994]. It is this fatty acid (oil) that can be extracted and converted into biodiesel.

Type of transesterification	Advantage	Disadvantage
Chemical catalysis	1-reaction condition can be well controlled	1-reaction temperature is relative high and the process is complex
	2-Large scale production	2-The later disposal process is complex
	3-The cost of the production process is cheap	3-The process need much energy
	4-The methanol produced in the process can be recycled	4-Need a installation for methanol recycle
	5-high conversion of the production	5-the waste water pollute the environment
Enzymatic catalyst	1-Moderate reaction condition	1-Limitation of enzyme in the conversion of short chain fatty acids
	2-The small amount of methanol required in the reaction	2-Chemicals arise in the process of production are poisons to enzyme
	3-Have no pollution to natural environment	
Supercritical fluid techniques	1-Easy to be controlled	1-High temperature and high pressure in the reaction condition leads to high coast for production and waste energy
	2-It is safe and fast	
	3-friendly to environment	

Table. 2. Types of transesterification catalysts

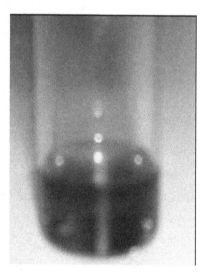

Fig. 4. Biodiesel from algae

Fig. 5 shows a schematic representation of the algal biodiesel value chain stages, starting with the selection of microalgae species depending on local specific conditions and the design and implementation of cultivation system for microalgae growth. Then, it follows the biomass harvesting, processing and oil extraction to supply the biodiesel production unit.

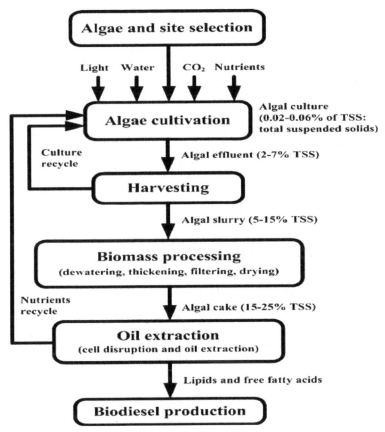

Fig. 5. Microalgae biodiesel value chain stages.

Algae's potential as a feedstock is dramatically growing in the biofuel market. Microalgae (to distinguish it from such macroalgae species as seaweed) have many desirable attributes as energy producers [Choe et al., 2002]:
- Algae is the most promising non-food source of biofuels,
- Algae has a simple cellular structure,
- a lipid-rich composition (40–80% in dry weight),
- a rapid reproduction rate,
- Algae can grow in salt water and harsh conditions,
- Algae thrive on carbon dioxide from gas- and coal-fired power Plants,
- Algae biofuel contains no sulfur, is non-toxic and highly biodegradable.

- The utilization of microalgae for biofuels production can also serve other purposes. Some possibilities currently being considered are listed below.
- Removal of CO_2 from industrial flue gases by algae bio-fixation [Wang et al., 2008], reducing the GHG emissions of a company or process while producing biodiesel. Wastewater treatment by removal of NH^+_4, NO^-_3, PO^{-3}_4, making algae to grow using these water contaminants as nutrients [Wang et al., 2008].
- After oil extraction the resulting algae biomass can be processed into ethanol, methane, livestock feed, used as organic fertilizer due to its high N:P ratio, or simply burned for energy cogeneration (electricity and heat) [Wang et al., 2008];
- Combined with their ability to grow under harsher conditions, and their reduced needs for nutrients, they can be grown in areas unsuitable for agricultural purposes independently of the seasonal weather changes, thus not competing for arable land use, and can use wastewaters as the culture medium, not requiring the use of freshwater.
- Depending on the microalgae species other compounds may also be extracted, with valuable applications in different industrial sectors, including a large range of fine chemicals and bulk products, such as fats, polyunsaturated fatty acids, oil, natural dyes, sugars, pigments, antioxidants, high-value bioactive compounds, and other fine chemicals and biomass [Raja et al., 2008].
- Because of this variety of high-value biological derivatives, with many possible commercial applications, microalgae can potentially revolutionize a large number of biotechnology areas including biofuels, cosmetics, pharmaceuticals, nutrition and food additives, aquaculture, and pollution prevention [Raja et al., 2008].

7. Environmental advantages of algal biofuels

In order to be a viable alternative energy source, a biofuel should provide a net energy gain, have environmental benefits, be economically competitive and be producible in large quantities without reducing food supplies [Hill, 2006]. In the subsections below we illustrate how the use of microalgae as feedstocks for biodiesel production can provide significant environmental benefits by reducing the land, pollutant and water footprints of biofuel production.

7.1 Advantages of biodiesel from algae oil (Table 3)
- Rapid growth rates
- Grows practically anywhere
- A high per-acre yield (7–31 times greater than the next best crop – palm oil)-
- No need to use crops such as palms to produce oil
- A certain species of algae can be harvested daily
- Algae biofuel contains no sulfur
- Algae biofuel is non-toxic
- Algae biofuel is highly bio-degradable
- Algae oil extracts can be used as livestock feed and even processed into ethanol
- High levels of polyunsaturates in algae biodiesel is suitable for cold weather climates
- Can reduce carbon emissions based on where it's grown

7.2 Disadvantages of biodiesel from algae oil
- Produces unstable biodiesel with many polyunsaturates
- Biodiesel performs poorly compared to it's mainstream alternative
- Relatively new technology

Type of organism	Advantage	Disadvantage
Microalgal oil	1-Fatty acid profile similar to vegetable oil	1-Most algal lipid have lower fuel value than diesel fuel
	2-Under certain condition it may be as high as 85% of the dry weight	2-The cost of cultivation is higher compared to common crop oil currently
	3-Short-time growth cycle	
	4-Composition is relative single in microalgae	
Bacteria oil	1-Fast growth rate	1-Most of bacteria can not yield lipids but complicated lipoid
Oleaginous yeast and mildews	1-Resource are abundant in the nature	1-Filteration and cultivation of yeasts and mildews with high-content are required
	2-High oil content in some species	2-Process of oils extracted is complex and new technology
	3-Short time growth cycle	3-The cost of cultivation is also higher compared to common crops currently
	4-Strong capability of growth in different cultivation on conditions	
Waste oils	1-The waste oil is cheap compared to crop oils	1-Conataing a lot of saturated fatty acids which is hard to converted to biodiesel by catalyst

Table 3. Advantage and disadvantage of algae as biodiesel source compared with bacteria, yeast and waste oils

8. Comparison between biodiesel production from algae and vegetables

Quantifying the land use changes associated with intensive biofuel feedstock production relies upon many assumptions [Chisti,. 2007], but it is clear that the accelerated cultivation of terrestrial plant biomass for biofuels will have an exceptionally large land footprint (Table 4). For example, the United States has the fourth largest absolute biodiesel potential of the 119 countries studied by Johnston and Holloway [Johnston, M. and Holloway, 2007]. However, recent work has suggested that the projected year 2016 demand for corn ethanol alone would require 43% of all U.S. land used for corn production in 2004 [Chisti,. 2007]. A related study concluded that the annual corn production needed to satisfy one half of all U.S. transportation fuel needs would require an area equivalent to more than eight times the U.S. land area that is presently used for crop production [Chisti,. 2007]. Other land-based crops would require less cropland, based on their oil content: oil palm (24% of current cropland area), coconut (54%), jatropha (77%), canola (122%) and soybean (326%) [Chisti,. 2007]. Moreover, recent work indicates that the ability of countries to grow terrestrial crops explicitly for the production of biofuels such as ethanol and biodiesel is significantly overestimated [Johnston, M. and Holloway, 2007], contributing to concerns that these biofuels are not feasible options for providing a significant fraction of global fuel demand.

Biodiesel feedstock	Area needed to meet global oil demand (10^6 hectans)	Area required as a percent of total global land	Area required as a percent of total arable global land
Cotton	15000	101	757
Soybean	10900	73	552
Mastard seed	8500	57	430
Sunflower	5100	34	258
Rapeseed/Canola	4100	27	207
Jatropha	2600	17	130[b]
Oil palm	820	5.5	41
Microalgae (10 g/m³/day, 30%TAG)	410	2.7	21[c]
Microalgae (50 g/m³/day, 50%TAG)	49	0.3	25[c]

[b] Jatropha is mainly grown on marginal land
[c] Assuring that microalgal ponds and bioreactors are located on non-arable land

Table 4. Comparison of estimated biodiesel production efficiencies from vascular plants and microalgae

9. The physical and chemical properties of biodiesel produced from algal cell

Analysis of the produced biodiesel from the promising alga *Dictyochloropsis splendida* (Table 5). showed that the unsaturated fatty acids percentage was increased in alga cultivated in nitrogen free media (0.0g/1 N) two times more than normal conditions (13.67, 4.81% respectively). However, the composition of fatty acids was different in these algae depending on its growth condition as showed in table 3. These results were in agreements with those reported by Wood (1974) relative to *Chlorophycean* species. Furthermore Ramos *et al*. (2009) reported that monounsaturated, polyunsaturated and saturated methyl esters were built in order to predict the critical parameters of European standard for any biodiesel, composition. The extent of unsaturation of microalgae oil and its content of fatty acids with more than four double bonds can be reduced easily by partial catalytic hydrogenation of the oil (Jang *et al*., 2005, Dijkstra, 2006). Concerning the fatty acids contents of the produced biodiesel from microalgae, Chisti (2007) reported in his review that, microalgal oils differ from vegetable oils in being quite rich in polyunsaturated fatty acids with four or more double bands (Belarbi *et al*., 2000) as eicosapentanoic acid (C20:5n-3) and docosahexaenoic acid (C22:6n-3) which occurred commonly in algal oils. The author added that, fatty acids and fatty acid methyl esters with four and more double bands are susceptible to oxidation during storage and this reduces their acceptability for use in biodiesel especially for vehicle use (European standard EN 14214 limits to 12%) while no such limitation exists for biodiesel intended for use as healing oil. In addition to the content of unsaturated fatty acids in the biodiesel also its iodine value (represented total unsaturation) must be taken in consideration (not exceeded 120 g iodine/100g biodiesel according to the European standard.

Fatty acids	*RT	Fatty acids percentage	
		Algae cultivated under normal conditions	Algae cultivated under free nitrogen media
C10:0 (Capric acid)	1.223	0.0	1.26
C14:0 (Myristic acid)	2.437	13.04	13.88
C16:0 (Palmitic acid)	2.860	81.14	69.59
C17:0 (Margeric acid)	3.240	1.01	1.21
C18:0 (stearic acid)	4.335	0.0	0.38
C18:1 (Oleic acid)	4.667	0.26	1.11
C18:2 (Linoleic acid)	5.333	4.39	12.14
C18:3 (linolenic acid)	6.948	0.15	0.42
Total saturated fatty acids		95.19	86.33
Total unsaturated fatty acids		4.81	13.67
TU/TS		0.05	0.16

*Retention time; TU/TS: total unsaturated/ total saturated fatty acids ratio.

Table 5. Analysis of fatty acids of the obtained biodiesel from the promising green microalgae *Dictyochloropsis* sp

10. Enhancement the biodiesel production from algae

Lipid productivity, the mass of lipid that can be produced per day, is dependent upon plant biomass production as well as the lipid content of this biomass. Algal biodiesel production will therefore be limited not only by the standing crop of microalgae, but also by its lipid content, which can vary from <1% to >50% dry weight [Shifrin, N.S. and Chisholm, 1980]. Given that a strong and predictable response of microalgal biomass to phosphorus enrichment has consistently been exhibited by freshwater ecosystems worldwide (Box 2), it can be expected that the volumetric lipid content (in mg L-1) of water contained in algal bioreactors should also in general increase with an increase in the total phosphorus content of the system, as has been reported for lakes by Berglund *et al.* [Berglund, 2001]. However, both the quantity and the quality of lipids produced will vary with the identity of the algal species that are present in the water, as well as with site-specific growth conditions. This variability probably reflects modifications in the properties of cellular membranes, and alterations in the relative rates of production and utilization of storage lipids [Roessler, 1990]. In the presence of moderate temperatures and sufficient light, many dozens of studies during the past several decades have revealed that algal lipid content is particularly sensitive to conditions of nutrient limitation . For example, silicon-starved diatoms can contain almost 90% more lipids than silicon-sufficient cells [Shifrin, N.S. and Chisholm, 1980]. However, silicon will be a growth-limiting nutrient only for the limited subset of microalgal species that have an absolute requirement of this element for their cellular growth. A stronger stimulation of lipid production occurs in response to conditions of nitrogen limitation, which potentially can occur in all known microalgae. Nitrogen-starved cells can contain as much as four times the lipid content of Nsufficient cells [Shifrin, N.S. and Chisholm, 1980], and maximizing the lipid production of pond bioreactors should therefore depend on their operators' ability to reliably and consistently induce N-limitation in the resident algal cells. Resource-ratio theory and the principles of ecological stoichiometry, provide additional new insights into the control of algal biomass and lipid production in pond bioreactors. the nutrient limitation status of microalgae can be directly controlled by regulating the ratio of nitrogen and phosphorus (N:P) supplied in the incoming nutrient feed: nitrogen limitation occurs at N:P supply ratios that lie below the optimal N:P ratio for microalgal growth, whereas phosphorus limitation occurs at ratios that exceed this ratio. A transition between N- and P-limitation of phytoplankton growth typically occurs in the range of N:P supply ratios between ca. 20:1 to ca. 50:1 by moles . Such shifts between N- and P-limitation have extremely important implications for algal biofuel production because diverse species of microalgae grown under nitrogen-limited conditions (i.e. low N:P supply ratios) can exhibit as much as three times the lipid content of cells grown under conditions of phosphorus limitation (high N:P supply ratios) . Both the total phosphorus concentration as well as the total nitrogen concentration in the nutrient feeds to pond bioreactors should therefore impact algal biodiesel production, because the N:P ratio of incoming nutrients will strongly influence algal biomass production as well as the cellular lipid content. Given the inverse relationship observed between N:P and cellular lipids , and the positive, hyperbolic relationship observed between N:P and microalgal biomass , we conclude that optimal lipid yields (in terms of mass of lipid produced per unit bioreactor volume per day) should occur at intermediate values of the N:P supply ratio. From the strong apparent interactions between the effects of nitrogen and carbon dioxide availability on microalgal lipids, we also conclude that the effects of N:P supply ratios on volumetric lipid production might be even greater if the bioreactors are simultaneously provided with supplemental CO_2 (cf. Figure 2).

Algal species	Chloroform/methanol (2:1, v/v)	Hexane/ether (1:1, v/v)
Jania rubens	4.4±0.12	2.8±0.04
Galaxaura oblongata	2.5±0.09	2.4±0.01
Gelidium latifolium	3.0±0.0	3.1±0.02
Asporagopsis taxiformis	4.1±0.08	3.4±0.05
Ulva lactuca	4.2±0.1	3.5±0.1
Colpomenia sinuosa	3.5±0.05	2.3±0.03
Dictyochloropsis splendida	12.5±0.23	2.4±0.14
Spirulina platensis	9.2±0.25	3.0±0.10
LSD	0.3261	0.3261

Each value is presented as mean of triplet treatments, LSD: Least different significantly at $P \le 0.05$ according to Duncan's multiple range test.

Table 6. Comparison between lipid percentage (%) produced by eight algal species using two different extraction system.

Eight algal species (4 *Rhodo*, 1 *chloro* and 1 *phaeophycean* macroalgae, 1 *cyanobacterium* and 1 green microalga) were used for the production of biodiesel using two extraction solvent systems (Hexane/ether (1:1, v/v)) and (Chloroform/ methanol (2:1, v/v)) Table 6. Biochemical evaluations of algal species were carried out by estimating biomass, lipid, biodiesel and sediment (glycerin and pigments) percentages. Hexane/ ether (1:1, v/v) extraction solvent system resulted in low lipid recoveries (2.3-3.5% dry weight) while; chloroform/methanol (2: 1, v/v) extraction solvent system was proved to be more efficient for lipid and biodiesel extraction (2.5 – 12.5% dry weight) depending on algae species (Table 7). The green microalga *Dictyochloropsis splendida* extract produced the highest lipid and biodiesel yield (12.5 and 8.75% respectively) followed by the cyanobacterium *Spirulina maxima* (9.2 and 7.5 % respectively). On the other hand, the macroalga (red, brown and green) produced the lowest biodieselyield. The fatty acids of *Dictyochloropsis splendida* Geitler biodiesel were determined using gas liquid chromatography. Lipids, biodiesel and glycerol production of *Dictyochloropsis splendida* Geitler (the promising alga) were markedly enhanced by either increasing salt concentration or by nitrogen deficiency (Table 8) with maximum production of (26.8, 18.9 and 7.9 % respectively) at nitrogen starvation condition. (Afify *et al.*, 2010)

Algal sp.	Lipid %	Biodiesel%	Sediment %	Biodiesel color
Jania rubens	4.4±0.12	0.25±0.01	4.2^a ± 0.05	Light brown
Galaxaura oblongata	2.5±0.09	2.06±0.02	0.08±0.0	Light green
Gelidium latifolium	3.0±0.0	1.3±0.0	1.6±0.01	yellow
Asporagopsis taxiformis	4.1±0.08	3.64^c ± 0.10	0.40±0.01	Dark green
Ulva lactuca	4.2±0.1	3.8±0.12	0.44±0.0	Light green
Colpomenia sinuosa	3.5±0.05	3.1±0.05	0.31±0.05	yellow
Dictyochloropsis splendida	12.5±0.23	8.75±0.24	3.75±0.08	colorless
Spirulina platensis	9.2±0.25	7.5±0.30	1.66±0.06	Light green
LSD	0.3261	0.3314	0.1786	

Each value is presented as mean of triplet treatments, LSD: Least different significantly at P ≤ 0.05 according to Duncan's multiple range test.

Table 7. Total lipid, biodiesel, sediment percentage and biodiesel color of eight algal species

Natural biotic communities in outdoor bioreactors require the external provision of potentially growth-limiting resources (e.g. light, carbon dioxide and the essential mineral nutrients N and P). These resources act as "bottom-up" regulators of the potential microalgal biomass that can be produced. Once harvested, the cellular lipids in this microalgal biomass can be extracted and processed to create biodiesel fuels. The lipid content of microalgal biomass is not constant, however, and can be influenced by many factors, including nitrogen:phosphorus supply ratios, light, CO_2 and the hydraulic residence time of the bioreactor. Moreover, natural assemblages of microalgae are taxonomically diverse: some species are small and can easily be consumed by herbivorous zooplankton. Undesirable grazing losses of edible microalgae (and their cellular lipids) to large-bodied zooplankton can be reduced by adding zooplanktivorous fish, which can greatly restrict large-bodied zooplankton growth via sizeselective predation ("top-down" regulation).

Sample culture conditions	Lipid content (%)	Biodiesel content (%)	Glycerol+ pigments content (%)	Biodiesel color
Control (2.5 g/l NaCl and 25g/l NaNO₃)	12.50±0.36	8.75±0.25	3.75±0.12	Colorless
NaCl stress				
5 g/l	14.50±1.2	8.90±0.62	5.60±0.18	Colorless
7.5 g/l	17.00±0.53	11.94±0.98	5.06±0.22	Light green
10 g/l	17.50±0.36	11.38±0.80	5.11±0.24	Light green
Nitrogen stress				
12.5 g/l	15.40±2.10	8.90±0.36	6.50±0.30	Yellowish green
6.25g/l	16.20±1.8	10.01±1.0	6.19±0.12	Light Yellow
0.0g/l	26.80± 2.12	18.90±1.2	7.9±0.50	Yellow
LSD	0.3643	0.1681	0.1431	

Each value is presented as mean of triplet treatments, LSD: Least different significantly at P ≤ 0.05 according to Duncan's multiple range test.

Table 8. Total lipid, biodiesel, sediment percentage and biodiesel color of Dictyochloropsis sp cultivated under stress

11. Wastewater nitrogen and phosphorous as microalgae nutrients

There is a unique opportunity to both treat wastewater and provide nutrients to algae using nutrient-rich effluent streams. By cultivating microalgae, which consume polluting nutrients in municipal wastewater, and abstracting and processing this resource, then the goals of sustainable fuel production and wastewater treatment can be combined (Andersen, 2005). Treated wastewater is rich in nitrogen and phosphorus, which if left to flow into waterways, can spawn unwanted algae blooms and result in eutrophication (Sebnem Aslan, 2006). These nutrients can instead be utilized by algae, which provide the co-benefit of producing biofuels and removing nitrogen and phosphorus as well as organic carbon (Mostafa and Ali, 2009). Wastewater treatment using algae has many advantages. It offers the feasibility to recycle these nutrients into algae biomass as a fertilizer and thus can offset treatment cost. Oxygen rich effluent is released into water bodies after wastewater treatment using algae (Becker, 2004).

Cyanobacteria strains (Anabaena flos aquae, Anabaena oryzae, Nostoc humifusum, Nostoc muscorum, Oscillatoria sp., Spirulina platensis, Phormedium fragile and Wollea saccata) and the green alga strain Chlorella vulgaris were obtained from the Microbiology Department, Soils, Water and Environment Res. Inst. (SWERI), Agric. Res., Center (ARC). Cyanobacteria strains were maintained in BG11 medium (Rippka et al., 1979) except Spirulina platensis which was cultivated in Zarrouk medium (Zarrouk, 1966). While, Bold medium (Nichols and Bold, 1965) was used for the green alga Chlorella vulgaris. Cultures were incubated in a growth chamber under continuous shaking (150 rpm) and illumination (2000 lux) at 25 ± 1 °C for 30 days. Shalaby et al. (2011). The effluent of the secondary treated sewage wastewater from Zenien Waste Water Treatment Plant (ZWWTP), Giza

Governorate, Egypt was used after filtered using glass microfiber filter to remove large particles and indigenous bacteria for the experiment and the chemical and physical parameters were analysis as reported by APHA (1998) Table (2). The supplementation of $NaNO_3$, K_2HPO_4 and $FeSO_4.7H_2O$ in amounts equal to those of the standard BG11, Bold and Zarrouk were used as basal media. The algal strains were grown in 500 ml Erlenmeyer flasks containing 200 ml of 100% effluent supplemented with basal nutrients and 100% effluent without basal nutrients with/without sterilization and the synthetic media (BG11, Bold and Zarrouk) were used as control. Two per cent algal inoculums were added to each flask. The experiment was conducted in triplicates and cultures were incubated at 25 °C ±1°C, under continuous shaking (150 rpm) and illumination (2000 lux) for 15 days. This work aimed to evaluate the laboratory cultivation of nine algal strains belonging to Nostocales and Chlorellales in secondary treated municipal domestic wastewater for biomass and biodiesel production as shown in Table (9 and 10).

Algal species	Total lipids	Biodiesel content	Glycerin + pigments	Color	pH
Nostoc muscorum	16.80±3.62	12.52±1.74	4.28±1.74	Brown	7.4±0.33
Anabaena flous aquae	5.50±0.58	4.00±0.41	1.50±0.41	Red	6.9±0.95
Chlorella vulgaris	12.50±1.20	8.8±0.16	3.70±0.16	Green	8.1±1.0
Oscillatoria sp	8.00±0.58	4.30±0.32	3.70±0.32	Yellow	7.5±0.85
Spirulina platensis	10.0±0.11	7.80±0.17	2.20±0.17	Light green	8.0±0.32
Anabaena oryzae	7.40±0.90	4.50±0.10	2.90±0.10	Orange	7.3±0.96
Wollea sp	6.30±1.31	3.90±0.60	2.40±0.60	Yellow	7.8±0.35
Nostoc humifusum	14.80±2.40	10.20±1.30	4.6±1.30	Yellowish brown	7.5±0.50
Phormedium sp	12.20±1.66	10.10±1.50	2.10±1.50	Dark brown	7.1±0.0
LSD	0.159	0.151	0.151		1.659

Each value is presented as mean of triplet treatments, LSD: Least different significantly at P ≤ 0.05 according to Duncan's multiple range tests.
T1: waste water without treatment; T2: waste water after sterilization; T3: waste water+ nutrients with sterilization T4: waste water+ nutrients without sterilization

Table 9. Total lipids, biodiesel, glycerine+pigments percentage and color, pH of biodiesel from different microalgae species cultivated in different waste water

Algal species	Optimal waste water treatment	Total lipids	Biodiesel	Glycerin + pigments
Nostoc muscorum	T3	12.50±2.65	7.40±0.74	5.10±0.74
Anabaena flous aquae	T3	7.40±0.95	5.00±0.61	2.40±0.61
Chlorella vulgaris	T3	13.20±1.87	8.50±1.74	4.70±1.74
Oscillatoria sp	T2	6.80±0.65	3.80±0.32	3.00±0.32
Spirulina platensis	T3	7.30±0.44	5.00±0.51	2.30±0.51
Anabaena oryzae	T4	8.00±0.16	4.70±0.12	3.30±0.12
Wollea sp	T3	7.20±1.32	4.00±0.22	3.23±0.22
Nostoc humifusum	T1	15.50±1.65	11.80±1.52	3.70±1.52
Phormedium sp	T2	11.60±0.88	8.40±0.65	3.20±0.65
LSD		0.159	0.159	0.152

Each value is presented as mean of triplet treatments, LSD: Least different significantly at P ≤ 0.05 according to Duncan's multiple range tests.
T1: waste water without treatment; T2: waste water after sterilization; T3: waste water+ nutrients with sterilization T4: waste water+ nutrients without sterilization

Table 10. Total lipids, biodiesel, glycerine+pigments percentage and color, pH of biodiesel from different microalgae species cultivated in different waste water

12. Economic importance

Compared to biofuels from agricultural crops, the amount of land required would be minimal. Trials in ideal conditions show that fast-growing micro-algae can yield 1800–2000 gallons/(acre - year) of oil—compare this with 50 gallons for soyabeans, 130 gallons for rapeseed and _650 gallons for palm oil. It can grow on fresh or brackish water on marginal land so that it does not compete with areas for agricultural cultivation. As Sean Milmo points out in his article in Oils and Fats International [Milmo, 2008]; oil from algae on 20–40 M acres of marginal land would replace the entire US supply of imported oil, leaving 450 M acres of fertile soil in the country entirely for food production. Biomass can also be harvested from marine algae blooms and algae can even be cultivated in sewage and water treatment plants. However, most estimates of algal fuel productivity estimate that with current production technologies algal diesel can be manufactured for, at best, $4.54 per gallon using high density photobioreactors. In order to compete economically with petroleum diesel costs – and not accounting for any potential subsidy scheme, which is a likely possibility – requires the reduction of these costs to near $1.81 per gallon relative to

2006 fuel prices. These cost reduction figures take into account the fact that materials input and refining of fuels (in this case the algae vegetable oil) account for roughly 71% of total at pump fuel cost [Chisti, 2007]. Algal biodiesel becomes even more plausible given the potential for GHG regulation in the near future. Since for every ton of algal biomass produced, approximately 1.83 tons of carbon dioxide is fixed while petroleum diesel carries a massive negat balance, the competitiveness of algae diesel increases as GHG externalities are taken into account. Given certain research objectives these cost reductions are achievable in the near future. The National Renewable Energy Laboratory (NREL) outlines many such research objects including: increasing photosynthetic efficiency of algae species for high lipid production, control of mechanisms of algae biofocculation, understanding the effects of non-steady-state operating conditions, and methods of species selection and control [Sheehan *et al.*, 1998].

13. The problems related with algae

Most problems with marine microalgae cultures are related to predation by various types of protozoans (e.g. zooflagellates, ciliates, and rhizopods). Other problem is the blooming of unwanted or toxic species such as the blue-green algae or dinoflagellates (red tides) that can result in high toxicity for consumers and even for humans. Examples are the massive development of green chlorococcalean algae, such as Synechocystis in freshwater, and also the development of Phaeodactylum in seawater that is undesirable for bivalve molluscs. [De Pauw *et al.*, 1984].

14. Other application of algae

Algae have mainly been used in west countries as raw material to extract alginates (from brown algae) and agar and carragenates (from red algae). Moreover, algae also contain multitude of bioactive compounds (phenolic compounds, alkaloids, plant acids, terpenoids and glycosides) that might have antioxidant, antibacterial, antiviral, anticarcinogenic, etc. properties. (Plaza, *et al.*, 2008).

15. References

Afify, AMM.; Shanab, SM.; Shalaby, EA. (2010). Enhancement of biodiesel production from different species of algae, grasas y aceites, 61 (4), octubre-diciembre, 416-422.

Antolin, G.; Tinaut, F. V.; Briceno, Y. (2002). Optimisation of biodiesel production by sunflower oil transesterification. Bioresour Technol ., 83:111–4.

Baruch, J.J. (2008). Combating global warming while enhancing the future. Technol. Soc. 30, 111–121.

Becker, E.W. in: J. Baddiley, *et al.* (Eds.), Microalgae: Biotechnology and Microbiology, Cambridge Univ. Press, Cambridge, NY, 1994, p. 178.

Benemann, J.R.; Koopman, B.L.; Weissman, J.C.; Eisenberg, D.M.; Oswald, W.J. (1978). An Integrated System for the Conversion of Solar Energy with Sewage-grown Microalgae, Report, Contract D(0-3)-34, U.S. Dept. of Energy, SAN-003-4-2.

Berglund, O. *et al.* (2001) The effect of lake trophy on lipid content and PCB concentrations in planktonic food webs. Ecology 82, 1078–1088.

Bligh EG, Dayer WJ. 1959. A rapid method for total lipid extraction and purification. Can. J. Biochem and Physiol., 37: 911-7.

Canakci, M.; Van Gerpen, J. (2003). A pilot plant to produce biodiesel from high free fatty acid feedstocks. Trans ASAE, 46: 945-54.

Chisti, Y. (2007). Biodiesel from microalgae. Biotechnol. Adv. 25, 294–306.

Chisti, Y. (2008). Biodiesel from microalgae beats bioethanol. Cell Press 26, 126–131.

Choe, S.H.; Jung, I.H.(2002). J. Ind. Eng. Chem. 8 (4): 297.

COM (2006) 34 final. An EU strategy for biofuels. Commission of the European Communities, Brussels, 8.2.2006.

Demirbas, A. (2007). Importance of biodiesel as transportation fuel. Energy Policy 35, 4661–4670.

De Pauw, N.; Morales, J.; Persoone, G. (1984). Mass culture of microalgae in aquaculture systems: progress and constraints. Hydrobiologia,116/117: 121-34.

Dewulf J, Van Langenhove H. Renewables-based technology: sustainability assessment. John Wiley & Sons, Ltd; 2006.

Dos Santos, M. D.; Guaratini, T.; Lopes, J. L. C.; Colepicolo, P. and Lopes, N. P. (2005). Plant cell and microalgae culture. In: Modern Biotechnology in Medicinal Chemistry and Industry. Research Signpost, Kerala, India.

European Environmental Agency (EEA) Report N85. Copenhagen, Denmark; 2007.

European Environmental Agency (EEA). Greenhouse gas emission trends and projections in Europe 2004: progress by the EU and its Member States towards achieving their Kyoto Protocol targets. Report N85. Copenhagen, Denmark; 2004.

Johnston, M. and Holloway, T. (2007) A global comparison of national biodiesel production potentials. Environ. Sci. Technol., 41: 7967–7973.

Hill, J. *et al.* (2006) Environmental, economic, and energetic costs and benefits of biodiesel and ethanol biofuels. Proc. Natl. Acad. Sci. U. S. A.

Hossain ABM, Salleh A. (2008). Biodiesel fuel production from algae as renewable energy. Am. J. Biochem. and Biotechn., 4(3): 250-254.

International Energy Agency (IEA). World Energy Outlook 2007. China and India Insights, Paris, France; 2007.

Kulkarni, M. G.; Dalai, A. K. (2006). Waste cooking oil – an economical source for biodiesel: a review. Ind Eng Chem Res., 45: 2901-13.

Krawczyk, T. (1996). Biodiesel: Alternative fuel makes inroads but hurdles remain. INFORM, 7: 801-829.

Laherrere J. Forecasting production from discovery. In: ASPO; 2005.

Lang, X.; Dalai, A. K.; Bakhashi, N. N.; Reaney, M. J. (2002). Preparation and characterization of biodiesels from various bio-oils. Biores Technol., 80: 53–62.

Lee-Saung, H.; Lee-Yeon, S.; Jung-Sang, H.; Kang-Sam, S. and Shin-Kuk, H. (2003). Antioxidant activities of fucosterol from the marine algae *Pelvetia siliquosa*. Archives of Pharmacal Research, 26: 719-722.

Lowenstein. J. (1986). The secret life of seaweeds. Oceans, 19: 72-75.

Ma, F.; Hanna, M. A. (1999). Bidiesel production: a review. Bioresour Technol., 70: 1–15.

Mata, T. M.; Martins, A. A.; Caetano, N. S. (2010). Microalgae for biodiesel production and other applications: A review. Renewable and Sustainable Energy Reviews 14, 217–232.

Mayer, A. M. S. and Hamann, M.T. (2004). Marine pharmacology in 2000: marine compounds with antibacterial, anticoagulant, antifungal, anti-inflammatory, antimalarial, antiplatelet, antituberculosis, and antiviral activities; affecting the cardiovascular, immune, and nervous system and other miscellaneous mechanisms of action. Mar. Biotechnol. 6: 37–52.

Milmo, S. (2008). Oil Fat Int. 24 (2): 22.

National Biodiesel Board. 2002. USA. Available in www.biodiesel.org/.

Naik, S.N., Meher, L.C., Sagar, D.V. (2006). Technical aspects of biodiesel production by transesterification – a review. Renew. Sust. Energy Rev. 10, 248–268.

Ormerod WG, Freund P, Smith A, Davison J. Ocean storage of CO2. IEA greenhouse gas R&D programme. UK: International Energy Agency; 2002.

Radich, A. (2006). Biodiesel performance, costs, and use. US Energy Information Administration. <http://www.eia.doe.gov/oiaf/analysispaper/biodiesel/index.html>.

Raja, R.; Hemaiswarya, S.; Kumar, N.A.; Sridhar, S,; Rengasamy, R. A. (2008). perspective on the biotechnological potential of microalgae. Critical Reviews in Microbiology, 34(2):77–88.

Renewable Fuel Agency (RFA). The Gallagher review of the indirect effects of biofuels production; 2008.

Richmond A. Handbook of microalgal culture: biotechnology and applied phycology. Blackwell Science Ltd; 2004.

Rittmann, B.E. (2008). Opportunities for renewable bioenergy using microorganisms. Biotechnol. Bioeng. 100, 203–212.

Roessler, P.G. (1990) Environmental control of glycerolipid metabolism in microalgae: commercial implications and future research directions. J. Phycol. 26, 393–399.

Scarlat N, Dallemand JF, Pinilla FG. Impact on agricultural land resources of biofuels production and use in the European Union. In: Bioenergy: challenges and opportunities. International conference and exhibition on bioenergy; 2008.

Schenk, P.M., Thomas-Hall, S.R., Stephens, E., Marx, U.C., Mussgnug, J.H., Posten, C., Kruse, O., Hankamer, B. (2008). Second generation biofuels: high-efficiency microalgae for biodiesel production. Bioenergy Res. 1, 20–43.

Shay, E.G. (1993). Diesel fuel from vegetable oils: status and opportunities. Biomass and Bioenergy, 4: 227–242.

Sheehan, J.; Cambreco, J.; Graboski, M.; Shapouri, H. (1998). An overview of biodiesel and petroleum diesel life cycles. US Department of agriculture and Energy Report, p. 1–35.

Shifrin, N.S. and Chisholm, S.W. (1980) Phytoplankton lipids: environmental influences on production and possible commercial applications. In Algae Biomass (Shelef, G. and Soeder, C.J., eds), pp. 627–645, Elsevier

Wang, B.; Li, Y.; Wu, N.; Lan, CQ. (2008). CO_2 bio-mitigation using microalgae. Applied Microbiology and Biotechnology, 79(5):707–18.

6

Getting Lipids for Biodiesel Production from Oleaginous Fungi

Maddalena Rossi, Alberto Amaretti, Stefano Raimondi and Alan Leonardi

University of Modena and Reggio Emilia,
Italy

1. Introduction

Biomass-based biofuel production represents a pivotal approach to face high energy prices and potential depletion of crude oils reservoirs, to reduce greenhouse gas emissions, and to enhance a sustainable economy (Zinoviev et al., 2010). Microbial lipids can represent a valuable alternative feedstock for biodiesel production, and a potential solution for a bio-based economy.

Nowadays, the production of biodiesel is based mostly on plant oils, even though animal fats, and algal oils can also be used. In particular, soybean, rapeseed, and palm oils are adopted as the major feedstock for biodiesel production. They are produced on agricultural land, opening the debate on the impact of the expansion of bioenergy crop cultures, which displace land from food production. Furthermore, their price restricts the large-scale development of biodiesel to some extent.

In order to meet the increasing demand of biodiesel production, other oil sources have been explored. Recently, the development of processes to produce single cell oil (SCO) by using heterotrophic oleaginous microorganisms has triggered significant attention (Azocar et al., 2010). These organisms accumulate lipids, mostly consisting of triacylglycerols (TAG), that form the storage fraction of the cell. The occurrence of TAG as reserve compounds is widespread among all eukaryotic organisms such as fungi, plants and animals, whereas it has only rarely been described in bacteria (Meng et al., 2009). In fact, bacteria generally accumulate polyhydroxyalkanoates as storage compound and only few bacterial species, belonging to the actinobacterial genera *Mycobacterium*, *Streptomyces*, *Rhodococcus* and *Nocardia* produce relevant amounts of lipids (Alvarez & Steinbuchel, 2002).

Among heterotrophic microorgansisms, oleaginous fungi, including both molds and yeasts, are increasingly been reported as good TAG producers. This chapter will focus on current knowledge advances in their metabolism, physiology, and in the result achieved in strain improvement, process engineering and raw material exploitation.

2. Ecology of oleaginous fungi

Oleaginous microorganisms are able to accumulate lipids above the 20% of their biomass, on dry basis. Several species of yeasts and filamentous fungi are regarded as oleaginous, since they have the capability to synthetize and accumulate high amounts of TAG within their cells, up to 70% of the biomass weight. These lipids have similar composition and energy value to plant and animal oils, but their production do not compete for food resources, in particular if it is based on inexpensive carbon sources, such as raw materials, by-products, and surplus. Furthermore, fungal SCO have a short process cycle, and their production is not subjected to seasonal and cyclical weather variations.

The study of oleaginous yeasts has a long history: their ability to accumulate lipids has been known from the 70s, but only in the last years the attention has been focused on exploitation of SCO for biodiesel production. The yeasts represent a part of the microbiota in all natural ecosystems, such as soils, freshwaters and marine waters, from the ocean surface to the deep sea. Widely distributed in the natural environment, they colonize also more extreme environments, such as low temperatures, low oxygen availabilities, and oceanic waters (Butinar et al., 2007). Approximately 1500 species of yeasts belonging to over 100 genera have been described so far (Satyanarayana & Kunze, 2010). Although the vast majority of yeasts are beneficial to human life, only a few are opportunistic human pathogens. As a whole, they play a pivoltal role in the food chain, and in the carbon, nitrogen and sulphur cycles. Among the huge number of species that have been described, only 30 are able to accumulate more than 25% of their dry weight as lipids (Beopoulos et al., 2009b).

Basidiomycetous yeasts strongly prevail among oleaginous yeasts, representing most of all the strains identified as lipid producers, even though some important oleaginous species have been identified among Ascomycota as well (e.g. *Yarrowia lipolytica*).

The most deeply investigated oleaginous yeasts belong to the genera *Yarrowia, Candida, Rhodotorula, Rhodosporidium, Cryptococcus,* and *Lypomyces* (Ageitos et al., 2011; Li et al., 2008; Rossi et al., 2009). *Yarrowia lipolytica,* previously referred to as *Candida lipolytica,* is a good candidate for single-cell oil production (Beopoulos et al., 2009a; Beopoulos et al., 2009b). *Yarrowia* are hemiascomycetous dimorphic fungi that belong to the order *Saccharomycetales.* They are able to degrade hydrophobic substrates such as n-paraffins and oils very efficiently and this physiological feature prompted the scientific community to explore several biotechnological applications (Bankar et al., 2009). The common habitats of these fungi are oil-polluted environments and foods such as cheese, yogurt, kefir, shoyu, meat, and poultry products. Despite *Y. Lipolytica* is distantly related to the conventional yeast *Saccharomyces cerevisiae,* the genome displays an expansion of protein families and genes involved in hydrophobic substrate (such as alkanes and lipids) utilization. Wild-type strains accumulate up to 38% of dry weight (DW) as lipids. Albeit the levels are lower than those of other oleaginous yeasts, it became a model organism because it can be subjected to genetic and metabolic engineering, having be developed a reliable and versatile system for disruption, cloning and expression of target genes.

Within the *Candida* genus, *Candida curvata* (Holdsworth & Ratledge, 1991) also referred as *Apiotrichum curvatum* and *Candida freyschussii* (Amaretti et al., 2011) synthetize and store significant amount of lipids. *Candida* comprises an extremely heterogeneous group of Ascomycota that can all grow with yeast morphology, classified in 150 heterogeneous species, among which only a minority have been implicated in human diseases, since approximately 65% of *Candida* species are unable to grow at 37°C, then they can not be successful pathogens or commensals of humans (Calderone, 2002). Therefore, most of the species can be exploited for biotechnological applications, despite of unwarranted negative public perceptions.

Lipomyces spp. present a great propensity to accumulate triacylglycerols. This genus belongs to the *Saccharomycetales* order and represents a unique branch in the evolution of the ascomycetes (van der Walt, 1992). *Lipomyces* are true soil inhabitants and have a worldwide distribution. The oleaginous species *Lipomyces starkeyi* has the capability to accumulate over 70% of its cell biomass as lipid under defined culture conditions, and can produce lipid on xylose, ethanol, and L-arabinose, or using a mixture of glucose and xylose (Zhao et al., 2008), as well as other wastes (Angerbauer et al., 2008).

Cryptococcus curvatus is a yeast with industrial potential as single-cell oil because it can grow and accumulate lipid on a very broad range of substrates. It requires minimal nutrients for growth, accumulating up to 60% of its cellular dry weight (DW) as intracellular lipid (Meesters et al., 1996; Zhang et al., 2011). Yeasts of the *Cryptococcus* genus are widely distributed in nature and may be isolated from various substrates such as air, soil, bird excreta, water, animal surfaces and mucosae, leaves, flowers, and decomposing wood. Most species are considered as free-living (non-symbiotic) and only a few have medical importance being responsible for disease in man and animals (*C. neoformans* and *C. gattii*). *C. curvatus* is recognized as an opportunistic pathogen of animals, including humans (Findley et al., 2009).

Species belonging to the genus *Rhodosporidium*, and to its asexual counterpart *Rhodotorula*, have been claimed as oleaginous yeasts. They belong to one of the three main lineages of the *Basidiomycota*, the *Pucciniomycotina*. *Rhodotorula* is a common environmental inhabitant. The synthesis of different commercially important natural carotenoids by yeast species belonging to the genus *Rhodotorula* has led to consider these microorganisms as a potential pigment sources. Within this genus, the mesophilic red yeast *Rhodotorula glutinis* is able to synthetize and store lipids also growing on glycerol, whereas the psychrophilic species *Rhodotorula glacialis*, that are not red yeasts, accumulates lipids in a range of temperature between -3 and 20°C (Amaretti et al., 2010). The red yeast *Rhodosporidium toruloides* is an oleaginous mesophilic species. *Rhodosporidium* are able to carry out a number of diverse biochemical reactions such as biodegradation of epoxides, biphenyls and oxiranes (Smit, 2004), biosynthesis of carotenoids (de Miguel et al., 1997) and other types of biotransformations, but a major biotechnological exploitation is associated to their ability to convert glycerol and lignocellulosic biowastes into lipids (Hu et al., 2009; Yu et al., 2011). Among the oily yeasts, two novel species of the anamorphic basidiomycetous genus *Trichosporon* have been recently identified (*T. cacaoliposimilis* and *T. oleaginosus*) (Gujjari et al., 2011), despite lipid accumulation has not yet explored in the perspective of biodiesel production. *Trichosporon* are basidiomycetous yeasts widely distributed in nature, consisting of soil- and water-associated species, predominantly found in environmental substrates, such as decomposing wood. They present distinct morphological characteristics of budding yeast cells and true mycelia that disarticulate to form arthroconidia. Some species are causative agents of diseases in man and cattle. They can occasionally belong to the gastrointestinal microbiota of humans as well as transiently colonize the skin and respiratory tract.

Exploitation of oleaginous filamentous fungi for biodiesel production has a more recent history, which, with few exceptions, derives from studies focused to poly-unsaturated fatty acid production (PUFA), such as arachidonic acid and γ-linolenic acid. The most relevant example of this biotechnological application is represented by exploitation of *Mortierella alpina* to produce oils containing n-1, n-3, n-4, n-6, n-7, and n-9 PUFAs (Sakuradani et al., 2009). Among the major lipid producers there is *Mucor circinelloides,* a zygomycete fungus, which is emerging as opportunistic pathogen in immunocompromised patients (Li et al., 2011). *M. circinelloides* has been used for the first commercial production of microbial lipids (Ratledge, 2004). Lipid accumulation in *M. circinelloides* has been extensively studied (Wynn et al., 2001), and its TAG have been proposed as feedstock for producing biodiesel by direct transformation of its lipids (Vicente et al., 2009). *M. circinelloides* represents an outstanding model within the *Zygomycota* phylum, based on the availability of an efficient transformation procedure (Gutierrez et al., 2011) and on the whole sequence of genome (http://genome.jgi-psf.org/Mucci2/Mucci2.home.html). Also the phylogenetically related *Umbelopsis isabellina* has emerging as a promising species to convert biomass residues to biodiesel precursors (Meeuwse et al., 2011a). To the best of our knowledge, limited are the attempts to get lipids with *Aspergillus oryzae* that, conversely, is extensively studied as lipase producer to carry out transesterification of TAG (Adachi et al., 2011).

3. Biochemistry of lipid accumulation

Lipid accumulation in oleaginous yeasts and molds has been demonstrated to occur when a nutrient in the medium (e.g. the nitrogen or the phosphorus source) becomes limited and the carbon source is present in excess. Nitrogen limitation is the most efficient condition for inducing lipogenesis. During the growth phase, nitrogen is necessary for the synthesis of proteins and nucleic acids, while the carbon flux is distributed among energetic and anabolic processes yielding carbohydrates, lipids, nucleic acids and proteins. When nitrogen gets limited, the growth rate slows down and the synthesis of proteins and nucleic acids tends to cease. In non-oleaginous species, the carbon excess remains unutilized or is converted into storage polysaccharides, while, in oleaginous species, it is preferentially channeled toward lipid synthesis, leading to the accumulation of TAG within intracellular lipid bodies (Ratledge & Wynn, 2002; Granger at al., 1993).

The biochemical pathway of lipid biosynthesis is not very different among eukaryotic organisms and does not differ in oleaginous and non-oleaginous fungi. The ability to accumulate high amounts of lipid depends mostly on the regulation the biosynthetic pathway and the supply of the precursors (i.e. acetyl-CoA, malonyl-CoA, and glycerol-3-phosphate) and the cofactor NADPH.

Most information were obtained from the model yeast *Saccharomyces cerevisiae* (Kohlwein, 2010), that does not accumulate lipids, and *Yarrowia lipolytica*, that represent a model for bio-oil production and is suitable for genetic manipulation (Beopoulos et al., 2009b).

3.1 Fatty acids biosynthesis and modifications

De novo synthesis of fatty acids (FA), the first step of lipid accumulation, is carried out in the cytosol by fatty acids synthetase (FAS) complex. In yeasts, FAS bears phosphopantheteine transferase activity to activate its acyl carrier protein (ACP) by loading the coenzyme pantothenate. FAS is a multimer of 6 α and 6 β subunits encoded by *fas2* and *fas1*, respectively, each subunit containing four functional domains. Therefore, FAS consists in a α6β6 molecular complex of 2.6 MDa with 48 functional centers that catalyze all reactions required for synthesis of fatty acids through cycles of multistep reactions. FAS firstly loads acetyl-CoA on its β-ketoacyl-ACP synthase (KS), then it exherts β-ketoacyl-ACP reductase (KR), β-hydroxyacyl-ACP dehydratase (DH), and enoyl-ACP reductase (EAR) activities. This set of reactions is repeated cyclically seven times to yield palmitoyl-ACP (Fig. 1) (Tehlivets et al, 2007).

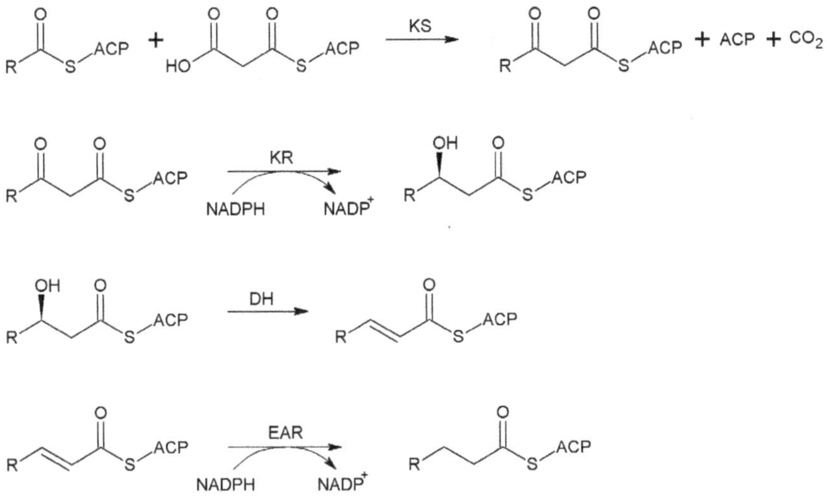

Fig. 1. Reactions occurring sequentially in fatty acid synthetase: condensation of acyl-ACP and malonyl-ACP mediated by KS, NADPH-dependent reduction of the keto group to a hydroxyl group by means of KR, dehydration to create a double bond with DH and reduction of the double bond by means of EAR. R = H, $CH_3(CH_2)_{2n;}$ $n_{max}=7$.

The biosynthesis of FA requires the constant supply of acetyl-CoA as initial biosynthetic unit and of malonyl-CoA as the elongation unit, supplying two carbons at each step. Non-oleaginous yeasts receive acetyl-CoA mostly from glycolysis. In oleaginous yeasts, acetyl-CoA is mostly provided by the cleavage in the cytosol of citrate, which accumulated as a consequence of nitrogen limitation (Ratledge, 2002) (Fig. 2). In fact, lipid accumulation by oleaginous fungi does not occur under balanced nutrient conditions.

In oleaginous yeasts, nitrogen limitation activates AMP-deaminase (Ratledge & Wynn, 2002), which supply ammonium to the nitrogen-starved cell. As a consequence, mitochondrial AMP concentration decreases, causing isocitrate dehydrogenase activity to drop. The TCA cycle is then blocked at the level of isocitrate, which accumulates and equilibrates with citrate through aconitase. Excess of citrate from TCA cycle is exported out of the mitochondrion via the malate/citrate antiport. Cytosolic ATP-citrate lyase (ACL) cleaves citrate to give oxaloacetate and acetyl-CoA (Fig. 2).

ACL represents one of the key enzymes that contribute to the oleaginous trait of yeasts, whereas its activity is negligible in non-oleaginous species. ACL is composed of two subunits, encoded by ACL1 and ACL2 and is negatively regulated by exogenous FA.

Malonyl-CoA is produced from acetyl-CoA by acetyl-CoA carboxylase (ACC) that condensate an acetyl-CoA unit with bicarbonate:

$$\text{Acetyl-CoA} + \text{HCO}_3^- + \text{ATP} \leftrightarrows \text{malonyl-CoA} + \text{ADP} + \text{Pi}$$

ACC is also a key enzyme in *de novo* FA synthesis, since *ACC1* mutants became FA auxotrophs or maintain low levels of ACC activity (Tehlivets et al., 2007). ACC1 undergoes allosteric activation by citrate. Furthermore the transcription of *FAS1*, *FAS2*, and *ACC1* is coordinately regulated, being negatively regulated by FA.

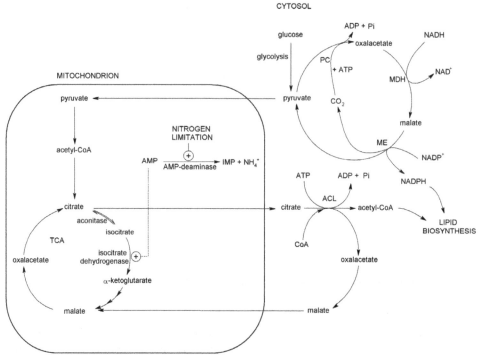

Fig. 2. Lipid biosynthesis from excess of citrate as a consequence of nitrogen limitation. Adapted from Ratledge, 2004.

Cytosolic NADPH is required for KR and EAR functions of FAS. For each elongation step of the acyl chain, two molecules of NADPH are required. One of the major sources of cytosolic NADPH are the pentose phosphate pathway and the transhydrogenase cycle, which transforms NADH into NADPH through the activity of pyruvate carboxylase (PC), malate dehydrogenase (MDH), and malic enzyme (ME), catalyzing the following reactions:

$$\text{Pyruvate} + CO_2 + \text{ATP} \leftrightarrows \text{oxaloacetate} + \text{ADP} + \text{Pi (PC)}$$

$$\text{Oxaloacetate} + \text{NADH} \leftrightarrows \text{Malate} + \text{NAD}^+ \text{ (MDH)}$$

$$\text{Malate} + \text{NADP}^+ \leftrightarrows \text{pyruvate} + CO_2 + \text{NADPH (ME)}$$

$$\text{NADH} + \text{NADP}^+ \leftrightarrows \text{NADPH} + \text{NAD}^+$$

ME has been found in several oleaginous fungi and it has been regarded as a key enzyme involved in lipid accumulation (Ratledge, 2002). In *Mortierella circinelloides*, overexpression of ME enhanced lipid accumulation (Zhang et al., 2011), whereas overexpression of the ME homologous in *Yarrowia lipolytica* did not result in yield improvement.

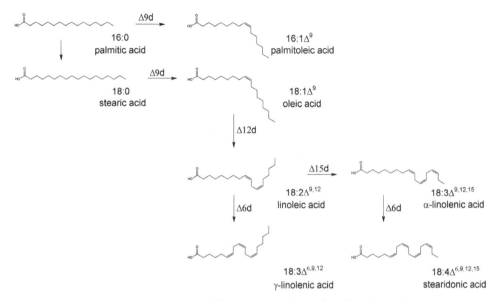

Fig. 3. Biosynthesis of poly-unsaturated fatty acid. Δ9d, Δ12d and Δ15d are the most common desaturases which are present in the endoplasmic reticulum (Ratledge 2004).

The final products of FAS are myristic or palmitic acids, depending on the yeast species. Reactions resulting in further elongation or desaturation occur in the endoplasmic reticulum (ER). Elongation reactions are catalyzed by elongases (such as malonyl-palmitoil transacylase, MPT) organized in a complex that requires malonyl-CoA provided by ACC. Desaturations are introduced by ER desaturases, hydrophobic membrane-bound proteins. The most common desaturases are Δ9, which inserts the first double bond onto palmitic and/or stearic acids, and Δ12, which catalyzes the insertion of the second unsaturation into oleic acid to produce linoleic acid. Δ6 and Δ15 desaturase activities have been recently described in in psychrophilic oleaginous yeasts, based on production of γ and α-linolenic acids, respectively (Fig. 3).

3.2 Biosynthesis of triacyl-glycerol

The fatty acyl-CoA produced by *de novo* synthesis are esterified with glycerol or sterols to produce triacyl-glycerol (TAG) and steryl-esters (SE), respectively. In oleaginous fungi, the neutral lipids SE and TAG are store inside the lipid bodies (LB). TAG are mostly formed by consecutive acylation of glycerol-3-phosphate (G3P), carried out by diverse acyl transferases. G3P is formed from glycerol by glycerol kinase or can be synthesized from dihydroxyacetone phosphate (DHAP) by G3P dehydrogenase, in a reversible reaction. *S. cerevisiae* can use both G3P and DHAP as acyl-group acceptor. The addition of the first acyl group leads to 1-acyl G3P, also named lysophosphatidic acid (LPA). LPA can also be formed by reduction of acyl-DHAP, carried out by a NADPH dependent reductase. A second acyltransferase loads an other acyl group, producing 1,2-diacyl G3P (phosphatidic acid, PA). Phosphate is removed from PA by phosphatidate phosphatase isoenzymes, generating diacylglycerol (DAG). DAG can be the direct precursor of TAG, or can be channeled toward phospholipids biosynthesis (Fig. 4).

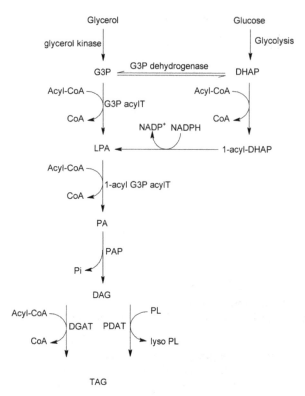

Fig. 4. *de novo* synthesis of TAG (adapted from Czabany et al., 2007)

The last step of *de novo* synthesis of TAG can be carried either by using diverse acyl donors, such as acyl-CoA or with phospholipids. In the former case, DAG acyl transferases (DGAT), which are integral proteins of the ER, can directly load the third Acyl-CoA. A DGAT enzyme is present in *S. cerevisiae* and *Y. lipolytica* and is mostly active during the stationary phase, although it is expressed also during the exponential phase. A second DGAT, more active during the exponential growth phase, has been identified in *Y. lipolytica*. In *S. cerevisiae* the phospholipid:DAG acyltransferase (PDAT) is localized in the ER, whereas in *Y. lipolytica* it is present both in the ER and in the surface of LB (Fig. 4).

3.3 Biogenesis of lipid bodies

In eukaryotes, neutral lipids are stored in specialized compartments known as lipids bodies (LB). They are assembled at a specialized subdomain of the ER where most biosynthetic enzymes and structural proteins are located (Waltermann et al., 2005). The neutral lipids do not fit among phospholipids and are thus deposited between the two leaflets of the membrane bilayer. However, substantial amounts of neutral lipids cannot be incorporated into membrane bilayer of ER and ongoing neutral lipid synthesis leads to the formation of a bud which buds off of the ER as a mature LB after reaching the critical size (Fig. 5).

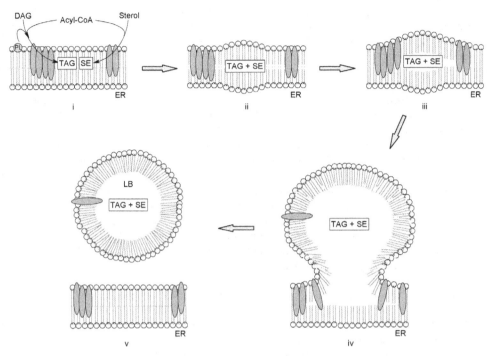

Fig. 5. Model of lipid bodies biogenesis from the membrane of the ER. TAG and SE accumulate between the two leaflets of the phospholipid bilayer (i to iii). The micro-droplet generated (iii, iv) evolve to lipid bodies (v) (Figure adapted from Czabany et al., 2007).

In most oleaginous yeasts, the neutral lipids of LB consist mostly of TAG (up to 90% or more) whereas a small fraction is represented by steryl esters. The presence of significant quantity of free fatty acids (FFA) within LP has been reported only for *Y. lipolytica*. In *S. cerevisiae*, which accumulates less than 15% lipids of its biomass, LB contain similar amounts of TAG and SE.

The core of LB, consisting of neutral lipids is surrounded by a phospholipid monolayer where several proteins are embedded. These proteins exert a key role in lipid metabolism, biosynthesis, and substrate trafficking. Upon requirement, storage lipids are mobilized from this compartment by triacylglycerol lipases and steryl ester hydrolases. The respective degradation products serve as energy sources and/or building blocks for membrane formation. In fact, FA hydrolyzed from TAG or SE are either channeled to the peroxisome, where β-oxidation takes place, or to phospholipid biosynthesis.

4. Metabolic engineering of oleaginous yeasts

The availability of genome data and genetic tools, such as the possibility to integrate homologous or heterologous genes, opened up the possibility to use metabolic engineering to understand the molecular mechanisms involved in lipid accumulation or to increase the yield of stored lipids in *S. cerevisiae* and *Y. lipolytica*. Whereas *S. cerevisiae* has been used mostly as a model to investigate and understand the lipid metabolism, in *Y. lipolytica* several attempts have been done in order to address the carbon flux toward TAG production and accumulation. Similar approaches are precluded to other oleaginous fungi since they lack genome information and the necessary tools for gene manipulation and strain improvement.

In *Y. lipolytica*, the role of glycerol-3-phosphate (G3P) in triacylglycerol (TAG) biosynthesis and accumulation has been investigated (Beopoulos et al., 2008). In this yeast G3P is formed from glycerol by the glycerol kinase encoded by *GUT1*, or it is synthetized from dihydroxyacetone phosphate (DHAP) by the G3P dehydrogenase (*GPD1*). The antagonist reaction, which produces DHAP from G3P, is carried out in competition by a second isoform of the G3P dehydrogenase, encoded by *GUT2*. In order to force the conversion of DHAP into G3P, the gene *GPD1* was over-expressed and the gene *GUT2* was deleted.

A diverse strategy to increase lipid accumulation was based on the disruption of the β-oxidative metabolism, through the deletion of the 6 *POX* genes (*POX1 to POX6*) that encode the peroxisomal acyl-coenzyme oxidases (Mlickowa et al., 2004a; Mlickowa et al., 2004b; Beopoulos et al., 2008). As a whole, the best results in terms of percentage of lipids per dry biomass, were reached coupling the increased level of G3P with the disactivation of the β-oxidation pathway (65%) (Dulermo et al., 2011).

Metabolic engineering strategies have been recently exploited to expand the range of substrates used by oleaginous fungi, also through functional expression of heterologous genes. Recently, it has been found that inulin is a good material for bio-productions (Chi et al., 2009). In order to make the oleaginous yeast *Y. lipolytica* able to accumulate lipids on inulin containing materials, the *Kluyveromyces marxianus* exo-inulinase gene (*INU1*) was heterologously expressed on a high copy plasmid (Zhao et al., 2010). The inulinase was efficiently secreted by *Y. lipolytica*, and inulin was hydrolyzed, assimilated and converted into TAG.

5. Cultivation condition of oleaginous yeasts

Lipid accumulation by oleaginous yeasts depends mostly on nutrient limitation conditions when excess carbon is present in the medium. Nutrient limitation prevents cells from being generated, while the carbon excess is converted into storage TAG. Published studies reports that phosphorus, magnesium, zinc, or iron limitation lead to lipid accumulation in model oleaginous yeasts (Hall & Ratledge, 1977; Beopoulos et al., 2009; Wu et al., 2010). However, nitrogen limitation is the most efficient form of nutrient limitation for lipogenesis induction, leading to the highest values of substrate/lipid conversion yield and lipid content within biomass (Hall & Ratledge, 1977; Wynn et al., 2001). Thus, nitrogen limitation is commonly used to induce lipogenesis in oleaginous fungi and the utilization of cultural media with appropriate C/N ratio is crucial to maximize lipid production.

Several studies focused on determining the optimal composition of cultural media for oleaginous fungi with the aim to optimize the performance of lipid-producing bioprocesses. The effect of the C/N ratio on lipid metabolism has been investigated for a number of oleaginous yeasts and molds, such as *Y. lipolytica* and many oleaginous species of *Rhodotorula, Candida, Apiotrichum/Cryptococcus, Mortierella* (Hall & Ratledge, 1977; Papanikolau et al., 2003; Granger et al., 1992; Wu et al., 2010; Park et al., 1990; Jang et al., 2005; Amaretti et al., 2010), and has been mathematically modeled for some of these organisms (Ykema et al., 1986; Granger et al., 1993; Economou et al., 2011). *Y. lipolytica* is the oleaginous microorganism for which information about the metabolic response to different C/N ratios is most abundant (Beopoulos et al., 2009a), particularly due to the availability of molecular tools for genetic engineering of this organism. Therefore, *Y. lipolytica* is regarded as a model organism for microbial oil production and the main traits of its metabolism can be used to give a general description of the metabolic response to different C/N ratios in the majority of oleaginous yeasts. With the increase of the C/N ratio, different metabolic behaviors were observed in *Y. lipolytica*: i) growth with mobilization of storage lipids, ii) growth of fat-free biomass, iii) growth with accumulation of lipids, and iv) growth with lipid accumulation and production of organic acids (Fig. 6).

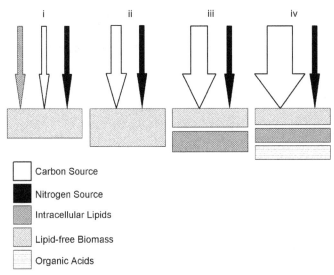

Fig. 6. Metabolic activity of oleaginous fungi (e.g. *Y. lipolitica*) as a function of carbon flow rate for a fixed nitrogen flow rate. Arrows indicate the consumption of nitrogen and carbon sources by the cells; squares indicate production rate. The dimension of arrows and squares is proportional to flow. (adapted from Beopoulos et al., 2009a)

If the medium is carbon limited or when the extracellular carbon supply gets exhausted, previously stored intracellular lipid can be mobilized and utilized by oleaginous microorganisms to sustain cells generation and production of lipid-free biomass (Park et al., 1990) (Fig. 6, i). If the medium is balanced and/or furnishes just the right amount of carbon flow to satisfy the growth need, balanced growth occurs without any accumulation of storage lipids (Fig. 6, ii). In conditions of carbon excess, a part of the carbon flow, which is proportional to nitrogen availability (Granger et al., 1993), is directed toward cells generation, whereas the carbon exceeding growth needs is channeled to the production of storage lipids (Fig. 6, iii). In some oleaginous fungi, the presence of a large carbon excess leads to the production of great amounts of organic acids, such as pyruvic acid and diverse TCA-cycle intermediates, at the expenses of lipid accumulation (Fig. 6, iv). In these latter conditions, *Y. lipolytica* produces citric acid (Levinson et al., 2007) but other oleaginous yeasts have never been reported to behave this way.

6. Batch, fed-batch and fermentation processes

Batch, fed batch, and continuous modes of culture have been developed to culture oleaginous microorganisms. Lipid production in batch cultures is carried out in a cultural medium with a high initial C/N ratio, the carbon source being present in an adequate excess with respect to the nitrogen source. In fact, in this condition, the flow of carbon utilization is limited only by the substrate uptake system of the cell, while the changes in nitrogen concentration determine the passage from a phase of balanced growth to a phase of lipid accumulation, causing the process to proceed through two phases. As nitrogen is consumed from the culture the C/N ratio tends to increase, but growth remains exponential and balanced until nitrogen is not the limiting substrate. During the growth phase, the carbon flow is mostly channeled to satisfy the growth need, therefore growth is balanced and lipid-free biomass is mostly produced (Fig. 6 ii). As nitrogen concentration becomes limiting, the growth rate and the carbon flow toward biomass generation decrease, while lipid production is triggered, resulting in a shift of microbial metabolism into the lipogenic phase (Fig. 6 iii, Fig. 7).

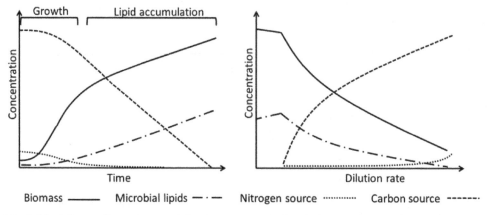

Fig. 7. Modeling and prediction of the timecourse of a batch fermentation (left) and the steady-state values of a continuous process (right) for microbial production of lipids. Axis are in arbitrary scales.

In batch cultures the initial C/N ratio of the cultural medium has a pivotal role in determining the bioprocess performance. In fact, both the rate and the yield of lipid production depend by the C/N ratio, which affects the duration of the exponential phase and the amount of biomass produced during growth. With a fixed carbon concentration, higher amounts of lipid-free biomass produced during the growth phase correspond to higher lipid production rates during the lipogenic phase, but to lower amounts of lipid content within cells and lipid/substrate conversion yields. Therefore, the initial C/N ratio needs to be optimized to maximize lipid productivity in batch cultures. The optimal C/N value is always high (e.g. in the range between 80 and 350 mol/mol) and strongly depends on the microorganism, the medium composition, the carbon source (e.g. glucose, glycerol, etc.), and the nitrogen source (e.g. diverse organic or inorganic sources). The minimal C/N ratio suitable for lipid accumulation can be estimated as $(Y_{X/S} \cdot q)^{-1}$, where $Y_{X/S}$ is the biomass/carbon source yield coefficient under conditions of carbon limitations (C-mol/C-mol) and q is the nitrogen/carbon content of biomass (N-mol/C-mol) (Ykema et al., 1986). However, it should be considered that extremely high C/N ratios may cause the production of unwanted byproducts, such as organic acids (Fig. 6 iv), or may lead to severe nitrogen deficiency, causing a rapid decrease in cells viability.

Unlike batch processes, in fed-batch mode, nutrients are fed into the bioreactor in a controlled manner, with the purpose to monitor and control the specific growth rate and the flows of nitrogen and carbon utilization. Through the judicious management of the feeding rate and composition, it is possible to control the C/N ratio within the culture and maintain the oleaginous microorganism in the optimal metabolic status, as appropriate, first for the growth phase, and later for the lipogenic phase. The lipogenic phase is the most extensive, corresponding to lipid production under nitrogen limitation, with constant C/N ratio, preventing loss of viability and acids production (Beopoulos et al., 2009a).

In continuous cultures, at the steady state, the assimilation of C and N sources and the microbial growth occur at constant rates, which ultimately depend by the dilution rate (D). The concentration of the substrates within the bioreactor is steady and depends by the dilution rate as well, the actual C/N ratio of the culture remaining constant unlike in batch cultures. Likewise in batch cultures, in continuous cultures the C/N ratio of the fresh medium needs to be higher than $(Y_{X/S} \cdot q)^{-1}$ to obtain some lipid accumulation (Ykema et al., 1986). However, at the steady-state with this medium, the C/N ratio within the culture is higher than in the fresh medium, due to nitrogen consumption. The extent of substrates utilization, and also the biomass and lipid concentrations are the highest at low D values and decrease with the increase of D (Fig. 7). While low D values promote lipid production and a more complete substrate utilization, on the contrary, the volumetric productivity of continuous processes is positively affected by the increase of the dilution rate (Ykema et al., 1986; Meeuwse et al., 2011b). Therefore, both the C/N content of the medium and the dilution rate need to be thoroughly tuned to maximize lipid productivity of continuous processes.

7. Substrates and raw material

The demand for the inexpensive production of biofuels has intensified due to increasing concerns of climate change, depletion of petroleum-based fuels, and environmental problems. In a market economy, corporations aim to maximize profit, seeking the most competitive feedstock. To produce single-cell oils for biodiesel production, the carbon source has necessarily to be cheap and available in large quantities. Therefore, while the first investigations on oleaginous fungi most commonly employed glucose as carbon source, nowadays the production of single-cell oils is predominantly addressed to transformation of raw materials, by-products and surplus.

Glucose is the carbon source most commonly employed for growth of oleaginous fungi and lipid production (Boulton & Ratledge, 1984; Hansson & Dostalek, 1986; Hassan et al., 1993; Heredia & Ratledge, 1988; Jacob, 1991; Jacob, 1992; Johnson et al., 1992; Li et al., 2007; Pan et al., 1986; Ratledge, 2004; Rau et al., 2005; Saxena et al., 2008; Zhao et al., 2008). High glucose concentrations enhance the carbon flow that is directed toward TAG production, thus improving lipid production in several yeasts. However, growth of some yeasts (e.g. *R. toruloides)* is inhibited by high concentration of glucose, (Li et al., 2007). Furthermore, in batch cultures, initial glucose concentration also affects the fatty acids composition of the lipids (Amaretti et al., 2010).

Carbon sources other than glucose, such as xylose (Chistopher et al., 1983; Heredia & Ratledge, 1988;), lactose (Christopher et al., 1983; Daniel et al., 1999;), arabinose, mannose (Hansson & Dostalek, 1986), mannitol (Hansson & Dostalek, 1986), ethanol (Chistopher et al., 1983; Eroshin & Krylova, 1983), have been also investigated in the 80s and 90s for the production of microbial lipids.

Albeit glucose is a very good carbon source for lipid production with oleaginous fungi, molasses, which carbohydrate fraction is mainly composed of sucrose, glucose, and fructose, do not represent a promising raw material for lipid production, since they are characterized by a high nitrogen content which delays the unbalanced growth, where number of cells can not augment anymore and lipids are accumulated (Johnson et al., 1995).

Carbons sources obtained from ligno-cellulosic biomasses represent one of the most important potential to produce biodiesel. In fact, several waste biomasses containing forest residues, agricultural residues, food wastes, municipal wastes, and animal wastes can be utilized for the production of lignocellulosic based microbial lipids. Microbial oil production from sulphuric acid treated rice straw hydrolysate (SARSH) by the yeast *Trichosporon fermentans* pointed out the difficulty to perform the process of lipid accumulation in presence of the inhibitory compounds released during hydrolysis, such as acetic acid, furfural, 5-hydroxymethylfurfural, and water soluble lignin (Huang et al., 2009). Selected strains were able to grow on xylose and glucose (Zhu et al., 2008), but the crude hydrolyzate did not result an optimal substrate for a high yield process of lipid production. Cellulose and hemicellulose are generally hardly hydrolyzed and assimilated by yeasts, while they can be degraded and used as carbon source by filamentous fungi. A screening of endophytic fungi from the oleaginous plants was the selection of strains belonging to the genera *Microsphaeropsis, Phomopsis, Cephalosporium, Sclerocystis* and *Nigrospora* that simultaneously accumulated lipids (21.3 to 35.0% of dry weight) and produced cellulase (Peng & Chen, 2007). Albeit these strains could be exploited as microbial oil producers by utilising straw as substrate, they have never been claimed again as a SCO producers on lingo-cellulosic biomass. Attempts to carry out lipid production in Solid State Fermentation (SSF) on wheat straw have been performed exploiting a cellulolytic strain of *Aspergillus oryzae* (Lin et al., 2010). This strain is able to use cellulose as substrate and accumulate lipids in a low cost fermentation system on this abundant cellulosic by-product.

Other complex matrices have been used, such as solids from wheat bran fermentation (Jacob 1991), sewage sludge (Angerbauer et al., 2008), wastewaters of animal fat treatment (Papanikolaou et al., 2002), whey derivatives (Ykema et al., 1989; Vamvakaki et al., 2010), olive oil mill wastewaters (Yousuf et al., 2010), and tomato waste hydrolysate (Fakas et al., 2008).

Nowadays, lipid production with oleaginous yeasts is focused on selection and development of yeasts as converters of glycerol into lipid for biodiesel production, since it is the major side-product of the biodiesel production process. The biotransformation of glycerol into TAG is therefore regarded as a promising way to decrease the cost of biodiesel process through simultaneous reutilization of its major byproduct. In general, for every 100 kg of biodiesel produced, approximately 10 kg of crude glycerol are created. Crude glycerol is a mixture of glycerol (65–85%, w/w), methanol, and soap, and contains macro elements such as calcium, potassium, magnesium, sulfur and sodium. In order to minimize unknown variables introduced through the use of crude glycerol, several studies to determine whether or not glycerol could be used as substrate or co-substrate for growth have been conducted using purified glycerol.

A deep characterization of lipid accumulation on glycerol has been carried out with *Yarrowia lipolytica*, that is able to metabolize several important industrial and agro-industrial by-products such as raw glycerol, producing large amounts of SCO and organic acids (Papanikolaou et al., 2003; Papanikolaou & Aggelis, 2002; Rymowicz et al., 2010; Rywinska et al., 2009). Biochemistry of lipid production on glycerol has been investigated in this organism: glycerol passes into the microbial cell by facilitated diffusion and the conversion is carried out via phosphorylation pathway, with direct phosphorylation to G3P and subsequent dehydrogenation. Recently, *Y. lipolytica* has been subjected to targeted and purposeful alteration of G3P shuttle pathway to better utilize glycerol for lipid production. In the genetically manipulated strains, lipid accumulation resulted from a complex interrelation between different processes in diverse cell compartments, such as lipid synthesis in the cytosol, location and storage in ER and LB, mobilization and degradation processes (Dulermo & Nicaud, 2011).

Pure glycerol supported growth and lipid accumulation of *Rhodotorula glutinis* and *Candida freyschussii* (Easterling et al., 2009; Amaretti et al., 2011), being used as sole carbon and energy source or in addition to xylose or glucose. The diverse composition of the medium affected not only the lipid/biomass yield, but also the TAG composition, in terms of ration of saturated, monounsaturated, and polyunsaturated fatty acids (Easterling et al., 2009).

Attempts to convert crude glycerol into lipids have been successfully performed exploiting the oleaginous yeast *Cryptococcus curvatus* (Liang et al., 2010). Different processes have been developed with very efficient yields and productivities. In a 12 days two-stage fed-batch where raw glycerol was fed, the biomass density and the lipid content reached 32.9 g/l and 52%, respectively. Methanol of crude glycerol did not pose a significant inhibitory effect even though it was existent in the bioreactor. Lipid accumulated by *C. curvatus* on glycerol presented high amount of monounsaturated fatty acid, turning out as excellent substrate for transformation into biodiesel.

8. Conclusions and perspectives

Oleaginous fungi, and particularly yeasts, are very efficient in the accumulation of intracellular TAG and it is expected that they will be exploited by the biofuel industry in the future. Nonetheless, the costs of microbial lipids are still too high in order to compete with plant oils for biodiesel manufacturing. Cheap carbon sources have necessarily to be used as carbon sources for the cultivation of these microorganisms and the performance of the bioprocess has to be further improved in terms of both the yield and the productivity. The exploration of the natural biodiversity is a promising strategy to identify novel oleaginous species that assimilate and get fat on agro-industrial residues, particularly the lingo-cellulosic biomass and crude glycerol from biodiesel industry. Further approaches combining genomic, transcriptomic, metabolomics, and lipidomic techniques will undoubtedly provide deeper information of lipid production by oleaginous fungi. A metabolic engineering approach is very promising, but it is still precluded for the most oleaginous species, for which genome disclosure has not been accomplished and genetic tools are not available yet.

9. References

Adachi, D., Hama, S., Numata, T., Nakashima, K., Ogino, C., Fukuda, H., & Kondo, A., (2011) Development of an *Aspergillus oryzae* whole-cell biocatalyst coexpressing triglyceride and partial glyceride lipases for biodiesel production. *Bioresource Technology*. Vol. 102, No. 12, pp. 6723-6729, ISSN 0960-8524.

Ageitos, J. M., Vallejo, J. A., Veiga-Crespo, P., & Villa TG. (2011) Oily yeasts as oleaginous cell factories. *Applied Microbiology and Biotechnology*. Vol. 90, No. 4, pp. 1219-1227, ISSN 0175-7598.

Alvarez, H. M. & Steinbuchel, A. (2002) Triacylglycerols in prokaryotic microorganisms. *Applied Microbiology and Biotechnology*. Vol. 60, No. 4, pp. 367-376, ISSN 0175-7598.

Amaretti, A., Raimondi, S., Sala, M., Roncaglia, L., De Lucia, M., Leonardi, A., & Rossi, M. (2010) Single cell oils of the cold-adapted oleaginous yeast *Rhodotorula glacialis* DBVPG 4785. *Microbial Cell Factories*. Vol. 23, No. 9, pp. 73, ISSN 1475-2859.

Amaretti, A., Raimondi, S., Leonardi, A., & Rossi, M. (2011) Lipid production from glycerol by *Candida Freyschusii*. *Proceedings of FEMS 2011 14th Congress of European Microbiologists*, pp. 126, Geneva, Switzerland, June 26-30, 2011

Angerbauer C, Siebenhofer M, Mittelbach M, Guebitz GM (2008) Conversion of sewage sludge into lipids by *Lipomyces starkeyi* for biodiesel production. *Bioresource Technology*. Vol. 99, No. 8, pp. 3051-3056, ISSN 0960-8524.

Azócar, L., Ciudad, G., Heipieper, H. J., & Navia, R. (2010) Biotechnological processes for biodiesel production using alternative oils. *Applied Microbiology and Biotechnology*. Vol. 88, No. 3, pp. 621-636, ISSN 0175-7598.

Bankar, A. V., Kumar, A. R., & Zinjarde, S. S. (2009) Environmental and industrial applications of Yarrowia lipolytica. *Applied Microbiology and Biotechnology*. Vol. 84, No. 5, pp. 847-865, ISSN 0175-7598.

Beopoulos, A., Mrozova, Z., Thevenieau, F., Le Dall, M. T., Hapala, I., Papanikolaou, S., Chardot, T., & Nicaud, J. M. (2008) Control of lipid accumulation in the yeast *Yarrowia lipolytica*. *Applied and Environmental Microbiology*. Vol. 74, No. 24, pp. 7779-7789, ISSN 0099-2240.

Beopoulos, A., Cescut, J., Haddouche, R., Uribelarrea, J. L., Molina-Jouve, C., & Nicaud, J. M. (2009a) *Yarrowia lipolytica* as a model for bio-oil production. *Progress in Lipid Research*. Vol. 48, No. 6, pp. 375-387, ISSN 0163-7827.

Beopoulos, A., Chardot, T., & Nicaud, J.M. (2009b) *Yarrowia lipolytica*: a model and a tool to understand the mechanisms implicated in lipid accumulation. *Biochimie*. Vol. 91, No. 6, pp. 692-696, ISSN 0300-9084.

Boulton, C. A. & Ratledge, C. (1984) *Cryptococcus terricolus*, an oleaginous yeast re-appraised. *Applied Microbiology and Biotechnology*. Vol. 20, No. 1, pp. 72-76, ISSN 0175-7598.

Butinar, L., Spencer-Martins, I., & Gunde-Cimerman, N. (2007) Yeasts in high Arctic glaciers: the discovery of a new habitat for eukaryotic microorganisms. *Antonie Van Leeuwenhoek*. Vol. 91, No. 3, pp. 277-289, ISSN 1572-9699.

Calderone, R. A. (2002) Candida and Candidiasis. Edited by R. A. Calderone, *American Society of Microbiology Press*. ISSN 1058-4838, Washington, DC, USA.

Chi, Z. M.,Chi, Z., Zhang, T., Liu, G. L.,& Yue,L. X. (2009) Inulinase expressing microorganisms and applications of inulinases. *Applied Microbiology and Biotechnology*. Vol. 82, No. 2, pp. 211-220, ISSN 0175-7598.

Christopher, T., Scragg, A. H., & Ratledge, C. (1983) A comparative study of citrate efflux from mitochondria of oleaginous and non oleaginous yeasts. *European Journal of Biochemistry* Vol. 130, No. 1, pp. 195-204, ISSN 0014-2956

Czabany, T., Athenstaedt, K., & Daum, G. (2007) Synthesis, storage and degradation of neutral lipids in yeast. *Biochimica et Biophysica Acta*. Vol. 1771, pp. 299-309. ISSN 0006-3002

Dahlqvist, A., Stahl, U., Lenman, M., Banas, A., Lee, M., Sandager, L., Ronne, H., & Stymne, S. (2009) Phospholipid:diacylglycerol acyltransferase: an enzyme that catalyzes the acyl-CoA-independent formation of triacylglycerol in yeast and plants. *Proceeding of the National Accademy of Science USA*. Vol. 6, No. 12, pp. 6487-6492, ISSN 0027-8424

Daniel, H. J., Otto, R. T., Binder, M., Reuss, M., & Syldatk, C. (1999) Production of sophorolipids from whey: development of a two-stage process with *Cryptococcus curvatus* ATCC 20509 and *Candida bombicola* ATCC 22214 using deproteinized whey concentrates as substrate. *Applied Microbiology and Biotechnology*. Vol. 51, No. 1, pp. 40–45, ISSN 0175-7598

de Miguel, T., Calo, P., Díaz, A., & Villa, T. G. (1997) The genus Rhodosporidium: a potential source of beta-carotene. *Microbiologia*. Vol. 13, No. 1, pp. 67-70, ISSN 0213-4101

Dulermo, T. & Nicaud, J. M. (2011) Involvement of the G3P shuttle and β-oxidation pathway in the control of TAG synthesis and lipid accumulation in *Yarrowia lipolytica*. *Metabolic engineering*. Vol. 40, No. 4, pp. 483-488, ISSN 1096-7176

Easterling, E. R., French, W. T., Hernandez, R., & Licha, M. (2009) The effect of glycerol as a sole and secondary substrate on the growth and fatty acid composition of Rhodotorula glutinis. *Bioresource Technology*. Vol. 100, No. 1, pp. 356-361. ISSN 0960-8524

Economou, C.N., Aggelis, G., Pavlou, S., & Vayenas, D.V. (2011) Modeling of single-cell oil production under nitrogen-limited and substrate inhibition conditions. *Biotechnology and Bioengineering*. Vol. 108, No. 5, pp. 1049-1055, ISSN 0006-3592.

Fakas, S., Certik, M., Papanikolaou, S., Aggelis, G., Komaitis, M., & Galiotou-Panayotou, M. (2008) Gamma-linolenic acid production by *Cunninghamella echinulata* growing on complex organic nitrogen sources. *Bioresource Technology*. Vol. 99, No. 13, pp. 5986-5990, ISSN 0960-8524

Findley, K., Rodriguez-Carres, M., Metin, B., Kroiss, J., Fonseca, A., Vilgalys, R., & Heitman, J. (2009) Phylogeny and phenotypic characterization of pathogenic Cryptococcus species and closely related saprobic taxa in the Tremellales. *Eukaryotic Cell*. Vol. 8, No. 3, pp. 353-361, ISSN 1535-9786

Gujjari, P., Suh, S. O., Coumes, K., Zhou, J. J. (2011) Characterization of oleaginous yeasts revealed two novel species: Trichosporon cacaoliposimilis sp. nov. and Trichosporon oleaginosus sp. nov. *Mycologia*. in press, ISSN 1557-2536.

Granger, L.M., Perlot, P., Goma, G., & Pareilleux, A. (1992) Kinetics of growth and fatty acid production of Rhodotorula glutinis. *Applied Microbiology and Biotechnology*. Vol. 37, No. 1, pp. 13-17, ISSN 0175-7598.

Granger, L.M., Perlot, P., Goma, G., & Pareilleux, A. (1993) Efficiency of fatty acid synthesis by oleaginous yeasts: Prediction of yield and fatty acid cell content from consumed C/N ratio by a simple method. *Journal of biochemical and microbiological technology and engineering*. Vol. 42, No. 10, pp. 1151-1156, ISSN 0006-3592.

Gutiérrez, A., López-García, S., & Garre, V. (2011) High reliability transformation of the basal fungus Mucor circinelloides by electroporation. *Journal of Microbiological Methods*. Vol. 84, No. 3, pp. 442-446, ISSN 0167-7012

Hall, M. J., & Ratledge, C. (1977) Lipid accumulation in an oleaginous yeast (Candida 107) growing on glucose under various conditions in a one- and two-stage continuous culture. *Applied and Environmental Microbiology*. Vol. 33, No. 3, pp. 577-584, ISSN 0099-2240.

Hansson, L. & Dostalek, M. (1986) Influence of cultivation conditions on lipid production by Cryptococcus albidus. *Applied Microbiology and Biotechnology*. Vol. 24, No. 1, pp. 12-18, ISSN 0175-7598.

Hassan, M., Blanc, P. J., Granger, L. M., Pareilleux, A., & Goma, G. (1993) Lipid production by an unsaturated fatty acid auxotroph of the oleaginous yeast *Apiotrichum curvatum* grown in single-stated continuous culture. *Applied Microbiology and Biotechnology*. Vol. 40, No. 4, pp. 483-488, ISSN 0175-7598.

Heredia, L. & Ratledge, C. (1988) Simultaneous utilization of glucose and xylose by *Candida curvata* D in continuous culture. *Biotechnology Letters*. Vol. 10, No. 1, pp. 25-30, ISSN 0141-5492.

Holdsworth, J. E. & Ratledge, C. (1991) Triacylglycerol synthesis in the oleaginous yeast *Candida curvata* D. *Lipids*. Vol. 26, No. 2, pp. 111-118, ISSN 0024-4201.

Hu, C., Zhao, X., Zhao, J., Wu, S., & Zhao, Z. K. (2009) Effects of biomass hydrolysis by-products on oleaginous yeast *Rhodosporidium toruloides*. *Bioresource Technology*. Vol. 100, No. 20, pp. 4843-4847, ISSN 0960-8524.

Huang, C., Zong, M. H., Wu, H., & Liu, Q. P. (2009). Microbial oil production from rice straw hydrolysate by *Trichosporon fermentans*. *Bioresource Technology*. Vol. 100, No. 19, pp. 4535-4538, ISSN 0960-8524

Jacob Z. (1991) Enrichment of wheat bran by *Rhodotorula gracilis* through solid-state fermentation. *Folia Microbiologica*. Vol. 36, No. 1, pp. 86–91, ISSN 1874-9356.

Jacob, Z. (1992) Linear growth and lipid synthesis in the oleaginous yeast *Rhodotorula gracilis*. *Folia Microbiologica*. Vol. 37, No. 2, pp. 117–121, ISSN 1874-9356.

Jang, H.D., Lin, Y.Y., & Yang, S.S. (2005) Effect of culture media and conditions on polyunsaturated fatty acids production by *Mortierella alpina*. *Bioresource Technology*. Vol. 96, No. 15, pp. 1633-1644, ISSN 0960-8524.

Johnson, V. W., Singh, M., Saini, V. S., Adhikari, D. K., SIsta, V. R., & Yadav, N. K. (1992) Effect of pH on lipid accumulation by an oleaginous yeast: *Rhodotorula glutinis* IIP-30. *World Journal of Microbiology and Biotechnology* Vol. 8, pp. 382–384, ISSN 0959-3993.

Johnson, V. W., Sigh, M., Saini V. S., Adhikari, D. K., Sista, V., & Yadav, N. K. (1995) Utilization of molasses for the production of fat by an oleaginous yeast, *Rhodotorula glutinis* IIP-30. *Journal of Industrial Microbiology and Biotechnology* Vol. 14, No. 1 pp. 1–4, ISSN 1476-5535

Kohlwein, S. D. (2010) Triacylglycerol homeostasis: insights from yeast. *Journal of Biological Chemistry*. Vol. 285, No. 21, pp. 15663-15667, ISSN 0021-9258.

Levinson, W.E., Kurtzman, C.P., & Min, T. (2007) Characterization of *Yarrowia lipolytica* and related species for citric acid production from glycerol. *Enzyme and Microbial Technology*. Vol. 41, No. 3, pp. 292-295, ISSN 0141-0229.

Liang, Y., Cui, Y., Trushenski, J., & Blackburn, J. W. (2010) Converting crude glycerol derived from yellow grease to lipids through yeast fermentation. *Bioresource Technology*. Vol. 101, No. 19, pp. 7581-7586, ISSN 0960-8524

Li, C. H., Cervantes, M., Springer, D. J., Boekhout, T., Ruiz-Vazquez, R. M., Torres-Martinez, S. R., Heitman, J., & Lee, S. C. (2011) Sporangiospore Size Dimorphism Is Linked to

Virulence of Mucor circinelloides. *PLoS Pathogen*. Vol. 7, No. 6, e1002086, ISSN 1553-7366

Li, Q., Du, W., & Liu D. (2008) Perspectives of microbial oils for biodiesel production. *Applied Microbiology and Biotechnology*. Vol. 80, No. 5, pp. 749-756, ISSN 0175-7598.

Li, Y., Zhao, Z., & Bai, F. (2007) High-density cultivation of oleaginous yeast *Rhodosporidium toruloides* Y4 in fed-bach culture. *Enzyme and Microbial Technology*. Vol. 41, No. 3, pp. 312-317, ISSN 0141-0229.

Lin, H., Cheng, W., Ding, H. T., Chen, X. Zhou, Q. F., Zhao, Y. H. (2010) Direct microbial conversion of wheat straw into lipid by a cellulolytic fungus of *Aspergillus oryzae* A-4 in solid-state fermentation. *Bioresource technology*. Vol. 101, No. 19, pp. 7556-7562, ISSN 0960-8524

Meesters, P. A. E. P., Huijberts, G. N. M., & Eggink, G. (1996) High-cell-density cultivation of the lipid accumulating yeast *Cryptococcus curvatus* using glycerol as a carbon source. *Applied Microbiology and Biotechnology*. Vol. 45, No. 5, pp. 749-756, ISSN 0175-7598.

Meeuwse, P., Tramper, J., & Rinzema, A. (2011a) Modeling lipid accumulation in oleaginous fungi in chemostat cultures: I. Development and validation of a chemostat model for *Umbelopsis isabellina*. *Bioprocess and Biosystems Engineering*. Epub ahead of print, ISSN 1615-7591.

Meeuwse, P., Tramper, J., & Rinzema A. (2011b) Modeling lipid accumulation in oleaginous fungi in chemostat cultures. II: Validation of the chemostat model using yeast culture data from literature. *Bioprocess and biosystems Engineering*. Epub ahead of print, ISSN 1615-7591.

Meng, X., , Yang, J., Xu, X., Zhang, L., Nie, Q., & Xian, M. (2009) Biodiesel production from oleaginous microorganisms. *Renewable Energy*. Vol. 34, No. 1, pp. 1-5, ISSN 0960-1481.

Mlickova, K., Luo, Y., d´Andrea, S., Pec, P., Chardot, T., & Nicaud, J.-M. (2004a). Acyl-CoA oxidase, a key step for lipid accumulation in the yeast *Yarrowia lipolytica*. *Journal of molecular catalysis. B, Enzymatic*. Vol. 28, No. 2-3, pp. 81-85 ISSN 1381-1177

Mlíckova, K., Roux, E., Athenstaedt, K., d'Andrea, S., Daum, G., Chardot, T., & Nicaud, J-M. (2004b). Lipid accumulation, lipid body formation, and acyl coenzyme A oxidases of the yeast *Yarrowia lipolytica*. *Applied and Environmental Microbiology*. Vol. 70, No. 7, pp. 3918-3924 ISSN 0099-2240

Pan, J. G., Kwak, N. Y., & Rhee, J. S. (1986) High density cell culture of *Rhodotorula glutinis* using oxygen-enriched air. *Biotechnology Letters*. Vol. 8, No. 10, pp. 715–718, ISSN 0141-5492

Papanikolaou, S., & Aggelis, G. (2002) Lipid production by *Yarrowia lipolytica* growing on industrial glycerol in a single-stage continuous culture. *Bioresource technology*. Vol 82, No. 1, pp. 43–49, ISSN 0960-8524

Papanikolaou, S., Chevalot, I., Komaitis, M., Marc, I., & Aggelis, G. (2002) Single cell oil production by *Yarrowia lipolytica* growing on an industrial derivative of animal fat in batch cultures. *Applied Microbiology and Biotechnology* Vol. 58, No. 3, pp. 308-312, ISSN 0175-7598

Papanikolaou, S., Muniglia, L., Chevalot, I., Aggelis, G., & Marc, I. (2003) Accumulation of a cocoa-butter-like lipid by *Yarrowia lipolytica* cultivated on agro-industrial residues. *Current Microbiology*. Vol. 46, No. 2, pp.124-130, ISSN 0343-8651.

Park, W.S., Murphy, P.A., & Glatz, B.A. (1990) Lipid metabolism and cell composition of the oleaginous yeast *Apiotrichum curvatum* grown at different carbon to nitrogen ratios. *Canadian Journal of Microbiology*. Vol. 36, No. 5, pp. 318-326, ISSN 0008-4166.

Peng, X. W. & Chen, H. Z. (2007) Microbial oil accumulation and cellulase secretion of the endophytic fungi from oleaginous plants. *Annals of Microbiology*. 57 Sakaiu,239-242, ISSN 1590-4261

Rajakumari, S., Grillitsch, K., & Daum, G. (2008) Synthesis and turnover of non-polar lipids in yeast. *Progress in Lipid Research*. Vol. 47, No. 3, pp. 157-171, ISSN 0163-7827

Ratledge, C. (2002) Regulation of lipid accumulation in oleaginous micro-organisms. *Biochemical Society Transactions*. Vol. 30, No. 6, pp. 1047-1050, ISSN 0300-5127

Ratledge, C., (2004) Fatty acid biosynthesis in microorganisms being used for single cell oil production. *Biochimie*. Vol. 86, No. 11, pp. 807–815, ISSN 0300-9084.

Ratledge, C. & Wynn, J. P. (2002) The biochemistry and molecular biology of lipid accumulation in oleaginous microorganisms. *Advances in Applied Microbiology*. Vol. 51, No. 1, pp. 1–44, ISSN 0065-2164.

Rymowicz, W., Fatykhova, A. R., Kamzolova, S. V., Rywińska, A., & Morgunov, I.G. (2010) Citric acid production from glycerol-containing waste of biodiesel industry by *Yarrowia lipolytica* in batch, repeated batch, and cell recycle regimes. *Applied Microbiology and Biotechnology*. Vol 87, No. 3, pp. 971-979. ISSN 0175-7598

Rywinska, A., Rymowicz, W., & Marcinkiewicz, M. (2010) Valorization of raw glycerol for citric acid production by *Yarrowia lipolytica* yeast. *Electronic Journal of Biotechnology*. Vol. 13, No. 4, ISSN 0717-3458

Rau, U., Nguyen, L. A., Roeper, H., Koch, H., & Lang, S. (2005) Fed-batch bioreactor production of mannosylerythritol lipids secreted by *Pseudozyma aphidis*. *Applied Microbiology and Biotechnology*. Vol. 80, No. :607–613, ISSN 0175-7598

Rossi, M., Buzzini, P., Cordisco, L., Amaretti, A., Sala, M., Raimondi, S., Ponzoni, C., Pagnoni, U. M., & Matteuzzi, D. (2009) Growth, lipid accumulation, and fatty acid composition in obligate psychrophilic, facultative psychrophilic, and mesophilic yeasts. *FEMS microbiology ecology*. Vol. 69, No. 3, pp. 363-372, ISSN 0168-6496

Sakuradani, E., Ando, A., Ogawa, J., & Shimizu, S. (2009) Improved production of various polyunsaturated fatty acids through filamentous fungus *Mortierella alpina* breeding. *Applied Microbiology and Biotechnology*. Vol. 84, No. 1, pp. 1-10. ISSN 0175-7598

Satyanarayana, T., & Kunze, G. (2010) Gatersleben, Germany (Eds.) Yeast Biotechnology: Diversity and Applications 746 pages ISBN: 1402082916

Saxena, V., Sharma, C. D., Bhagat, S. D., Saini, V. S., & Adhikari, D. K. (2008) Lipid and fatty acid biosynthesis by *Rhodotorula minuta*. *Journal of the American Oil Chemists Society*. Vol. 75, No. 4, pp. 501–505, ISSN 0003-021X.

Smit, M. S. (2004) Fungal epoxide hydrolases: new landmarks in sequence-activity space. *Trends in biotechnology*. Vol. 22, No. 3, pp. 123-129, ISSN 0167-7799

Tehlivets, O., Scheuringer, K., & Kohlwein, S. D. (2007) Fatty acid synthesis and elongation in yeast. *Biochimica et Biophysica Acta*. Vol. 1171, No. 3, pp. 255-270, ISSN 0006-3002

van der Walt, J. P. (1992) The Lipomycetaceae, a model family for phylogenetic studies. *Antonie Van Leeuwenhoek*. Vol. 62, No. 4, pp. 247-250, ISSN 0003-6072.

Vamvakaki, A. N., Kandarakis, I., Kaminarides, S., Komaitis, M. & Papanikolaou, S. (2010) Cheese whey as a renewable substrate for microbial lipid and biomass production by Zygomycetes *Engineering in Life Sciences* Vol. 10, No. 4, pp. 348–360, ISSN 1618-0240

Vicente, G., Bautista, L. F., Rodríguez, R., Gutiérrez, F. J., Sádaba, I., Ruiz-Vázquez, R. M., Torres-Martínez, S., & Garre, V. (2009) Biodiesel production from biomass of an oleaginous fungus. *Biochemical Engineering Journal.* Vol. 48, No. 1, pp. 22–27, ISSN 1369-703X

Wang, S. L., Sun, J. S., Han, B. Z., & Wu, X. Z. (2007) Optimization of beta-carotene production by *Rhodotorula glutinis* using high hydrostatic pressure and response surface methodology. *Journal of Food Science.* Vol. 72, No. 8, pp. M325-329, 1911, ISSN 0022-1147

Wu, S., Hu, C., Jin, G., Zhao, X., & Zhao, Z.K. (2010) Phosphate-limitation mediated lipid production by *Rhodosporidium toruloides*. *Bioresource Technology.* Vol. 101, No. 15, pp. 6124-6129, ISSN 0960-8524

Wynn, J.P., Hamid, A. A., Li, Y, & Ratledge, C. (2001) Biochemical events leading to the diversion of carbon into storage lipids in the oleaginous fungi *Mucor circinelloides* and *Mortierella alpina*. *Microbiology.* Vol. 147, No. 10, pp. 2857-2864, ISSN 1350-0872.

Ykema, A., Verbree, E.C., Verseveld, H.W., & Smit, H. (1986) Mathematical modelling of lipid production by oleaginous yeasts in continuous cultures. *Antonie van Leeuwenhoek.* Vol. 52, No. 6, pp. 491-506, ISSN:1572-9699

Ykema, A., Kater, M. M. & Smit H. (1989) Lipid production in whey permeate by an unsaturated fatty acid mutant of the oleaginous yeast *Apiotrichum curvatum*. *Biotechnology Letters.* Vol.11, No. 7, pp. 477-482, ISSN 0141-5492

Yousuf, A., Sannino, F., Addorisio, V., Pirozzi, D. (2010) Microbial conversion of olive oil mill wastewaters into lipids suitable for biodiesel production. *Journal of agricultural and food chemistry.* Vol. 11 No. 58(15), pp. 8630-8635, ISSN 0021-8561

Yu, X., Zheng, Y., Dorgan, K. M., Chen, S. (2011) Oil production by oleaginous yeasts using the hydrolysate from pretreatment of wheat straw with dilute sulfuric acid. *Bioresource Technology.* Vol. 102, No. 10, pp. 6134-6140, ISSN 0960-8524

Zhang, Y., Adams, I. P., & Ratledge, C. (2007) Malic enzyme: the controlling activity for lipid production? Overexpression of malic enzyme in *Mucor circinelloides* leads to a 2.5-fold increase in lipid accumulation. *Microbiology.* Vol. 153, No. 7, pp. 2013-2025, ISSN 1350-0872

Zhang, J., Fang, X., Zhu, X. L., Li, Y., Xu, H. P., Zhao, B. F., Chen, L. & Zhang, X. D. (2011) Microbial lipid production by the oleaginous yeast *Cryptococcus curvatus* O3 grown in fed-batch culture. *Biomass and Bioenergy.* Vol. 35, No. 5, pp. 1906-1911, ISSN 0961-9534

Zhao, C. H., Cuim W., Lium X. Y., Chi, Z. M., & Madzak, C. (2010) Expression of inulinase gene in the oleaginous yeast *Yarrowia lipolytica* and single cell oil production from inulin-containing materials. *Metabolic engineering.* Vol. 12, No. 6, pp. 510-517, ISSN 1096-7176

Zhao, X., Kong, X., Hua, Y., Feng, B., Zhao, Z. K. (2008) Medium optimization for lipid production through co-fermentation of glucose and xylose by the oleaginous yeast *Lipomyces starkeyi*. *European Journal of Lipid Science and Technology.* Vol. 110, No. 5, pp. 405–412, ISSN 1438-9312

Zhu, L. Y., Zong, M. H., & Wu, H, (2008) Efficient lipid production with *Trichosporon fermentans* and its use for biodiesel preparation. *Bioresource Technology*. Vol. 99, No. 16, pp. 7881-7885, ISSN 0960-8524

Zinoviev, S., Müller-Langer, F., Das, P., Bertero, N., Fornasiero, P., Kaltschmitt, M., Centi, G., & Miertus, S. (2010) Next-generation biofuels: Survey of emerging technologies and sustainability issues. *Chemistry & sustainability, energy & materials*. Vol. 3, No. 10, pp. 1106-1133, ISSN 1864-564X

7

Microalgae as Feedstocks for Biodiesel Production

Jin Liu[1], Junchao Huang[2,3] and Feng Chen[3,4]
[1]Department of Applied Sciences and Mathematics,
Arizona State University, Polytechnic Campus, Mesa,
[2]Kunming Institute of Botany, Chinese Academy of Sciences,
[3]School of Biological Science, The University of Hong Kong, Hong Kong,
[4]Institute for Food & Bioresource Engineering, College of Engineering,
Peking University, Beijing,
[1]USA
[2,3,4]China

1. Introduction

Fossil-based fuels including oil, coal and gas play a pivotal role in modern world energy market. These fossil fuels, according to world energy outlook 2007, will remain the major sources of energy and are expected to meet about 84% of energy demand in 2030. However, fossil fuels are non-renewable and will be finally diminished. It has been recently estimated that the global oil, coal and gas last only approximately for 35, 100 and 37 years respectively, based on a modified Klass model (Shafiee & Topal, 2009). In order to sustain a stable energy supply in the future, it is necessary to develop other sources of energy, e.g., renewable energy. Renewable energy is derived from natural processes that are replenished constantly, including hydropower, wind power, solar energy, geothermal energy, biodiesel, etc. An estimated $150 billion was invested in renewable energy worldwide in 2009, around 2.5 times of the 2006 investment (Figure 1).

It is well known that transport is almost totally dependent on petroleum-based fuels, which will be depleted within 40 years. An alternative fuel to petrodiesel must be technically feasible, easily available, economically competitive, and environmentally acceptable (Demirbas, 2008). Biodiesel is such a candidate fuel for powering the transport vehicles. Biodiesel refers to a biomass-based diesel fuel consisting of long-chain alkyl (methyl, propyl or ethyl) esters. In addition to being comparable to petrodiesel in most technical aspects, biodiesel has the following distinct advantages over petrodiesel (Knothe, 2005a):

1. derived from renewable domestic resources, thus reducing dependence on and preserving petroleum;
2. biodegradable and reduced exhaust emissions, being environment-friendly;
3. higher flash point, being safer for handling and storage; and
4. excellent lubricity.

Like petrodiesel, biodiesel operates in compression ignition engines. Biodiesel is miscible with petrodiesel in all ratios. Currently, the blends of biodiesel and petrodiesel instead of net biodiesel have been widely used in many countries and no engine modification is required (Singhania et al., 2008). These blends of biodiesel with petrodiesel are usually denoted by acronyms, for example B20 which indicates a blend of 20% biodiesel with petrodiesel (Knothe, 2005a).

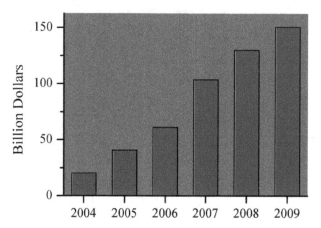

Fig. 1. Global investment in renewable energy, 2004-2009. Adapted from REN21 (2010)

The global markets for biodiesel are entering a period of rapid and transitional growth. In the year 2007, there were only 20 nations producing biodiesel for the needs of over 200 nations; by the year 2010, more than 200 nations become biodiesel producing nations and suppliers (Thurmond, 2008). Global biodiesel production has massively increased to 16.6 billion liters per year over the last nine years (Figure 2). Much of the growth is happening in just three countries: the United States, Brazil and Germany, which together account for over half of biodiesel (Checkbiotech, 2009). The International Energy Agency's report suggests that world production of biodiesel could top 25 million tons per year by 2012 if the recent trends continue.

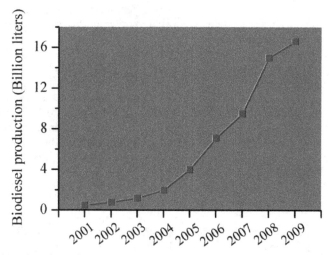

Fig. 2. Global biodiesel production, 2001-2009. Adapted from REN21 (2010)

Biodiesel can be produced from a variety of feedstocks, including plant oils, animal fats and waste oils as well as microalgae (Demiras, 2008). Each feedstock has its advantages and disadvantages in terms of oil content, fatty acid composition, biomass yield and geographic distribution. Depending on the origin and quality of feedstocks, changes may be required for the production process of biodiesel.

The use of plant oils as biodiesel feedstocks has been long recognized and well documented in numerous studies (Abdullah et al., 2009; de Oliveira et al., 2005; Graef et al., 2009; Hawash et al., 2009; Hill et al., 2006; Jain & Sharma, 2010; Nakpong & Wootthikanokkhan, 2010; Patil & Deng, 2009; Rashid & Anwar, 2008; Sahoo & Das, 2009; Saka & Kusdiana, 2001). These feedstocks include the oils from soybean, rapeseed, palm, canola, peanut, cottonseed, sunflower and safflower. Based on the geographic distribution, soybean is the primary source for biodiesel in USA, palm oil is used as a significant biodiesel feedstock in Malaysia and Indonesia, and rapeseed is the most common base oil used in Europe for biodiesel production (Demiras, 2008). The vast majority of these plants are also used for food and feed production, which means that possible food versus fuel conflicts are present. Thus, the use of these plant oils as feedstocks for biodiesel seems insignificant for the developing countries which are importers of edible oils (Meher et al., 2008). In addition to these edible oils, various non-edible, tree-borne oils from jatropha, karanja, jojoba and neem are the potential biodiesel feedstocks (Jain & Sharma, 2009; Meher et al., 2008; Sahoo & Das, 2009). Jatropha and karanja are two oilseed plants that are not widely exploited due to the presence of toxic components in the oils. In India, they are popularly used as biodiesel feedstocks.

In addition to the plant oils, animal fats and waste oils are the potential sources for commercial biodiesel production (Thompson et al., 2010). Among these feedstocks, tallow, lard, yellow grease and waste cooking oils have received most interest (Banerjee et al., 2009; Canakci, 2007; da Cunha et al., 2009; Dias et al., 2009; Diaz-Felix et al., 2009; Oner & Altun, 2009; Phan & Phan, 2008). However, animal fats and waste oils usually contain large amounts of free fatty acids, which can be as high as 41.8% (Canakci, 2007). Free fatty acids cannot be directly converted to biodiesel in alkali-catalyzed transesterificatoin but react with alkali to form soaps that inhibit the separation of biodiesel from glycerin and wash water fraction (Huang et al., 2010). A two-step process was developed for these high fatty acid feedstocks: acid-catalyzed pretreatment and alkali-catalyzed transesterificaton. Because animal fats and waste oils have relatively high level of saturation (Canakci, 2007), the biodiesel from these sources exhibits poor cold flow properties.

Microalgae represent a wide variety of aquatic photosynthetic organisms with the potential of producing high biomass and accumulating high level of oil. The production of biodiesel from microalgal oil has long been recognized and been evaluated in response to the United States Department of Energy for research in alternative renewable energy (Sheehan et al., 1998). Currently, the commercialization of algae-derived biodiesel is still in its infancy stage. Using microalgae as biodiesel feedstocks has received unprecedentedly increasing interest, including but not restricted to microalgal strain selection and genetic engineering, mass cultivation for biomass production, lipid extraction and analysis, transesterification technologies, fuel properties and engine tests (Abou-Shanab et al., 2011; Brennan & Owende, 2010; Demirbas, 2009; Greenwell et al., 2010; Miao & Wu, 2006; Pruvost et al., 2011; Radakovits et al., 2010; Rodolfi et al., 2009; Ross et al., 2008; Sydney et al., 2011). Considering their unique characteristics, microalgae have been considered as the most promising feedstock of biodiesel that has the potential to displace fossil diesel (Chisti, 2007). This review mainly focuses on the potential of using microalgae as biodiesel feedstocks, biodiesel production pipeline, and possibility of employing genetic engineering for improving microalgal productivity.

2. Potential of using microalgae as biodiesel feedstocks

Microalgae represent a large and diverse group of prokaryotic or eukaryotic photosynthetic microorganisms that are in unicellular or multicellular form. Examples of prokaryotic microorganisms are cyanobacteria (commonly referred to as blue-green algae) that are closely related to Gram-negative bacteria and eukaryotic ones are for example green microalgae and diatoms (Graham et al., 2009). Microalgae can be found in a wide range of environmental conditions, including water, land, and even unusual environments such as snow and desert soils (Lee, 2008). It is estimated that there are more than 50,000 species around the world, among which only about 30,000 have been studied and analyzed (Mata et al., 2010). Extensive collections of microalgae have been established by researchers in different countries, including the Freshwater Microalgae Collection of University of Coimbra (Portugal), the Collection of the Goettingen University (Germany), the Provasoli-Guillard National Center for Culture of Marine Phytoplankton (CCMP, USA), the University of Texas Algal Culture Collection (USA), the CSIRO collection of Living Microalgae (CCLM, Australia), the National Institute for Environmental Studies Collection (NIES, Japan), the American Type Culture Collection (ATCC, USA), and the Freshwater Algae Culture Collection of Institute of Hydrobiology (China). Together more than 10,000 microalgal strains are available to be selected for use in a broad range of applications, for example, as biodiesel feedstocks.

The use of microalgae for biodiesel production has long been recognized and its potential has been widely reported by many research studies recently (Abou-Shanab et al., 2011; Afify et al., 2010; Ahmad et al., 2011; Cheng et al., 2009; Damiani et al., 2010; Gouveia et al., 2009; Liu et al., 2010; Rodolfi et al., 2009; Yoo et al., 2010). Microalgae reproduce themselves autotrophically using CO_2 from air and light through photosynthesis. Compared with higher plants, microalgae exhibit higher photosynthetic efficiency and grow much faster, finishing an entire growth cycle within a few days (Christi, 2007). Typical growth rates are presented in Figure 3 as the doubling time for each microalgal species. A low doubling time corresponds to a high specific growth rate. Microalgae double themselves with an average time of 26 h, and some can even reproduce within 8 h. Moreover, they can be adapted to grow in a broad range of environmental conditions, suggesting the possibility of finding species best suited to local environments which is not suitable for cultivating oil plants (e.g. palm, soybean and rapeseed).

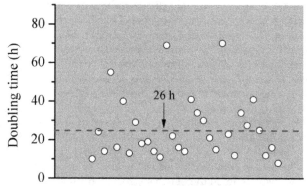

Fig. 3. Doubling time for some microalgal species. The dash line indicates the average value. $T_d=\ln(2)/\mu$, T_d, doubling time, μ, specific growth rate.

In addition to growth rate, lipid content is another important factor to assess the potential of microalgae for biodiesel production. Over the past few decades, thousands of algae and cyanobacterial species have been screened for high lipid production, and numerous oleaginous species have been isolated and characterized. The lipid contents of these oleaginous algae are species- and/or strains-dependent, vary greatly, and may reach as high as 68% of dry weight, as shown in Table 1. Generally, microalgae synthesize a low content of lipids under nutrient replete conditions (Figure 4), with membrane lipids (e.g., phospholipids and glycolidips) being the main components; whereas under stress conditions such as nitrogen deficiency, a great increase in total lipids was observed (Figure 4) with neutral lipids in particular triacylglycerols (TAGs) being the dominant components (Hu, 2004). TAGs are considered to be superior to phospholipids or glycolipids for biodiesel feedstocks because of their higher percentage of fatty acids and lack of phosphate (Pruvost et al., 2009). Unlike higher plants in which individual classes of lipids may be synthesized and localized in a specific cell, tissue or organ, algae produce these different lipids in a single cell (Hu et al., 2008b). The synthesized TAGs are deposited in lipid bodies located in cytoplasm of algal cells (Damiani et al., 2010; Rabbani et al., 1998).

Algal species	Culture conditions	Lipid content (%)	biomass productivity (g/L/day)	Lipid productivity (mg/L/day)	References
Chlorophyta					
Botryococcus braunii	Phototrophic	9.5-13.5	0.02-0.04	2.6-4.5	Chinnasamy et al., 2010
Botryococcus braunii	Phototrophic	17.85	0.346		Órpez et al., 2009
Botryococcus braunii	Phototrophic	24	0.077	21	Yoo et al., 2010
Botryococcus sp.	Phototrophic	15.8-35.9	0.14-0.22	21.3-46.9	Yeesang and Cheirsilp, 2011
Chlamydomonas reinhardtii	Mixotrophic	12.2-46	0.21-0.36	29-95	Li et al., 2010a
Chlorella ellipsoidea	Phototrophic	32	0.07	22.4	Abou-Shanab et al., 2011
Chlorella ellipsoidea	Phototrophic	15-43		11.4	Yang et al., 2011
Chlorella protothecoides	Heterotrophic	48.1-63.8	1.02-1.73	3432-6293	De la Hoz Siegler et al., 2011
Chlorella protothecoides	Heterotrophic	49	1.2	586.8	Gao et al., 2010
Chlorella saccharophila	Phototrophic	12.9-18.1	0.02	2.7-4.2	Chinnasamy et al., 2010
Chlorella sorokiniana	Phototrophic	19.3	0.23	44.7	Rodolfi et al., 2009
Chlorella sp.	Phototrophic	33.9	0.528	178.8	Chiu et al., 2008
Chlorella sp.	Phototrophic	22.4-66.1	0.08-0.34	51-124	Hsieh and Wu., 2009
Chlorella sp.	Phototrophic	34.1 [a]	0.053	22	Matsumoto et al., 2010
Chlorella sp.	Phototrophic	18.7	0.23	42.1	Rodolfi et al., 2009
Chlorella vulgaris	Phototrophic	20-42	0.21-0.35	44-147	Feng et al., 2011
Chlorella vulgaris	Phototrophic, Mixotrophic, heterotrophic	21-38	0.01-0.26	4-54	Liang et al., 2009
Chlorella vulgaris	Phototrophic	19.2	0.17	32.6	Rodolfi et al., 2009
Chlorella vulgaris	Phototrophic	26-52		11.6-13.2	Widjaja et al., 2009
Chlorella vulgaris	Phototrophic	35	0.117	41	Yeh et al., 2010
Chlorella zofingiensis	Heterotrophic	52	0.72	374.4	Liu et al., 2010

Algal species	Culture conditions	Lipid content (%)	biomass productivity (g/L/day)	Lipid productivity (mg/L/day)	References
Chlorella zofingiensis	Phototrophic	25.8	0.136	35.1	Liu et al., 2011
Chlorococcum sp.	Phototrophic	19.3	0.28	53.7	Rodolfi et al., 2009
Choricystis minor	Phototrophic	21-59.3	0.35	82	Sobczuk and Chisti, 2010
Dunaliella tertiolecta	Phototrophic	12.2-15.2	0.03-0.04	4.0-4.6	Chinnasamy et al., 2010
Dunaliella tertiolecta	Phototrophic	16.7	0.12	20	Gouveia and Oliveira, 2009
Haematococcus pluvialis	Phototrophic	15.6-34.9			Damiani et al., 2010
Micractinium pusillum	Phototrophic	24	0.108	25.7	Abou-Shanab et al., 2011
Neochloris oleabundans	Phototrophic	19-56	0.03-0.15	10.7-38.8	Gouveia et al., 2009
Neochloris oleabundans	Phototrophic	7-40.3	0.31-0.63	38-133	Li et al., 2008
Ourococcus multisporus	Phototrophic	52	0.045	23.3	Abou-Shanab et al., 2011
Parietochloris incisa	Phototrophic	18-34 [a]	0.23-0.47	46-160	Solovchenko et al., 2008
Pseudochlorococcum sp.	Phototrophic	24.6-52.1	0.234-0.76	53-350	Li et al., 2011a
Scenedesmus obliquus	Phototrophic	21-58	0.08-0.09	19-43.3	Abou-Shanab et al., 2011
Scenedesmus obliquus	Phototrophic	17.7	0.09	15.9	Gouveia and Oliveira, 2009
Scenedesmus obliquus	Phototrophic	12-38.9	0.20-0.29	35.1-78.7	Ho et al., 2010
Scenedesmus obliquus	Phototrophic, Mixotrophic	12.6-58.3	0.51	270	Mandal and Mallick, 2009
Scenedesmus rubescens like	Phototrophic	11.3-27 [a]	0.44-0.54	108-133	Lin and Lin, 2011
Scenedesmus quadricauda	Phototrophic	18.4	0.19	35.1	Rodolfi et al., 2009
Scenedesmus sp.	Phototrophic	22-53	0.08	20.3	Xin et al., 2010
Scenedesmus sp.	Phototrophic	18	0.203	39	Yoo et al., 2010
Scenedesmus sp.	Phototrophic	21.1	0.26	53.9	Rodolfi et al., 2009
Tetraselmis chui	Phototrophic	17.3-23.5	1-2.6	235-450	Araujo et al., 2011
Tetraselmis sp.	Phototrophic	8.7-33	0.21	22.86	Huerlimann et al., 2010
Tetraselmis suecica	Phototrophic	8.5-12.9	0.28-0.32	27-36.4	Rodolfi et al., 2009
Tetraselmis tetrathele	Phototrophic	29.2-30.3	3.1-4.4	905-1333	Araujo et al., 2011
Bacillariophyceae					
Chaetoceros calcitrans	Phototrophic	39.8	0.04	17.6	Rodolfi et al., 2009
Chaetoceros gracilis	Phototrophic	15.5-60.3	3.4-3.7	530-2210	Araujo et al., 2011
Chaetoceros muelleri	Phototrophic	11.7-25.3	1.2-2.7	1404-6831	Araujo et al., 2011
Chaetoceros muelleri	Phototrophic	33.6	0.07	21.8	Rodolfi et al., 2009
Cylindrotheca closterium	Phototrophic	17-30			Pruvost et al., 2011
Navicula sp.	Phototrophic	47.6 [a]	0.055	26.4	Matsumoto et al., 2010
Nitzschia cf. pusilla	Phototrophic	48	0.065	31.4	Abou-Shanab et al., 2011
Nitzschia laevis	Heterotrophic	12.8	2.02	258.6	Chen et al., 2008

Algal species	Culture conditions	Lipid content (%)	biomass productivity (g/L/day)	Lipid productivity (mg/L/day)	References
Nitzschia sp.	Phototrophic	32	0.013		Moazami et al., 2011
Phaeodactylum tricornutum	Phototrophic	18.7	0.24	44.8	Rodolfi et al., 2009
Skeletonema costatum	Phototrophic	21.1	0.08	17.4	Rodolfi et al., 2009
Skeletonema sp.	Phototrophic	31.8	0.09	27.3	Rodolfi et al., 2009
Thalassiosira pseudonana	Phototrophic	20.6	0.08	17.4	Rodolfi et al., 2009
Eustigmatophyceae					
Ellipsoidion sp.	Phototrophic	27.4	0.17	47.3	Rodolfi et al., 2009
Monodus subterraneus	Phototrophic	12.9-15 [a]	0.34-0.49	47.5-67.5	Khozin-Goldberg and Cohen, 2006
Monodus subterraneus	Phototrophic	16.1	0.19	30.4	Rodolfi et al., 2009
Nannochloropsis oculata	Phototrophic	22.8-23	2.4-3.4	547.2-782	Araujo et al., 2011
Nannochloropsis oculata	Phototrophic	26.2-30.7	0.37-0.50	84-151	Chiu et al., 2009
Nannochloropsis oculata	Phototrophic	7.9-15.9	0.06-0.13	9.1-16.4	Converti et al., 2009
Nannochloropsis sp.	Phototrophic	52	0.0465		Moazami et al., 2011
Nannochloropsis sp.	Phototrophic	23.1-37.8	0.06	20	Huerlimann et al., 2010
Nannochloropsis sp.	Phototrophic	28.7	0.09	25.8	Gouveia and Oliveira, 2009
Nannochloropsis sp.	Phototrophic	21.6-35.7	0.17-0.21	37.6-61	Rodolfi et al., 2009
Others					
Aphanothece microscopica	Heterotrophic	7.1-15.3	0.26-0.44	30-50	Queiroz et al., 2011
Crypthecodinium Cohnii	Heterotrophic	19.9	2.24	444.9	Couto et al., 2010
Isochrysis galbana	Phototrophic	24.6	0.057	14.02	Lin et al., 2007
Isochrysis sp.	Phototrophic	23.5-34.1	0.09	20.95	Huerlimann et al., 2010
Isochrysis sp.	Phototrophic	22.4-27.4	0.14-0.17	37.8	Rodolfi et al., 2009
Pavlova lutheri	Phototrophic	35.5	0.14	50.2	Rodolfi et al., 2009
Pavlova salina	Phototrophic	30.9	0.16	49.4	Rodolfi et al., 2009
Pavlova viridis	Phototrophic	24.8-32			Li et al., 2005
Pleurochrysis carterae	Phototrophic	9.7-12	0.03-0.04	2.7-4.4	Chinnasamy et al., 2010
Porphyridium cruentum	Phototrophic	9.5	0.37	34.8	Rodolfi et al., 2009
Rhodomonas sp.	Phototrophic	9.5-20.5	0.06	6.19	Huerlimann et al., 2010
Schizochytrium limacinum	Heterotrophic	50.3 [a]	3.48	1750	Ethier et al., 2011
Schizochytrium mangrovei	Heterotrophic	68 [a]	2.44	1659	Fan et al., 2007
Spirulina maxima	Phototrophic	4.1	0.21	8.6	Gouveia and Oliveira, 2009
Thalassiosira weissflogii	Phototrophic	6.3-13.2	0.5-1.5	31.5-198	Araujo et al., 2011

[a] Total fatty acid content

Table 1. Lipid content and productivity of various microalgal species.

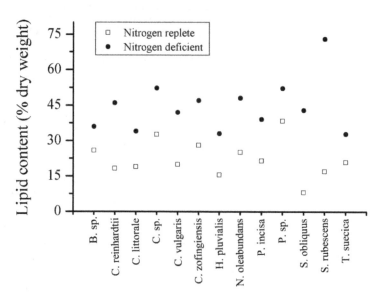

Fig. 4. Lipid content under nitrogen replete (open squares) and nitrogen deficient (filled circles) conditions for *Chlorophyta*. *B.* sp., *Botryococcus* sp. (Yeesang and Cheirsilp, 2011); *C. reinhardtii*, *Chlamydomonas reinhardtii* (Li et al., 2010); *C. littorale*, *Chlorocuccum littorale* (Ota et al., 2009); *C.* sp., *Chlorella* sp. (Hsieh and Wu, 2009); *C. vulgaris*, *Chlorella vulgaris* (Feng et al., 2011); *C. zofingiensis*, *Chlorella zofingiensis* (Liu et al., 2010); *H. pluvialis*, *Haematococcus pluvialis* (Damiani et al 2010); *N. oleabundans*, *Neochloris oleabundans* (Gouveia et al., 2009); *P. incisa*, *Parietochloris incisa* (Solovchenko et al., 2010); *P.* sp., *Pseudochlorococcum* sp. (Li et al., 2011); *S. obliquus*, *Scenedesmus obliquus* (Mandal and Mallick, 2009); *S. rubescens*, *Scenedesmus rubescens* (Mandal and Mallick, 2009); *T. suecica*, *Tetraselmis suecica* (Rodolfi et al., 2009).

The important properties of biodiesel such as cetane number, viscosity, cold flow, oxidative stability, are largely determined by the composition and structure of fatty acid esters which in turn are determined by the characteristics of fatty acids of biodiesel feedstocks, for exmaple carbon chain length and unsaturation degree (Knothe, 2005b). Fatty acids are either in saturated or unsaturated form, and the unsaturated fatty acids may vary in the number and position of double bones on the acyl chain. Based on the number of double bones, unsaturated fatty acids are clarified into monounsaturated fatty acids (MUFAs) and polyunsaturated fatty acids (PUFAs). The fatty acid profile of a great many algal species has been investigated and is shown in Table 2. The synthesized fatty acids in algae are commonly in medium length, ranging from 16 to 18 carbons, despite the great variation in fatty acid composition. Specifically, the major fatty acids are C16:0, C18:1 and C18:2 or C18:3 in green algae, C16:0 and C16:1 in diatoms and C16:0, C16:1, C18:1 and C18:2 in cyanobacteria. It is worthy to note that these data are obtained from algal species under specific conditions and vary greatly when algal cells are exposed to different environmental or nutritional conditions such as temperature, pH, light intensity, or nitrogen concentration (Guedes et al 2010; James et al., 2011; Sobczuk & Chisti, 2010; Tatsuzawa et al., 1996). Generally, saturated fatty esters possess high cetane number and superior oxidative stability; whereas unsaturated, especially

Fatty acids / Algal species	C12:0	C14:0	C15:0	C16:0	C16:1	C16:2	C16:3	C16:4	C17:0	C18:0	C18:1	C18:2	C18:3	C18:4	C20:0	C20:4	C20:5	C22:5	C22:6	Refs
Chlorophyta																				
Botryococcus braunii				29.5	3.4				1	44.9	21.2									Yoo et al., 2010
Botryococcus sp.		3.95	1.56	30.04	0.94				1.54	12.02	37.68	5.01	7.35		0.63					Yeesang and Cheirsilp, 2011
Chlamydomonas reinhardtii				30.7	3	1.8	1.6	2.7		3.2	27.2	18.3	11		0.5					James et al., 2011
Chlorella ellipsoidea		2		26						4	40	23				5				Abou-Shanab et al 2011
Chlorella protothecoides				14.3	1				0.32	2.7	71.6	9.7								Cheng et al 2009
Chlorella pyrenoidosa		0.7		17.3	0.8	7	9.3			1.2	3.3	18.5	41.8							D'Oca et al 2011
Chlorella sorokiniana				25.4	3.1	10.7	4.1			1.4	12.4	34.4	7.1							Chen and Johns, 1991
Chlorella sp.	3.78		5.24	16.1	10.88	9.79			4.74	4.35	8.45	14.36	18.79							Li et al., 2011b
Chlorella vulgaris				24	2.1					1.3	24.8	47.8								Yoo et al., 2010
Chlorella zofingiensis				22.62	1.97	7.38	1.94	0.22		2.09	35.68	18.46	7.75	0.49						Liu et al., 2010
Chlorocuccum littorale				20.9	5.6			14.4		29.7	7.2	22.2								Ota et al., 2009
Choricystis minor				36					0.4	12.3	31.2	9.9	3.8	1.9						Sobczuk and Chisti, 2010
Dictyochloropsis splendida		13.88		69.59					1.21	0.38	1.11	12.14	0.42							Afify et al 2010
Dunaliella tertiolecta				26.4		2.3	1.27			0.6	16.8	13.1	39.6							Chen et al 2011
Haematococcus pluvialis	0.21	1.25		22.5	0.64				0.19	3.15	19.36	26.9	17.04		0.2	0.89	0.57			Damiani et al 2010
Micractinium pusillum				33	1						31	17	18							Abou-Shanab et al 2011
Neochloris oleabundans				23.3	0.6	1.6	2.4		0.2	4.5	43	17.8	5.8							Levine et al., 2011
Neochloris sp.		5.22		29.4					5.2	6.6	17.5	23.6	12.6							Moazami et al., 2011
Ourococcus multisporus		2		19	1					5	26	11	36							Abou-Shanab et al 2011
Parietochloris incise				9.1	0.7	0.6				2.1	15.1	9.3	1.6	1.2		58.9				Khozin-Goldberg et al., 2002
Scenedesmus obliquus		1.48		21.8	5.95	3.96	0.68	0.43		0.45	17.93	21.74	3.76	0.21						Gouveia and Oliveira, 2009
Scenedesmus sp.				36.3	4					2.7	25.9	31.1								Yoo et al., 2010
Tetraselmis sp.		0.6		27.8						0.9	28.2	9.3	23.9	3.7		0.9	3.4			Huerlimann et al., 2010
Bacillariophyceae																				
Chaetoceros sp.		23.6		9.2	36.5	6.9	2.6		2	3		1.4	0.6			4.1	8		1	Renaud et al., 2002
Cyclotella cryptica		1.4		15.2	10.7					3.9	1.2	3.5					9.7		1.7	Pahl et al., 2010
Navicula sp.				45	52.7					0.6	1.1	0.6								Matsumoto et al., 2010
Nitzschia cf. pusilla		6		31	57	0.27					6									Abou-Shanab et al 2011
Nitzschia laevis		16.9		28.5	23.9					0.7	5.1	3.4	4.1			5	11.7			Chen et al 2008
Nitzschia sp.		9	3.5	37.4					4.6	5.3	16.9	11.6								Moazami et al., 2011
Cyanobacteria																				
Nostoc commune				23.5	22.5					5.6	21.1	14.1								Pushparaj et al., 2008
Nostoc flagelliforme		0.65		21.27	14.91					6.2	22.59	15.03	19.35							Liu et al., 2005
Spirulina				49.2	5.9					1.7	2.9	22.7	17.5							Chaiklahan et al 2008
Spirulina maxima		0.34		40.16	9.19		0.42	0.16		1.18	5.43	17.89	18.32	0.08	0.06					Gouveia and Oliveira, 2009
Synechocystis PCC6803				52	3				1	3	9	29	3							Wada and Murata, 1990
Eustigmatophyceae																				
Monodus subterraneus		3.3		19.8	34.3					9.7	9	0.8	0.7			2.8	15.5			Khozin-Goldberg and Cohen, 2006
Nannochloropsis oculata				62						11	5	8	15							Converti et al 2009
Nannochloropsis sp.				23.4						7.14	45.4	11.7	12.2							Moazami et al., 2011

Fatty acids / Algal species	C12:0	C14:0	C15:0	C16:0	C16:1	C16:2	C16:3	C16:4	C17:0	C18:0	C18:1	C18:2	C18:3	C18:4	C20:0	C20:4	C20:5	C22:5	C22:6	Refs
Prymnesiophyceae																				
Isochrysis galbanan		19.3		18.1						29.5	2.6	3.6	13.8					4.1	7.5	Lin et al., 2007
Isochrysis sp.		8.9	0.4	13.7	5.1					0.2	22.8	2.3	4.8	22.5		0.1	0.6	1.7	12.7	Huerlimann et al., 2010
Pavlova lutheri		5.54		19	31.46					1.11	2.55	4.46	5.37	6.63			16.07		7.8	Guedes et al 2010
Pavlova viridis		19.9		13.9	16.1												21.2		8.7	Hu et al 2008a
Pavlova viridis		10.34		17.3	17.87					3.16	1.33	2.48	2.23				10.46		14.78	Li et al., 2005
Rhodophyta																				
Porphyridium cruentum				14.5	8.5					10.5	14				10.8	6.1			10.5	Oh et al., 2009
Others																				
Crypthecodinium cohnii	2.9	13.4		22.9	0.4					2.6	7.6							0.5	49.5	Couto et al., 2010
Glossomastrix chrysoplasta		22		4.4	4					6.6	3.9					5.5	39.2	13.3		Kawachi et al., 2002
Rhodomonas sp.		7.8	0.4	19.7	1.5					3	8.4	3	29.8	11.7		0.6	8.6	1.7	3	Huerlimann et al., 2010
Schizochytrium limacinum		3.96		54.61						3.86								6.47	31.09	Ethier et al 2011

Table 2. Fatty acid composition of various algal species (% of total fatty acids)

polyunsaturated, fatty esters have improved low-temperature properties (Knothe, 2008). In this regard, it is suggested that the modification of fatty esters, for example the enhanced proportion of oleic acid (C18:1) ester, can provide a compromise solution between oxidative stability and low-temperature properties and therefore promote the quality of biodiesel (Knothe, 2009). Thus, microalgae with high oleic acid are suitable for biodiesel production.

Currently the commercial production of biodiesel is mainly from plant oils and animal fats. However, the plant oil derived biodiesel cannot realistically meet the demand of transport fuels because large arable lands are required for cultivation of oil plants, as demonstrated in Table 3. Based on the oil yield of different plants, the cropping area needed is calculated and expressed as a percentage of the total U.S. cropping area. If soybean, the popular oil crop in United States is used for biodiesel production to meet the existing transport fuel need, 5.2 times of U.S. cropland will need to be employed. Even the high-yielding oil plant palm is planted as the biodiesel feedstock, more than 50% of current U.S. arable lands have to be occupied. The requirement of huge arable lands and the resulted conflicts between food and oil make the biodiesel from plant oils unrealistic to completely replace the petroleum derived diesel in the foreseeable future. It is another case, however, if microalgae are used to produce biodiesel. As compared with the conventional oil plants, microalgae possess significant advantages in biomass production and oil yield and therefore the biodiesel productivity. In terms of land use, microalgae need much less than oil plants, thus eliminating the competition with food for arable lands (Table 3).

In addition to biodiesel, microalgae can serve as sources of other renewable fuels such as biogas, bioethanol, bio-oil and syngas (Chisti, 2008; Demirbas, 2010; Mussgnug et al., 2010). Moreover, microalgal biomass contains significant amounts of proteins, carbohydrates and other high-value compounds that can be potentially used as feeds, foods and pharmaceuticals (Chisti, 2007). Thus, integrating the production of such co-products with biofuels will provide new insight into improving the production economics of microalgal biodiesel. Microalgae can also be used for sequestration of carbon dioxide from industrial flue gases and wastewater treatment by removal of nutrients (Chinnasamy et al 2010; Fulke et al., 2010; Levine et al., 2011; Yang et al., 2011). Coupled with these environment-beneficial approaches, the production potential of microalgae derived biodiesel is desirable.

Feedstocks	Oil content (% dry weight)	Oil yeild (L/ha year)	Land area needed (M ha)[a]	Percentage of existing US cropping area[a]
Corn	44	172	3480	1912
Hemp	33	363	1650	906
Soybean	18	636	940	516
Jatropha	28	741	807	443
Camelina	42	915	650	357
Canola	41	974	610	335
Sunflower	40	1070	560	307
Castor	48	1307	450	247
Palm oil	36	5366	110	60.4
Microalgae (low oil content)	30	58,700	10.2	5.6
Microalgae (medium oil content)	50	97,800	6.1	3.4
Microalgae (high oil content)	70	136,900	4.4	2.4

[a] For meeting all transport fuel needs of the United States. Adapted from Chisti, 2007 and Mata et al., 2010.

Table 3. Comparison of microalgae with other biodiesel feedstocks.

3. Biodiesel production from microalgae

The biodiesel production from microalgal oil shares the same processes and technologies as those used for other feedstocks derived oils. However, microalgae are microorganisms living essentially in liquid environments and thus have particular cultivation, harvesting, and downstream processing techniques for efficient biodiesel production. The microalgal biodiesel production pipeline is schematically presented in Figure 5, including strain selection, mass culture, biomass harvesting and processing, oil extraction and transesterification.

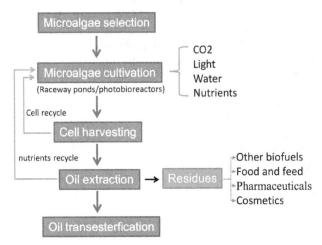

Fig. 5. Microalgal biodiesel production pipeline

3.1 Microalgae selection

There are more than 50,000 microalgal species around the world. Selection of an ideal species is of fundamental importance to the success of algal biodiesel production. Theoretically, an ideal species should own the following desirable characteristics: rapid growth rate, high oil content, wide tolerance of environmental conditions, CO2 tolerance and uptake, large cell size, easy of disruption, etc. However, it is unlikely for a single species to excel in all above mentioned characteristics. Thus, prioritization is required. Commonly, fast-growing strains with high oil content are placed on the priority list for biodiesel production. Fast growth makes sure the high biomass productivity and reduces the contamination risk owing to out-competition of slower growers. High oil content helps increase the process yield coefficient and reduce the cost of downstream extraction and purification. The selected species should be suitable for mass cultivation under local geographic and climatic conditions, for example, the inland prefers freshwater algae while the coastal place desires marine algal species. Ease of harvesting is an often-overlooked criterion and should be taken into account. Algal biomass harvest requires significant capital and accounts for up to 30% of total biomass production cost (Molina Grima et al., 2003). Therefore, it is desirable to choose algal species with properties that simplify harvesting, including large cell size, high specific gravity and autofloculation potential (Griffiths & Harrison, 2009). These properties can greatly influence the process economics for biodiesel production from algae. An additional algal characteristic is the suitability of lipids for biodiesel production; for example, neutral lipids in particular TAG are superior to polar lipids (phospholipids and glycolipids) for biodiesel and C18:1 has advantages over other fatty acids for improving biodiesel quality (Knothe, 2009).

3.2 Microalgae cultivation
3.2.1 Factors affecting algal lipids and fatty acids

Microalgae require several things to grow, including a light source, carbon dioxide, water, and inorganic salts. The lipid content and fatty acid composition are species/strain-specific and can be greatly affected by a variety of medium nutrients and environmental factors. Carbon is the main component of algal biomass and accounts for ca 50% of dry weight. CO_2 is the common carbon source for algal growth. But some algal species are also able to utilize organic carbon sources, for example sugars and glycerol (Easterling et al., 2009; Liu et al., 2010). Sugars particularly glucose are preferred and can be used to boost production of both algal biomass and lipids (Liu et al., 2010). Nitrogen is an important nutrient affecting lipid metabolism in algae. The influence of nitrogen concentration on lipid and fatty acid production has been investigated in numerous algal species. Nitrate was suggested to be superior to other nitrogen sources such as urea and ammonium for algal lipid production (Li et al., 2008). Generally, low concentration of nitrogen in the medium favors the accumulation of lipids particularly TAGs and total fatty acids. But in some cases, nitrogen starvation caused decreased synthesis of lipids and fatty acids (Saha et al., 2003). Nitrogen concentration also affects algal fatty acid composition. For example, in cyanobacteria, increased levels of C16:0 and C18:1 and decreased C18:2 levels were observed in response to nitrogen deprivation (Piorreck & Pohl, 1984). In the marine alga *Pavlova viridis*, nitrogen depletion resulted in an increase in saturated, monounsaturated fatty acids and C22:6 (n-3) contents (Li et al., 2005). Nitrogen starvation brought about a strong increase in the proportion of C20:4 (n-6) in the green algal *Parietochloris incisa* (Solovchenko et al., 2008). Similar to nitrogen, silicon is a key nutrient that affects lipid metabolism of diatoms, and can promote the accumulation of neutral lipids as well as of saturated and monounsaturated fatty acids when depleted from culture medium (Roessler, 1988). Other types of nutrient deficiency include phosphorus and sulfur limitations are also able to enhance lipid accumulation in algae (Khozin-Goldberg & Cohen, 2006; Li et al., 2010b; Sato et al., 2000). These types of nutrient deficiency, however, do not always lead to elevated overall lipid production, because they at the same time exert negative effect on algal growth and contribute to the reduced algal biomass production that compromises the enhanced lipid yield resulting from increased lipid content. Therefore, the manipulation of these nutrients needs to be optimized to induce lipid accumulation while maintaining algal growth for maximal production of lipids. Iron is a micro-nutrient required in a tiny amount for ensuring algal growth. Within a certain range of concentrations, high concentrations of iron benefit algal growth as well as cellular lipid accumulation and thus the overall lipid yield in the green alga *Chlorella vulgaris* (Liu et al., 2008).

Among the environmental factors, light is an important one that has a marked effect on the lipid production and fatty acid composition in algae (Brown et al., 1996; Damiani et al., 2010; Khotimchenko & Yakovleva, 2005; Napolitano, 1994; Sukenik et al., 1989; Zhekisheva et al., 2002, 2005). Generally, low light intensity favors the formation of polar lipids such as the membrane lipids associated with the chloroplast; whereas high light intensity benefits the accumulation of neutral storage lipids in particular TAGs. In *H. pluvialis*, for example, high light intensity resulted in a great increase of both neutral and polar lipids, but the increase extent of neutral lipids was much greater than that of polar lipids, leading to the dominant proportion of neutral lipids in the total lipids (Zhekisheva et al., 2002, 2005). Although the effect of light intensity on fatty acid composition differs among the algal species and/or strains, there is a general trend that the increase of light intensity contributes to the enhanced proportions of saturated and monounsaturated fatty acids and the concurrently the reduced proportion of polyunsaturated fatty acids (Damiani et al., 2010; Sukenik et al., 1989; Zhekisheva et al., 2002, 2005). Temperature is another important environmental factor that affects profiles of algal lipids and fatty acids. In response to temperature shift, algae commonly alter the physical properties and thermal responses of membrane lipids to maintain fluidity and function of membranes (Somerville, 1995). In general, increased temperature causes increased fatty acid saturation and at the same time decreased fatty acid unsaturation. For example, C14:0, C16:0, C18:0 and C18:2 increased and C18:3 (n-3), C18:4, C20:5 and C22:6 decreased in *Rhodomonas* sp., and C16:0 increased and C18:4 decreased in *Cryptomonas* sp. when temperature increased (Renaud et al., 2002). As for the effect of temperature on cellular lipid content, it differs in a species-dependent manner. In response to increased temperature, algae may show an increase (Boussiba et al., 1987), no significant change or even a decrease (Renaud et al., 2002) in lipid contents. Other environmental factors such as salinity, pH and dissolved O_2 are also important and able to affect algal lipid metabolism.

In addition to the nutritional and environmental factors, growth phase and aging of the culture affect algal lipids and fatty acids. Commonly, algae accumulate more lipids at stationary phase than at logarithmic phase (Bigogno et al., 2002; Mansour et al., 2003). Associated with the growth phase transition from logarithmic to stationary phase, increased proportions of C16:0 and C18:1 and decreased proportions of PUFAs are often observed. Besides, it is suggested that algal lipids and fatty acids can be greatly affected by cultivation modes. Algae growing under heterotrophic mode usually produce more lipids in particular TAG and higher proportion of C18:1 than under photoautotrophic mode (Liu et al., 2011).

3.2.2 Raceway ponds and photobioreactors

Currently, the commonly used culture systems for large-scale production of algal biomass are open ponds and enclosed photobioreactors. An open pond culture system usually consists of a series of raceways-type of ponds placed outdoors. In this system, the shallow pond is usually about one foot deep; algae are cultured under conditions identical to their natural environment. The pond is designed in a raceway configuration, in which a paddle wheel provides circulation and mixing of the algal cells and nutrients (Chisti, 2007). The raceways are typically made from poured concrete, or they are simply dug into the earth and lined with a plastic liner to prevent the ground from soaking up the liquid. Compared with photobioreactors, open ponds cost less to build and operate, and are more durable with a large production capacity. However, the open pond system has its intrinsic disadvantages including rapid water loss due to evaporation, contamination with unwanted algal species as well as organisms that feed on algae, and low biomass productivity. In addition, optimal culture conditions are difficult to maintain in open ponds and recovering the biomass from such a dilute culture is expensive.

Unlike open ponds, enclosed photobioreactors are flexible systems that can be employed to overcome the problems of evaporation, contamination and low biomass productivity encountered in open ponds (Mata et al., 2010). These systems are made of transparent materials with a large surface area-to-volume ratio, and generally placed outdoors using natural light for illumination. The tubular photobioreactor is the most widely used one, which consists of an array of straight transparent tubes aligned with the sun's rays (Chisti, 2007). The tubes are generally no more than 10 cm in diameter to maximize sunlight penetration. The medium broth is circulated through a pump to the tubes, where it is exposed to light for photosynthesis, and then back to a reservoir. In some photobioreactors, the tubes are coiled to form what is known as a helical tubular photobioreactor. Artificial illumination can be used for photobioreactor. But it adds to the production cost and thus is used for the production of high value products instead of biodiesel feedstock. The algal biomass is prevented from settling by maintaining a highly turbulent flow within the reactor using either a mechanical pump or an airlift pump (Chisti, 2007). The result of photosynthesis will generate oxygen. The oxygen levels will accumulate in the closed photobioreactor and inhibit the growth of algae. Therefore, the culture must periodically be returned to a degassing zone, an area where the algal broth is bubbled with air to remove the excess oxygen. In addition, carbon dioxide must be fed into the system to provide carbon source and maintain culture pH for algal growth. Photobioreactors require cooling during daylight hours and temperature regulation in night hours. This may be done through heat exchangers located either in the tubes themselves or in the degassing column.

Table 4 shows the comparison between open ponds and photobioreactors for microalgae cultivation.

Photobioreactors have obvious advantages over open ponds: offer better control, prevent contamination and evaporation, reduce carbon dioxide losses and allow to achieve higher biomass productivities. However, enclosed photobioreactors cost high to build and operate and the scale-up is difficult, limiting the number of large-scale commercial systems operating globally to high-value production runs (Greenwell et al., 2010). In this context, a hybrid photobioreactor-open pond system is proposed: using photobioreactors to produce contaminant-free inoculants for large open ponds.

Culture systems	Open ponds	Enclosed bioreactors
Contamination control	Difficult	Easy
Contamination risk	High	Reduced
Sterility	None	Achievable
Process control	Difficult	Easy
Species control	Difficult	Easy
Mixing	Very poor	Uniform
Operation regime	Batch or semi-continuous	Batch or semi-continuous
Area/volume ration	Low	High
Algal cell density	Low	High
Investment	Low	Hight
Operation cost	Low	High
Light utilization efficiency	Poor	High
Temperature control	difficult	More uniform temperature
Productivity	Low	High
Hydrodynamic stress on algae	Very low	Low-high
Evaporation of growth medium	High	Low
Gas transfer control	Low	High
O_2 inhibition	< bioreactors	Great problem
Scale-up	Difficult	Difficult

Table 4. Comparison of open ponds and photobioreactors for microalgae cultivation (Mata et al., 2010)

3.3 Biomass harvesting and concentration

Algal harvesting is the concentration of diluted algal suspension into a thick algal paste, with the aim of obtaining slurry with at least 2–7% algal suspension on dry matter basis. Biomass harvest is a very challenging process and may contribute to 20-30% of the total biomass production cost (Molina Grima et al., 2003). The most common harvesting methods include sedimentation, filtration, centrifugation, sometimes with a pre-step of flocculation or flocculation-flotation. Flocculation is employed to aggregate the microalgal cells into larger clumps to enhance the harvest efficiency by gravity sedimentation, filtration, or centrifugation (Molina Grima et al., 2003). The selection of a harvesting process for a particular strain depends on size and properties of the algal strain. The selected harvest method must be able to handle a large volume of algal culture broth.

Filtration is the most commonly used method for harvesting algal biomass. The process can range from micro-strainers to pressure filtration and ultra-filtration systems. Vacuum filtration is feasible for harvesting large microalgae such as *Coelastrum proboscideum* and *Spirulina platensis* but unsuitable for recovering small size algal cells such as *Scenedesmus, Dunaliella,* or *Chlorella* (Molina Grima et al., 2003). Membrane-based microfiltration and ultrafiltration have also been used for harvesting algal cells for some specific application purposes, but overall, they are more expensive. Centrifugation is an accelerated sedimentation process for algae harvesting. Generally, centrifugation has high capital and operation costs, but its efficiency is much higher than natural sedimentation. Because of its high cost, centrifugation as an algae harvesting method is usually considered only feasible for high value products rather than biofuels.

3.4 Biomass processing for oil extraction

After harvesting, chemicals in the biomass may be subject to degradation induced by the process itself and also by internal enzyme in the algal cells. For example, lipase contained in the cells can rapidly hydrolyze cellular lipids into free fatty acids that are not suitable for biodiesel production. Therefore, the harvested biomass need be processed rapidly. Drying is a major step to keep the quality of the oil. In addition, the solvent-based oil extraction can be difficult when wet biomass is used. Various drying methods such as sun drying, spray drying, freeze drying, and drum drying can be used for drying algal biomass (Mata et al., 2010). Due to the high water content of algal biomass, sun-drying is not a very effective method for algal powder production. Spray drying and freeze drying are rapid and effective, but also expensive and not economically feasible for biofuel production. Because of the high energy required, drying is considered as one of the main economical bottlenecks in the entire process.

There are several approaches for extracting oil from the dry algal biomass, including solvent extraction, osmotic shock, ultrasonic extraction and supercritical CO_2 extraction. Oil extraction from dried biomass can be performed in two steps, mechanical crushing followed by solvent extraction in which hexane is the main solvent used. For example, after the oil extraction using an expeller, the leftover pulp can be mixed with cyclohexane to extract the remaining oil. The oil dissolves in the cyclohexane and the pulp is filtered out from the solution. These two stages are able to extract more than 95% of the total oil present in the algae. Oil extraction from algal cells can also be facilitated by osmotic shock or ultrasonic treatment to break the cells. Osmotic shock is a sudden reduction in osmotic pressure causing cells to rupture and release cellular components including oil. The algae lacking the cell wall are suitable for this process. In the ultrasonic treatment, the collapsing cavitation bubbles near to the cell walls cause cell walls to break and release the oil into the solvent. Supercritical CO_2 is another way for efficient extraction of algal oil, but the high energy demand is a limitation for commercialization of this technology (Herrero et al., 2010).

3.5 Oil transesterification

Algal oil contained in algal cells can be converted into biodiesel through transesterification. Transesterification is a chemical conversion process involving reacting triglycerides of vegetable oils or animal fats catalytically with a short-chain alcohol (typically methanol or ethanol) to form fatty acid esters and glycerol (Figure 6). This reaction occurs stepwise with the first conversion of triglycerides to diglycerides and then to monoglycerides and finally to glycerol. The complete transesterification of 1 mol of triglycerides requires 3 mol of alcohol, producing 1 mol of glycerol and 3 mol of fatty esters. Considering that the reaction is reversible, large excess of alcohol is used in industrial processes to ensure the direction of fatty acid esters. Methanol is the preferred alcohol for industrial use because of its low cost, although other alcohols like ethanol, propanol and butanol are also commonly used.

$$
\begin{array}{lcccc}
CH_2\text{--}OOC\text{--}R_1 & & R_1\text{--}COO\text{--}R & & CH_2\text{--}OH \\
| & & & & | \\
CH\text{--}OOC\text{--}R_2 & +\ 3ROH \xrightleftharpoons{Catalyst} & R_2\text{--}COO\text{--}R & + & CH\text{--}OH \\
| & & & & | \\
CH_2\text{--}OOC\text{--}R_3 & & R_3\text{--}COO\text{--}R & & CH_2\text{--}OH \\
\text{Triglyceride} & \text{Alcohol} & \text{Esters} & & \text{Glycerol}
\end{array}
$$

Fig. 6. Transesterification of oil to biodiesel. R_{1-3} indicates hydrocarbon groups.

In addition to heat, a catalyst is needed to facilitate the transesterification. The transesterification of triglycerides can be catalyzed by acids, alkalis or enzymes. Acid transesterification is considered suitable for the conversion of feedstocks with high free fatty acids but its reaction rate is low (Gerpen, 2005). In contrast, alkali-catalyzed transesterification has a much higher reaction rate, approximately 4000 times faster than the acid-catalyzed one (Fukuda et al., 2001). In this context, alkalis (sodium hydroxide and potassium hydroxide) are preferred as catalysts for industrial production of biodiesel. The use of lipases as transesterification catalysts has also attracted much attention as it produces high purity product and enables easy separation from the byproduct glycerol (Ranganathan et al., 2008). However, the cost of enzyme is still relatively high and remains a barrier for its industrial implementation. In addition, it has been proposed that biodiesel can be prepared from oil via transesterification with supercritical methanol (Demirbas, 2002).

4. Genetic engineering of microalgae

4.1 Microalgal lipid biosynthesis
Although lipid metabolism, in particular the biosynthesis of fatty acids and TAG, is poorly understood in algae, it is generally recognized that the basic pathways for fatty acid and TAG biosynthesis are similar to those demonstrated in higher plants.

Algae synthesize the *de novo* fatty acids in the chloroplast using a single set of enzymes. A simplified schedule for saturated fatty acid biosynthesis is shown in Figure 7. Acetyl-CoA is the basic building block of the acyl chain and serves as a substrate for acetyl CoA

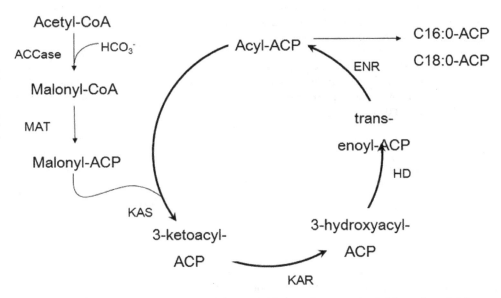

Fig. 7. Simplified overview of saturated fatty acid biosynthesis in algal chloroplast. ACCase, acetyl-CoA carboxylase; ACP, acyl carrier protein; CoA, coenzyme A; ENR, enoyl-ACP reductase; HD, 3-hydroxyacyl-ACP dehydratase; KAR, 3-ketoacyl-ACP reductase; KAS, 3-ketoacyl-ACP synthase; MAT, malonyl-CoA:ACP transacylase.

carboxylation and as well as a substrate for the initial condensation reaction. The formation of malonyl CoA from acetyl CoA is generally regarded as the first reaction of fatty acid biosynthesis, which is catalyzed by acetyl CoA carboxylase (ACCase). The malonyl group of malonyl CoA is transferred to a protein co-factor, acyl carrier protein (ACP), resulting in the formation of malonyl ACP that enters into a series of condensation reactions with acyl ACP (or acetyl CoA) acceptors. The first condensation reaction is catalyzed by 3-ketoayl ACP synthase III (KAS III), forming a four-carbon product. KAS I and KAS II catalyze the subsequent condensations. After each condensation, the 3-ketoacyl-ACP product is reduced, dehydrated, and reduced again, by 3-ketoacyl-ACP reductase, 3-hydroxyacyl-ACP dehydratase, and enoyl-ACP reductase, respectively, to form a saturated fatty acid. To produce an unsaturated fatty acid, a double bond is introduced onto the acyl chain by the soluble enzyme stearoyl ACP desaturase (SAD). Unlike plants, some algae produce long-chain acyl ACPs (C_{20}-C_{22}) that derive from the further elongation and/or desaturation of C_{18}. The fatty acid elongation is terminated when the acyl group is released from ACP by an acyl-ACP thioesterase that hydrolyzes the acyl ACP and produces free fatty acids or by acyltransferases that transfer the fatty acid from ACP to glycerol-3-phosphate or monoacylglycerol-3-phosphate. These released fatty acids serve as precursors for the synthesis of cellular membranes and neutral storage lipids like TAG.

It has been proposed that the biosynthesis of TAG occurs in cytosol via the direct glycerol pathway (Figure 8). Generally, acyl-CoAs sequentially react with the hydroxyl groups in glycerol-3-phosphate to form phosphatidic acid via lysophosphatidic acid. These two reactions are catalyzed by glycerol-3-phospate acyl transferase and lysophosphatidic acid acyl transferase respectively. Dephosphorylation of phosphatidic acid results in the release of DAG which accepts a third acyl from CoA to form TAG. This final step is catalyzed by diacylglycerol acyltransferase, an enzymatic reaction that is unique to TAG synthesis. In addition, an alternative pathway that is independent of acyl-CoA may also be present in algae for TAG biosynthesis (Dahlqvist et al., 2000). This pathway employs phospholipids as acyl donors and diacylglycerols as the acceptors and might be activated when algal cells are exposed to stress conditions, under which algae usually undergo rapid degradation of the photosynthetic membranes and concurrent accumulation of cytosolic TAG-enriched lipid bodies (Hu et al., 2008b).

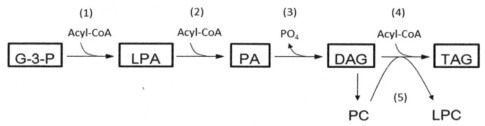

Fig. 8. Simplified illustration of the TAG biosynthesis in algae. DAG, diacylglycerol; LPA, lysophosphatidic acid; LPC, lysophosphatidylcholine; G-3-P, glycerol-3-phosphate; PA, phosphatidic acid; PC, phosphatidylcholine; TAG, triacylglycerol. (1) glycerol-3-phosphate acyl transferase, (2) lysophosphatidic acid acyl transferase, (3) phosphatidic acid phosphatase, (4) diacylglycerol acyl transferase, and (5) phospholipid:diacylglycerol acyltransferase.

4.2 Genetic engineering of microalgal lipids

Genetic engineering is a feasible and complimentary approach to increase algal productivity and improve the economics of algal biodiesel production. This has long been recognized but it seems that so far little progress has been made. The lack of full or near-full genome sequences and robust transformation systems makes genetic engineering of algae lag much behind that of bacteria, fungi and higher eukaryotes. Although certain algal species have been reported for efficient transformation, it proves to be difficult to produce stable transformants of algae. Currently, sophisticated genetic engineering whereby several genes are concurrently down-regulated or overexpressed is only really applicable to the green alga *Chlamydomonas reinhardtii*. This situation, however, is likely to change because of the growing scientific and commercial interest in other algal species that are of great potential for industrial applications.

Understanding the algal lipid biosynthesis is of great help to engineer algal lipid production. Although lipid metabolism in algae is not as fully understood as that in higher plants, they have similar lipid biosynthetic pathway as mentioned above. Theoretically, overexpression of the genes involved in fatty acid synthesis is able to increase lipid accumulation, in that fatty acids required as precursors for lipid biosynthesis are produced in excess. However, overexpressoin of the native ACCase, the rate-limiting enzyme catalyzing the first committed step of fatty acid biosynthesis in many organisms, could not increase the lipid production in diatom (Dunahay et al., 1995). It is possible that under high flux conditions through ACCase, the condensing enzymes or other factors may begin to limit fatty acid synthesis rate. Therefore, more complete control may come from certain transcription factors that can increase expression of the entire pathway. Another feasible approach of increasing cellular lipid contents is to inhibit metabolic pathways that lead to other carbon storage compounds, such as starch. Starch synthesis shares common carbon precursors with lipid synthesis in algae. Blocking starch synthesis is able to redirect carbon flux to lipid biosynthetic pathway, resulting in overproduction of fatty acids and thus total lipids (Li et al., 2010a). Neutral lipids in particular TAG surpass other lipids for biodiesel production, attracting the interest of enhancing cellular TAG contents through genetic engineering. Overexpression of genes involved in TAG assembly, e.g., glycerol-3-phosphate acyltransferase, lysophosphatidic acid acyltransferase, or diacylglycerol acyltransferase, all significantly increase TAG production in plants. Such strategies may also be applicable to algae for enhancing TAG levels. Commonly, algae produce larger amounts of lipids under unfavorable conditions than logarithmic growing condition. Enhancing lipid biosynthesis through genetic engineering, therefore, is likely to reduce algal proliferation and biomass production. In this context, the use of inducible promoters could overcome the problem because the transgenic expression can only be activated when a high cell density is achieved. The important properties of biodiesel such as cetane number, viscosity, cold flow, oxidative stability, are largely determined by the composition and structure of fatty acid esters which in turn are determined by the characteristics of fatty acids of biodiesel feedstocks, for example carbon chain length and unsaturation degree (Knothe, 2005b). Thus, the genetic modification of algal fatty acid composition is of also great interest. Generally, saturated fatty esters possess high cetane number and superior oxidative stability; whereas unsaturated, especially polyunsaturated fatty esters have improved low-temperature properties (Knothe, 2008). In this regard, it is suggested that the modification of fatty esters, for example the enhanced proportion of oleic acid (C18:1) ester, can provide a compromise solution between oxidative stability and low-temperature properties and therefore promote the quality of biodiesel (Knothe, 2008, 2009). Oleic acid is converted to linoleic acid (C18:2) in a single desaturation step, catalyzed by a Δ12 desaturase enzyme encoded by the *FAD2* gene. Inactivation of this desaturation step can greatly increase the proportion of oleic acid in soybean and may represent a possible strategy for elevated accumulation of oleic acid in algae.

Genetic engineering can also be used potentially to improve tolerance of algae to stress factors such as temperature, salinity and pH. These improved attributes will allow for the cost reduction in algal biomass production and be beneficial for growing selected algae under extreme conditions that limit the proliferation of invasive species. Photoinhibition is another technical challenge to be addressed by genetic engineering. When the light intensities exceed the value for maximum photosynthetic efficiency, algae show photoinhibition, a common phenomenon for phototrophy under which the growth rate slows down. Engineered algae with a higher threshold of light inhibition will significantly improve the economics of biodiesel production.

Engineering algae for biodiesel production is currently still in its infancy. Significant advances have only been achieved in the genetic manipulation of some model algae. It is likely that many of these advances can be extended to industrially important algal species in the future, making it possible to use modified algae as cell factories for commercial biodiesel production. Nevertheless, many challenges yet remain open and should be addressed before profitable algal biodiesel become possible.

5. Conclusion and perspectives

Microalgae have the potential for the production of profitable biodiesel that can eventually replace petroleum based fuel. Algal-biodiesel production, however, is still too expensive to be commercialized as no algal strains are available possessing all the advantages for achieving high yields of oil via the economical open pond culturing system. Current studies are still limited to the selection of ideal microalgal species, optimization of mass cultivation, biomass harvest and oil extraction processes, which contribute to high costs of biodiesel production from microalgae. Future cost-saving efforts for algal-biofuel production should focus on the production technology of oil-rich algae via enhancing algal biology (in terms of biomass yield and oil content) and culture-system engineering coupled with advanced genetic engineering strategies and utilization of wastes. In addition to oils, microalgae also contain large amounts of proteins, carbohydrates, and other nutrients or bioactive compounds that can be potentially used as feeds, foods and pharmaceuticals. Integrating the production of such co-products with biodiesel is an appealing way to lowering the cost of algal-biofuel production.

6. Acknowledgment

This work was supported by a grant from Seed-Funding Programme for Basic Research of the University of Hong Kong.

7. References

Abdullah, A.Z.; Salamatinia, B.; Mootabadi, H.& Bhatia, S. (2009). Current status and policies on biodiesel industry in Malaysia as the world's leading producer of palm oil. *Energy Policy*, 37, 5440-5448.

Abou-Shanab, R.A.I.; Hwang, J.-H.; Cho, Y.; Min, B.& Jeon, B.-H. (2011). Characterization of microalgal species isolated from fresh water bodies as a potential source for biodiesel production. *Applied Energy*, In Press, DOI: 10.1016/j.apenergy.2011.01.060.

Afify, A.; Shalaby, E.A.& Shanab, S.M.M. (2010). Enhancement of biodiesel production from different species of algae. *Grasas Y Aceites*, 61, 416-422.

Ahmad, A.L.; Yasin, N.H.M.; Derek, C.J.C.& Lim, J.K. (2011). Microalgae as a sustainable energy source for biodiesel production: A review. *Renewable and Sustainable Energy Reviews*, 15, 584-593.

Araujo, G.S.; Matos, L.J.B.L.; Gonçalves, L.R.B.; Fernandes, F.A.N.& Farias, W.R.L. (2011). Bioprospecting for oil producing microalgal strains: Evaluation of oil and biomass production for ten microalgal strains. *Bioresource Technology*, In Press, DOI: 10.1016/j.biortech.2011.01.089.

Banerjee, A.& Chakraborty, R. (2009). Parametric sensitivity in transesterification of waste cooking oil for biodiesel production--A review. *Resources, Conservation and Recycling*, 53, 490-497.

Bigogno, C.; Khozin-Goldberg, I.; Boussiba, S.; Vonshak, A.& Cohen, Z. (2002). Lipid and fatty acid composition of the green oleaginous alga *Parietochloris incisa*, the richest plant source of arachidonic acid. *Phytochemistry*, 60, 497-503.

Boussiba, S.; Vonshak, A.; Cohen, Z.; Avissar, Y.& Richmond, A. (1987). Lipid and biomass production by the halotolerant microalga *Nannochloropsis salina*. *Biomass*, 12, 37-47.

Brennan, L.& Owende, P. (2010). Biofuels from microalgae--A review of technologies for production, processing, and extractions of biofuels and co-products. *Renewable and Sustainable Energy Reviews*, 14, 557-577.

Brown, M.R.; Dunstan, G.A.; Norwood, S.J.& Miller, K.A. (1996). Effects of harvest stage and light on the biochemical composition of the diatom *Thalassiosira pseudonana*. *Journal of Phycology*, 32, 64-73.

Canakci, M. (2007). The potential of restaurant waste lipids as biodiesel feedstocks. *Bioresource Technology*, 98, 183-190.

Chaiklahan, R.; Chirasuwan, N.; Loha, V.& Bunnag, B. (2008). Lipid and fatty acids extraction from the cyanobacterium *Spirulina*. *Scienceasia*, 34, 299-305.

Checkbiotech (2009) Massive increase in global biofuel production. Available from: http://bioenergy.checkbiotech.org/news/massive_increase_global_biofuel_produc tion.

Chen, C.-Y.; Yeh, K.-L.; Aisyah, R.; Lee, D.-J.& Chang, J.-S. (2011). Cultivation, photobioreactor design and harvesting of microalgae for biodiesel production: A critical review. *Bioresource Technology*, 102, 71-81.

Chen, F.& Johns, M. (1991). Effect of C/N ratio and aeration on the fatty acid composition of heterotrophic *Chlorella sorokiniana*. *Journal of Applied Phycology*, 3, 203-209.

Chen, G.-Q.; Jiang, Y.& Chen, F. (2008). Salt-induced alterations in lipid composition of diatom *Nitzshia laevis* (Bacillariophyceae). *Journal of Phycology*, 44, 1309-1314.

Cheng, Y.; Zhou, W.; Gao, C.; Lan, K.; Gao, Y.& Wu, Q. (2009). Biodiesel production from Jerusalem artichoke (*Helianthus Tuberosus L.*) tuber by heterotrophic microalgae *Chlorella protothecoides*. *Journal of Chemical Technology & Biotechnology*, 84, 777-781.

Chinnasamy, S.; Bhatnagar, A.; Hunt, R.W.& Das, K.C. (2010). Microalgae cultivation in a wastewater dominated by carpet mill effluents for biofuel applications. *Bioresource Technology*, 101, 3097-3105.

Chisti, Y. (2007). Biodiesel from microalgae. *Biotechnology Advances*, 25, 294-306.

Chisti, Y. (2008). Biodiesel from microalgae beats bioethanol. *Trends in Biotechnology*, 26, 126-31.

Chiu, S.Y.; Kao, C.Y.; Chen, C.H.; Kuan, T.C.; Ong, S.C.& Lin, C.S. (2008). Reduction of CO_2 by a high-density culture of *Chlorella sp* in a semicontinuous photobioreactor. *Bioresource Technology*, 99, 3389-3396.

Chiu, S.-Y.; Kao, C.-Y.; Tsai, M.-T.; Ong, S.-C.; Chen, C.-H.& Lin, C.-S. (2009). Lipid accumulation and CO_2 utilization of *Nannochloropsis oculata* in response to CO_2 aeration. *Bioresource Technology*, 100, 833-838.

Converti, A.; Casazza, A.A.; Ortiz, E.Y.; Perego, P.& Del Borghi, M. (2009). Effect of temperature and nitrogen concentration on the growth and lipid content of *Nannochloropsis oculata* and *Chlorella vulgaris* for biodiesel production. *Chemical Engineering and Processing: Process Intensification*, 48, 1146-1151.

Couto, R.M.; Simoes, P.C.; Reis, A.; Da Silva, T.L.; Martins, V.H.& Sanchez-Vicente, Y. (2010). Supercritical fluid extraction of lipids from the heterotrophic microalga *Crypthecodinium cohnii*. *Engineering in Life Sciences*, 10, 158-164.

da Cunha, M.E.; Krause, L.C.; Moraes, M.S.A.; Faccini, C.S.; Jacques, R.A.; Almeida, S.R.; Rodrigues, M.R.A.& Caramao, E.B. (2009). Beef tallow biodiesel produced in a pilot scale. *Fuel Process Technology*, 90, 570-575.

Dahlqvist, A.; Stahl, U.; Lenman, M.; Banas, A.; Lee, M.; Sandager, L.; Ronne, H.& Stymne, H. (2000). Phospholipid : diacylglycerol acyltransferase: An enzyme that catalyzes the acyl-CoA-independent formation of triacylglycerol in yeast and plants. *Proceedings of the National Academy of Sciences*, 97, 6487-6492.

Damiani, M.C.; Popovich, C.A.; Constenla, D.& Leonardi, P.I. (2010). Lipid analysis in *Haematococcus pluvialis* to assess its potential use as a biodiesel feedstock. *Bioresource Technology*, 101, 3801-3807.

De la Hoz Siegler, H.; Ben-Zvi, A.; Burrell, R.E.& McCaffrey, W.C. (2011). The dynamics of heterotrophic algal cultures. *Bioresource Technology*, In Press, DOI: 10.1016/j.biortech.2011.01.081.

de Oliveira, D.; Di Luccio, M.; Faccio, C.; Dalla Rosa, C.; Bender, J.; Lipke, N.; Amroginski, C.; Dariva, C.& de Oliveira, J. (2005). Optimization of alkaline transesterification of soybean oil and castor oil for biodiesel production. *Applied Biochemistry and Biotechnology*, 122, 553-560.

Demirbas, A. (2002). Biodiesel from vegetable oils via transesterification in supercritical methanol. *Energy Conversion and Management*, 43, 2349-2356.

Demirbas, A. (2008). *Biodiesel-a realistic fuel alternative for diesel engines*. Springer - Verlag, London.

Demirbas, A. (2009). Progress and recent trends in biodiesel fuels. *Energy Conversion and Management*, 50, 14-34.

Demirbas, A. (2010). Use of algae as biofuel sources. *Energy Conversion and Management*, 51, 2738-2749.

Dias, J.M.; Alvim-Ferraz, M.C.M.& Almeida, M.F. (2009). Production of biodiesel from acid waste lard. *Bioresource Technology*, 100, 6355-6361.

Diaz-Felix, W.; Riley, M.R.; Zimmt, W.& Kazz, M. (2009). Pretreatment of yellow grease for efficient production of fatty acid methyl esters. *Biomass and Bioenergy*, 33, 558-563.

D'Oca, M.G.M.; Viêgas, C.V.; Lemões, J.S.; Miyasaki, E.K.; Morón-Villarreyes, J.A.; Primel, E.G.& Abreu, P.C. (2011). Production of FAMEs from several microalgal lipidic extracts and direct transesterification of the *Chlorella pyrenoidosa*. *Biomass and Bioenergy*, In Press, DOI: 10.1016/j.biombioe.2010.12.047.

Dunahay, T.G.; Jarvis, E.E.& Roessler, P.G. (1995). Genetic transformation of the diatoms *Cyclotella cryptica* and *Navicula saprophila*. *Journal of Phycology*, 31, 1004-1012.

Easterling, E.R.; French, W.T.; Hernandez, R.& Licha, M. (2009). The effect of glycerol as a sole and secondary substrate on the growth and fatty acid composition of *Rhodotorula glutinis*. *Bioresource Technology*, 100, 356-361.

Ethier, S.; Woisard, K.; Vaughan, D.& Wen, Z. (2011). Continuous culture of the microalgae Schizochytrium limacinum on biodiesel-derived crude glycerol for producing docosahexaenoic acid. *Bioresource Technology*, 102, 88-93.

Fan, K.-W.; Jiang, Y.; Faan, Y.-W.& Chen, F. (2007). Lipid characterization of Mangrove thraustochytrid - *Schizochytrium mangrovei*. *Journal of Agricultural and Food Chemistry*, 55, 2906-2910.

Feng, Y.; Li, C.& Zhang, D. (2011). Lipid production of Chlorella vulgaris cultured in artificial wastewater medium. *Bioresource Technology*, 102, 101-105.

Fukuda, H.; Kondo, A.& Noda, H. (2001). Biodiesel fuel production by transesterification of oils. *Journal of Bioscience and Bioengineering*, 92, 405-416.

Fulke, A.B.; Mudliar, S.N.; Yadav, R.; Shekh, A.; Srinivasan, N.; Ramanan, R.; Krishnamurthi, K.; Devi, S.S.& Chakrabarti, T. (2010). Bio-mitigation of CO_2, calcite formation and simultaneous biodiesel precursors production using *Chlorella* sp. *Bioresource Technology*, 101, 8473-8476.

Gao, C.; Zhai, Y.; Ding, Y.& Wu, Q. (2010). Application of sweet sorghum for biodiesel production by heterotrophic microalga *Chlorella protothecoides*. *Applied Energy*, 87, 756-761.

Gerpen, J.V. (2005). Biodiesel processing and production. *Fuel Process Technology*, 86, 1097-1107.

Gouveia, L.; Marques, A.; da Silva, T.& Reis, A. (2009). *Neochloris oleabundans* UTEX #1185: a suitable renewable lipid source for biofuel production. *Journal of Industrial Microbiology and Biotechnology*, 36, 821-826.

Gouveia, L.& Oliveira, A. (2009). Microalgae as a raw material for biofuels production. *Journal of Industrial Microbiology and Biotechnology*, 36, 269-274.

Graef, G.; LaVallee, B.J.; Tenopir, P.; Tat, M.; Schweiger, B.; Kinney, A.J.; Gerpen, J.H.V.& Clemente, T.E. (2009). A high-oleic-acid and low-palmitic-acid soybean: agronomic performance and evaluation as a feedstock for biodiesel. *Plant Biotechnology Journal*, 7, 411-421.

Graham, L.E.; Wilcox, L.W.& Graham, J. (2009). *Algae. 2nd ed.* Benjamin Cummings, San Francisco, CA.

Greenwell, H.C.; Laurens, L.M.L.; Shields, R.J.; Lovitt, R.W.& Flynn, K.J. (2010). Placing microalgae on the biofuels priority list: a review of the technological challenges. *Journal of The Royal Society Interface*, 7, 703-726.

Griffiths, M.& Harrison, S. (2009). Lipid productivity as a key characteristic for choosing algal species for biodiesel production. *Journal of Applied Phycology*, 21, 493-507.

Guedes, A.; Meireles, L.; Amaro, H.& Malcata, F. (2010). Changes in lipid class and fatty acid composition of cultures of *Pavlova lutheri*, in response to light intensity. *Journal of the American Oil Chemists' Society*, 87, 791-801.

Hawash, S.; Kamal, N.; Zaher, F.; Kenawi, O.& Diwani, G.E. (2009). Biodiesel fuel from Jatropha oil via non-catalytic supercritical methanol transesterification. *Fuel*, 88, 579-582.

Herrero, M.; Mendiola, J.A.; Cifuentes, A.& Ibáñez, E. (2010). Supercritical fluid extraction: Recent advances and applications. *Journal of Chromatography A*, 1217, 2495-2511.

Hill, J.; Nelson, E.; Tilman, D.; Polasky, S.& Tiffany, D. (2006). Environmental, economic, and energetic costs and benefits of biodiesel and ethanol biofuels. *Proceedings of the National Academy of Sciences*, 103, 11206-11210.

Ho, S.-H.; Chen, W.-M.& Chang, J.-S. (2010). *Scenedesmus obliquus* CNW-N as a potential candidate for CO2 mitigation and biodiesel production. *Bioresource Technology*, 101, 8725-8730.

Hsieh, C.-H.& Wu, W.-T. (2009). Cultivation of microalgae for oil production with a cultivation strategy of urea limitation. *Bioresource Technology*, 100, 3921-3926.

Hu, C.; Li, M.; Li, J.; Zhu, Q.& Liu, Z. (2008a). Variation of lipid and fatty acid compositions of the marine microalga *Pavlova viridis* (Prymnesiophyceae) under laboratory and outdoor culture conditions. *World Journal of Microbiology and Biotechnology*, 24, 1209-1214.

Hu, Q. (2004). Environmental effects on cell composition. in: *Handbook of microalgal culture*, (Ed.) Richmond, A., Blackwell. Oxford, pp. 83-93.

Hu, Q.; Sommerfeld, M.; Jarvis, E.; Ghirardi, M.; Posewitz, M.; Seibert, M.& Darzins, A. (2008b). Microalgal triacylglycerols as feedstocks for biofuel production: perspectives and advances. *Plant Journal*, 54, 621-639.

Huang, G.; Chen, F.; Wei, D.; Zhang, X.& Chen, G. (2010). Biodiesel production by microalgal biotechnology. *Applied Energy*, 87, 38-46.

Huerlimann, R.; de Nys, R.& Heimann, K. (2010). Growth, lipid content, productivity, and fatty acid composition of tropical microalgae for scale-Up production. *Biotechnology and Bioengineering*, 107, 245-257.

Jain, S.& Sharma, M.P. (2010). Prospects of biodiesel from Jatropha in India: A review. *Renewable and Sustainable Energy Reviews*, 14, 763-771.

James, G.O.; Hocart, C.H.; Hillier, W.; Chen, H.; Kordbacheh, F.; Price, G.D.& Djordjevic, M.A. (2011). Fatty acid profiling of Chlamydomonas reinhardtii under nitrogen deprivation. *Bioresource Technology*, 102, 3343-3351.

Kawachi, M.; Inouye, I.; Honda, D.; O'Kelly, C.J.; Bailey, J.C.; Bidigare, R.R.& Andersen, R.A. (2002). The Pinguiophyceae classis nova, a new class of photosynthetic stramenopiles whose members produce large amounts of omega-3 fatty acids. *Phycological Research*, 50, 31-47.

Khozin-Goldberg, I.; Bigogno, C.; Shrestha, P.& Cohen, Z. (2002). Nitrogen starvation induces the accumulation of arachidonic acid in the freshwater green alga *Parietochloris incisa* (trebuxiophyceae). *Journal of Phycology*, 38, 991-994.

Khozin-Goldberg, I.& Cohen, Z. (2006). The effect of phosphate starvation on the lipid and fatty acid composition of the fresh water eustigmatophyte *Monodus subterraneus*. *Phytochemistry*, 67, 696-701.

Knothe, G. (2005b). Dependence of biodiesel fuel properties on the structure of fatty acid alkyl esters. *Fuel Process Technology*, 86, 1059-1070.

Knothe, G. (2005a). Introduction: what is biodiesel. in: *The biodiesel handbook*, (Eds.) Knothe, G.;Gerpen, J.V.&Krahl, J., AOCS Press. Champaign, pp. 1-3.

Knothe, G. (2008). "Designer" biodiesel: Optimizing fatty ester composition to improve fuel properties. *Energy & Fuel*, 22, 1358-1364.

Knothe, G. (2009). Improving biodiesel fuel properties by modifying fatty ester composition. *Energy and Environmental Science*, 2, 759-766.

Lee, R.E. (2008). *Phycology. Forth ed.* Cambridge University Press, Cambridge.

Levine, R.B.; Costanza-Robinson, M.S.& Spatafora, G.A. (2011). *Neochloris oleoabundans* grown on anaerobically digested dairy manure for concomitant nutrient removal and biodiesel feedstock production. *Biomass and Bioenergy*, In Press, DOI: 10.1016/j.biombioe.2010.08.035.

Li, M.; Gong, R.; Rao, X.; Liu, Z.& Wang, X. (2005). Effects of nitrate concentration on growth and fatty acid composition of the marine microalga *Pavlova viridis* (Prymnesiophyceae). *Annals of Microbiology*, 55, 51-55.

Li, X.; Hu, H.-Y.; Gan, K.& Sun, Y.-X. (2010a). Effects of different nitrogen and phosphorus concentrations on the growth, nutrient uptake, and lipid accumulation of a freshwater microalga *Scenedesmus* sp. *Bioresource Technology*, 101, 5494-5500.

Li, Y.; Chen, Y.-F.; Chen, P.; Min, M.; Zhou, W.; Martinez, B.; Zhu, J.& Ruan, R. (2011b). Characterization of a microalga Chlorella sp. well adapted to highly concentrated municipal wastewater for nutrient removal and biodiesel production. *Bioresource Technology*, In Press, DOI: 10.1016/j.biortech.2011.01.091.

Li, Y.; Han, D.; Hu, G.; Sommerfeld, M.& Hu, Q. (2010b). Inhibition of starch synthesis results in overproduction of lipids in *Chlamydomonas reinhardtii*. *Biotechnology and Bioengineering*, 107, 258-268.

Li, Y.; Han, D.; Sommerfeld, M.& Hu, Q. (2011a). Photosynthetic carbon partitioning and lipid production in the oleaginous microalga *Pseudochlorococcum* sp. (Chlorophyceae) under nitrogen-limited conditions. *Bioresource Technology*, 102, 123-129.

Li, Y.; Horsman, M.; Wang, B.; Wu, N.& Lan, C. (2008). Effects of nitrogen sources on cell growth and lipid accumulation of green alga *Neochloris oleoabundans*. *Appl Microbiol Biotechnol*, 81, 629-636.

Liang, Y.; Sarkany, N.& Cui, Y. (2009). Biomass and lipid productivities of *Chlorella vulgaris* under autotrophic, heterotrophic and mixotrophic growth conditions. *Biotechnology Letters*, 31, 1043-1049.

Lin, Q.& Lin, J. (2011). Effects of nitrogen source and concentration on biomass and oil production of a *Scenedesmus rubescens* like microalga. *Bioresource Technology*, 102, 1615-1621.

Lin, Y.-H.; Chang, F.-L.; Tsao, C.-Y.& Leu, J.-Y. (2007). Influence of growth phase and nutrient source on fatty acid composition of *Isochrysis galbana* CCMP 1324 in a batch photoreactor. *Biochemical Engineering Journal*, 37, 166-176.

Liu, J.; Huang, J.; Fan, K.W.; Jiang, Y.; Zhong, Y.; Sun, Z.& Chen, F. (2010). Production potential of *Chlorella zofingienesis* as a feedstock for biodiesel. *Bioresource Technology*, 101, 8658-8663.

Liu, J.; Huang, J.; Sun, Z.; Zhong, Y.; Jiang, Y.& Chen, F. (2011). Differential lipid and fatty acid profiles of photoautotrophic and heterotrophic Chlorella zofingiensis: Assessment of algal oils for biodiesel production. *Bioresource Technology*, 102, 106-110.

Liu, X.-J.; Jiang, Y.& Chen, F. (2005). Fatty acid profile of the edible filamentous cyanobacterium *Nostoc flagelliforme* at different temperatures and developmental stages in liquid suspension culture. *Process Biochemistry*, 40, 371-377.

Liu, Z.-Y.; Wang, G.-C.& Zhou, B.-C. (2008). Effect of iron on growth and lipid accumulation in *Chlorella vulgaris*. *Bioresour Technol*, 99, 4717-4722.

Mandal, S.& Mallick, N. (2009). Microalga *Scenedesmus obliquus* as a potential source for biodiesel production. *Applied Microbiology and Biotechnology*, 84, 281-291.

Mansour, M.P.; Volkman, J.K.& Blackburn, S.I. (2003). The effect of growth phase on the lipid class, fatty acid and sterol composition in the marine dinoflagellate, *Gymnodinium* sp. in batch culture. *Phytochemistry*, 63, 145-153.

Mata, T.M.; Martins, A.A.& Caetano, N.S. (2010). Microalgae for biodiesel production and other applications: A review. *Renewable and Sustainable Energy Reviews*, 14, 217-232.

Matsumoto, M.; Sugiyama, H.; Maeda, Y.; Sato, R.; Tanaka, T.& Matsunaga, T. (2010). Marine diatom, *Navicula* sp. strain JPCC DA0580 and marine green alga, *Chlorella* sp. strain NKG400014 as potential sources for biodiesel production. *Applied Biochemistry and Biotechnology*, 161, 483-490.

Meher, L.C.; Naik, S.N.; Naik, M.K.& Dalai, A.K. (2008). Biodiesel production using karanja (*Pongamia pinnata*) and jatropha (*Jatropha curcas*) seed oil. in: *Handbook of plant-based biofuels*, (Ed.) Pandey, A., CRC Press. Boca Raton, FL pp. 255-266.

Miao, X.& Wu, Q. (2006). Biodiesel production from heterotrophic microalgal oil. *Bioresource Technology*, 97, 841-846.

Moazami, N.; Ranjbar, R.; Ashori, A.; Tangestani, M.& Nejad, A.S. (2011). Biomass and lipid productivities of marine microalgae isolated from the Persian Gulf and the Qeshm Island. *Biomass and Bioenergy*, In Press, DOI: 10.1016/j.biombioe.2011.01.039.

Molina Grima, E.; Belarbi, E.H.; Acién Fernández, F.G.; Robles Medina, A.& Chisti, Y. (2003). Recovery of microalgal biomass and metabolites: process options and economics. *Biotechnology Advances*, 20, 491-515.

Mussgnug, J.H.; Klassen, V.; Schlüter, A.& Kruse, O. (2010). Microalgae as substrates for fermentative biogas production in a combined biorefinery concept. *Journal of Biotechnology*, 150, 51-56.

Nakpong, P.& Wootthikanokkhan, S. (2010). High free fatty acid coconut oil as a potential feedstock for biodiesel production in Thailand. *Renewable Energy*, 35, 1682-1687.

Napolitano, G.E. (1994). The relationship of lipids with light and chlorophyll measurements in freshwater algae and periphyton. *Journal of Phycology*, 30, 943-950.

Oh, S.H.; Han, J.G.; Kim, Y.; Ha, J.H.; Kim, S.S.; Jeong, M.H.; Jeong, H.S.; Kim, N.Y.; Cho, J.S.; Yoon, W.B.; Lee, S.Y.; Kang, D.H.& Lee, H.Y. (2009). Lipid production in *Porphyridium cruentum* grown under different culture conditions. *Journal of Bioscience and Bioengineering*, 108, 429-434.

Oner, C.& Altun, S. (2009). Biodiesel production from inedible animal tallow and an experimental investigation of its use as alternative fuel in a direct injection diesel engine. *Applied Energy*, 86, 2114-2120.

Órpez, R.; Martínez, M.E.; Hodaifa, G.; El Yousfi, F.; Jbari, N.& Sánchez, S. (2009). Growth of the microalga *Botryococcus braunii* in secondarily treated sewage. *Desalination*, 246, 625-630.

Ota, M.; Kato, Y.; Watanabe, H.; Watanabe, M.; Sato, Y.; Smith Jr, R.L.& Inomata, H. (2009). Fatty acid production from a highly CO_2 tolerant alga, *Chlorocuccum littorale*, in the presence of inorganic carbon and nitrate. *Bioresource Technology*, 100, 5237-5242.

Pahl, S.L.; Lewis, D.M.; Chen, F.& King, K.D. (2010). Heterotrophic growth and nutritional aspects of the diatom *Cyclotella cryptica* (Bacillariophyceae): Effect of some environmental factors. *Journal of Bioscience and Bioengineering*, 109, 235-239.

Patil, P.D.& Deng, S. (2009). Optimization of biodiesel production from edible and non-edible vegetable oils. *Fuel*, 88, 1302-1306.

Phan, A.N.& Phan, T.M. (2008). Biodiesel production from waste cooking oils. *Fuel*, 87, 3490-3496.

Piorreck, M.& Pohl, P. (1984). Preparatory experiments for the axenic mass-culture of microalgae. 2. Formation of biomass, total protein, chlorophylls, lipids and fatty-acids in green and blue green-algae during one growth-phase. *Phytochem*, 23, 217-223.

Pruvost, J.; Van Vooren, G.; Cogne, G.& Legrand, J. (2009). Investigation of biomass and lipids production with *Neochloris oleoabundans* in photobioreactor. *Bioresource Technology*, 100, 5988-5995.

Pruvost, J.; Van Vooren, G.; Le Gouic, B.; Couzinet-Mossion, A.& Legrand, J. (2011). Systematic investigation of biomass and lipid productivity by microalgae in photobioreactors for biodiesel application. *Bioresource Technology*, 102, 150-158.

Pushparaj, B.; Buccioni, A.; Paperi, R.; Piccardi, R.; Ena, A.; Carlozzi, P.& Sili, C. (2008). Fatty acid composition of Antarctic cyanobacteria. *Phycologia*, 47, 430-434.

Queiroz, M.I.; Hornes, M.O.; da Silva-Manetti, A.G.& Jacob-Lopes, E. (2011). Single-cell oil production by cyanobacterium Aphanothece microscopica Nägeli cultivated heterotrophically in fish processing wastewater. *Applied Energy*, In Press, DOI: 10.1016/j.apenergy.2010.12.047.

Rabbani, S.; Beyer, P.; Lintig, J.v.; Hugueney, P.& Kleinig, H. (1998). Induced beta -carotene synthesis driven by triacylglycerol deposition in the unicellular alga *Dunaliella bardawil*. *Plant Physiology*, 116, 1239-1248.

Radakovits, R.; Jinkerson, R.E.; Darzins, A.& Posewitz, M.C. (2010). Genetic Engineering of Algae for Enhanced Biofuel Production. *Eukaryotic Cell*, 9, 486-501.

Ranganathan, S.V.; Narasimhan, S.L.& Muthukumar, K. (2008). An overview of enzymatic production of biodiesel. *Bioresource Technology*, 99, 3975-3981.

Rashid, U.& Anwar, F. (2008). Production of biodiesel through optimized alkaline-catalyzed transesterification of rapeseed oil. *Fuel*, 87, 265-273.

REN21. (2010) Renewables Global Status Report. Available from: http://www.ren21.net/Portals/97/documents/GSR/REN21_GSR_2010_full_revised%20Sept 2010.pdf.

Renaud, S.M.; Thinh, L.-V.; Lambrinidis, G.& Parry, D.L. (2002). Effect of temperature on growth, chemical composition and fatty acid composition of tropical Australian microalgae grown in batch cultures. *Aquaculture*, 211, 195-214.

Rodolfi, L.; Zittelli, G.C.; Bassi, N.; Padovani, G.; Biondi, N.; Bonini, G.& Tredici, M.R. (2009). Microalgae for oil: Strain selection, induction of lipid synthesis and outdoor mass cultivation in a low-cost photobioreactor. *Biotechnology and Bioengneering*, 102, 100-112.

Roessler, P.G. (1988). Changes in the activities of various lipid and carbohydrate biosynthetic enzymes in the diatom *Cyclotella cryptica* in response to silicon deficiency. *Archives of Biochemistry and Biophysics*, 267, 521-528.

Ross, A.B.; Jones, J.M.; Kubacki, M.L.& Bridgeman, T. (2008). Classification of macroalgae as fuel and its thermochemical behaviour. *Bioresource Technology*, 99, 6494-6504.

Saha, S.K.; Uma, L.& Subramanian, G. (2003). Nitrogen stress induced changes in the marine cyanobacterium *Oscillatoria willei* BDU 130511. *FEMS Microbiolgy Ecolology*, 45, 263-72.

Sahoo, P.K.& Das, L.M. (2009). Process optimization for biodiesel production from Jatropha, Karanja and Polanga oils. *Fuel*, 88, 1588-1594.

Saka, S.& Kusdiana, D. (2001). Biodiesel fuel from rapeseed oil as prepared in supercritical methanol. *Fuel*, 80, 225-231.

Sato, N.; Hagio, M.; Wada, H.& Tsuzuki, M. (2000). Environmental effects on acidic lipids of thylakoid membranes. in: *Recent advances in biochemistry of plant lipids*, (Eds.) Harwood, J.L.&Quinn, P.J., Portland Press. London, pp. 912-914.

Shafiee, S.& Topal, E. (2009). When will fossil fuel reserves be diminished? *Energy Policy*, 37, 181-189.

Sheehan, J.; Dunahay, T.; Benemann, J.& Roessler, P. 1998. A look back at the U.S. Department of Energy's aquatic species programme - Biodiesel from algae, Report No. NREL/TP-580-24190; National Renewable Energy Laboratory: Golden, CO.

Singhania, R.R.; Parameswaran, B.& Pandey, A. (2008). Plant-based bioufuels: an introduction. in: *Handbook of plant-based biofuels*, (Ed.) Pandey, A., CRC Press. Boca Raton, FL, pp. 3-12.

Sobczuk, T.M.& Chisti, Y. (2010). Potential fuel oils from the microalga Choricystis minor. *Journal of Chemical Technology and Biotechnology*, 85, 100-108.

Solovchenko, A.; Khozin-Goldberg, I.; Didi-Cohen, S.; Cohen, Z.& Merzlyak, M. (2008). Effects of light intensity and nitrogen starvation on growth, total fatty acids and arachidonic acid in the green microalga *Parietochloris incisa*. *Journal of Applied Phycology*, 20, 245-251.

Somerville, C. (1995). Direct tests of the role of membrane lipid composition in low-temperature-induced photoinhibition and chilling sensitivity in plants and cyanobacteria. *Proceedings of the National Academy of Sciences*, 92, 6215-8.

Sukenik, A.; Carmeli, Y.& Berner, T. (1989). Regulation of fatty acid composition by irradiance level in the eustigmatophyte *Nannochloropsis* sp. *Journal of Phycology*, 25, 686-692.

8

Eco-Physiological Barriers and Technological Advances for Biodiesel Production from Microalgae

Simrat Kaur[1], Mohan C. Kalita[2], Ravi B. Srivastava[3] and Charles Spillane[1]
[1]Genetics and Biotechnology Laboratory, Botany and Plant Science,
C306 Aras de Brun, National University of Ireland Galway,
[2]Department of Biotechnology, Gauhati University, Assam,
[3]Defence Institute of High Altitude Research, Defence Research & Development
Organisation, Leh (Jammu and Kashmir),
[1]Ireland
[2,3]India

1. Introduction

The combination of diminishing fossil fuel reserves (peak oil) and increasing prices of diesel provide a challenge to the majority of nations in terms of national energy security and ensuring sustainable energy supplies. Such pressing challenges have provided the impetus for an acceleration of renewable energy research to identify novel and innovative liquid biofuels for the future (IEA 2011). Any such liquid biofuels from renewable resources will need to have a lower environmental footprint than fossil fuel derived liquid biofuels in order to meet key sustainability criteria (Nuffield 2011). The most abundant available natural renewable resource on planet earth is solar energy. Photosynthetic organisms such as plants, algae (macro- and micro-algae) and some bacteria have been selected through evolution to convert solar energy to storable forms of energy. Such photosynthetic organisms can constitute a renewable resource which can effectively harvest and convert solar energy to a variety of energy-dense biofuels. In the case of microalgae, at least US$ 300 million has been committed to facilitate phycology research on bioprospecting microalgal diversity and evaluation of the feasibility of different microalgal species and strains for biofuels production (Sheehan et al., 1998). The use of oil crops such as palm, soy, and oilseed rape as feedstocks for biodiesel production has provided the basis for the first generation of biofuels. However, the cultivation of plants on arable land for biofuel production can compete with the use of the same land for food and animal feed production – the so called "food vs fuel" land-use competition dilemma. In addition, biofuel crops also have significant water and nutrient requirements which can adversely affect their sustainability when Life Cycle Analysis (LCA) is conducting across their value chain. For instance, microalgal production systems reliant on dwindling freshwater supplies will face sustainability problems if they are scaled up (Wigmosta et al., 2011). As a result microalgal systems based on saltwater or waste water are likely to be more sustainable. One approach being pursued for circumventing the 'food vs fuel' dilemma associated with first generation biofuel involves biological processing of lignocellulosic biomass as the basis for development of second generation biofuel systems. The development of commercial-scale efficient conversion technologies for exploitation of biological wastes as an obvious source for biofuel generation is a major focus of research and development efforts associated with this generation of biofuel. One of the third generation biofuels under development aims to exploit the photosynthetic capability of microalgae for the conversion of solar energy into energy-dense biomass. A major advantage of microalgae over the use of crop biomass for biodiesel production is the lower land area requirements for production of an equivalent amount of fuel. As understanding of microalgal genomes and biochemistry increases, opportunities are emerging for development of fourth generation biofuels where metabolic engineering of microbes leads to more effective domestication of microalgae for biofuel production. However, at present the commercial production of biofuels from microalgae is limited by a lack of effective systems for biomass production, harvesting, extraction, and recovery of oils that can economically integrate all operational units from growth through to biofuel product recovery. In this chapter, we discuss the limitations of individual operational units in the context of efforts underway to establish fourth generation microalgal biofuels that are economically and environmentally sustainable.

2. Microalgae biomass generation

2.1 Open cultivation

Large scale microalgal biomass production can be achieved either through open pond cultivation under natural sunlight or under the controlled conditions of a photobioreactor. In the USA, the history of mass production of microalgae dates back at least to 1953 with the production of *Scenedesmus* species in Washington. Many systems for cultivating microalgae on a large scale have been suggested in many countries including the USA, Germany, Japan, Israel, the UK, the Czech Republic and others. Typically, microalgae are first grown in inorganic nutrients and then, in a second phase, are cultivated is done using waste water streams.

Commercial cultivation of microalgae can be done in a range of different ways including (a) open cultivation using natural sunlight, (b) closed cultivation using natural light and (c) closed cultivation using artificial light (in photobioreactors). Each of these systems has advantages and disadvantages, and the choice of system depends on the degree of parameter control needed to produce the desired product and on the value of the end-product (Apt and Behrens, 1999). The most commonly used artificial open pond systems consist of large shallow ponds, tanks, circular ponds and raceway ponds (Ugwu et al., 2008). The construction and operation costs of such open cultivation systems are considerably less but are challenging to operate on a year round basis due to seasonal climatic variations. While open pond culture is cheaper than culture in closed photobioreactors (Borowitzka 1999), it is currently limited to a relatively small number of microalgal species. Rectangular ponds with a paddle wheel (raceway ponds) are the most widely used for the production of *Spirulina* sp., *Dunaliella salina* and *Haematococcus* sp. and currently represent the most efficient design for the large scale culture of most species of microalgae. Individual ponds are tyically up to 1 ha in area, with an average depth of about 20- 30 cm (Andersen 2005). The need to provide adequate light to the algal cells and maintaining an adequate water depth for mixing of the microalgae are important considerations for determining the pond depth. The diurnal natural light cycle results in the exposure of microalgae to limiting, saturating and over-saturating light conditions. High irradiances throughout the year and moderate temperatures are optimal for outdoor microalgae cultivation. For example, the geographical location of southern Spain with an average of 10-12 hours of sunlight per day, and a mean solar irradiance ranging from 400 µmol photons $m^{-2}s^{-1}$ during winter to 1800 µmol photons $m^{-2}s^{-1}$ is considered highly suitable for outdoor cultivation of microalgae. The maximal areal productivity of microalgae in outdoor conditions ranges from 20 to 30 $gm^{-2}d^{-1}$ (Cuaresma et al., 2011). To date, light-to-biomass conversion efficiencies of 1-4 % have been achieved for microalgae grown in conventional open pond cultivation systems. Because the scaling-up of microalgal biomass production in open raceway ponds is relatively easy, such systems are primarily considered for commercial applications. However, differences in weather variables such as solar irradiance, rainfall, and temperature significantly affect prospects for open cultivation of microalgae at different geographical locations. Temperature influences the rate of various reactions of photosynthesis (Raven, 1988). Therefore, microalgae exhibit an optimal growth within a narrow temperature range and die above a certain threshold temperature (Béchet et al., 2011). In addition, temperature is an important factor that affect the rate of evaporation from shallow algal ponds. In addition to changing the physical environment of open ponds, rainfall can lead to microbial contamination that inhibits microalgal growth (Hase et al., 2000).

The paddle wheels installed in open ponds are used to circulate the water, while compressed air can be introduced into the bottom of a pond to agitate the water, bringing microalgae from the lower levels upwards. Raceway channels are typically built in concrete or compacted earth, and are often lined with white plastic. During daylight, the microalgal culture is fed continuously in front of the paddlewheel where the flow begins. The biomass is harvested behind the paddlewheel, on completion of the circulation loop. The paddlewheel operates continuously to prevent sedimentation and flocculation (Chisti, 2007). The largest raceway-based biomass production facility currently occupies an area of 440,000 m^2 (Spolaore et al., 2006). This facility is owned by Earthrise Nutritional (www.earthrise.com) and is used to produce cyanobacterial biomass for food. In India, Pary Nutraceuticals (part of the Chennai-based Murugappa group) has been focusing on microalgal research and development and are commercially producing *Spirulina* for nutraceuticals.

2.2 Closed cultivation systems for microalgae

Several different closed systems using natural sunlight have been described for microalgae (Richmond et al. 1993, Molina Grima et al., 1995, Spektorova et al., 1997). In such systems, microalgae are grown in transparent glass or plastic vessels, and the vessels are placed under natural illumination. A higher surface to volume ratio is provided, so microalgal cell densities are often higher than in open ponds. However, these systems are also subject to variations in light intensity and temperature that make cultivation reproducibility problematic. In addition, removal of oxygen from the culture and the provision of adequate temperature control (especially if energy is required for cooling) pose a major problem with such closed systems. Large scale indoor cultivation using highly-controlled photobioreactors or fermentors have also been used successfully for microalgal biomass production. The wide range of different types of closed photobioreactors (PBRs) include vertical-column, flat-plate and tubular PBRs (Ugwu et al., 2008). These provide the ability to control and optimize culture parameters, and as a result such photobioreactors are suitable for culturing many different species of microalgae. The basic features which must be considered when designing a photobioreactor are: source of light, churning rate of algae (to avoid biomass sedimentation and for uniform availability of nutrients and light), material for construction, CO_2 supply, and removal of O_2, pH and temperature control (Kaur et al., 2010). Whether, closed or open systems will be optimal for commercial cultivation of different species (or strains) of microalgae is difficult to determine. However, it is clear that photobioreactors will play a critical role to feed open ponds with a high-cell-density unialgal inoculum (Cheng et al., 2011).

3. Thermodynamic efficiency of photosynthesis in microalgae

Photosynthesis is a chemical reaction governed by the laws of thermodynamics. Assuming a microalgal cell as 'boundary' and the process of photosynthesis as 'system', then according to the law of thermodynamics, the two kinds of work associated with this chemical reaction are electrical work and work of expansion. In a biological system such as microalgae, the production of ATP derived by the transfer of charges across the biofluidic membranes can be called electrical work. The growth or increase in size of cell and cellular components (including oil bodies) is the 'work of expansion'. In the very familiar photosynthetic reaction (Albarrán-Zavala & Angulo-Brown, 2007);

$$6CO_2 + 12\,H_2O \longrightarrow C_6H_{12}O_6 + 6H_2O + 6O_2$$

$$\Delta G^0 = 2880.31 \text{ kJ/mol}_{C6H12O6} \qquad \text{at } \lambda = 680 \text{ nm}$$

3.1 Photosynthetic conversion efficiency

In outdoor cultivation systems, the microalgal biomass productivity derived through photosynthesis depends on the solar energy input. The estimated yearly average solar energy density, including both direct beam radiation and diffuse scattered radiation is 10,038 MJ/m^2year. To account for non-sunny weather conditions, a more realistic theoretical maximum solar energy density is obtained after reducing this value by 10%, which corresponds to a value of 9034 MJ/m^2year. However, the actual value will exhibit temporal and spatial variation depending on the geographical location and will generally be lower (Cooney et al., 2011). The fraction of the solar energy spectrum (SEarth ~ 9034 MJ/m^2year) is further reduced by 45% to calculate the value of photosynthetically active radiation (PAR) that supports photosynthesis (SEarthPAR ~ 4065 MJ/m^2year). PAR is expressed in terms of photon flux as it reaches surface of microalgal cells in the form of photons, the energy of which varies inversely with the wavelength. The upper theoretical limit for the average PAR spectrum photon flux energy ($E_{MaxAvePAR}$) is 0.2253 MJ/mol that corresponds to $\lambda 531$ nm (green) (Weyer et al., 2009). Hence, the available photon flux reaching the earth surface and which is available for photosynthesis is calculated by the following formula (Cooney et al., 2011):

$$PF_{PAR} = \frac{S_{EarthPAR} \sim 4065 \text{ MJ} / m^2 / \text{year}}{E_{MaxAvePAR} \sim 0.2253 \text{ MJ} / \text{mol photon}} = 18{,}043 \text{ moles photons} / m^2 \text{year}$$

3.2 Maximum theoretical photosynthetic efficiency

The most cited values for maximum photosynthetic efficiencies in microalgae are in the range of 17-23% (Gordon & Polle 2007, Zemke et al., 2010). Cooney and coworkers (2011) illustrate how to obtain the maximum theoretical value of photosynthetic efficiency from the available PF_{PAR} of 18,043 moles photons/m^2 year. The theoretical maximum is calculated by considering both photon transmission (η_{PTE}) and photon conversion efficiencies (η_{PUE}) as 100%. Thus, according to the following equation:

$$\text{Photons utilized} = PF_{PAR} \times \eta_{PTE} \times \eta_{PUE} = 18,043 \text{ moles photons/}m^2 \text{ year} \times 1 \times 1$$

In a microalgal cell, these photons power the photosynthetic production of carbohydrates (CH_2O) which have an average energy content of 0.4825MJ/mole (Weyer et al., 2009). On average 10 photons are required to derive one mole of CH_2O. Hence, the total energy consumed during the photosynthetic conversion reaction is obtained as follows (Cooney et al 2011).

$$E_{CARB} = \frac{\left(18,043 \text{ moles photons } / \ m^2 \text{year} \times 1 \times 1\right). \left(0.4825MJ \ / \ \text{mole}\right)}{10} = 871 \ MJ \ / \ m^2 \text{year}$$

The estimated total photosynthetic efficiency during the conversion of PAR to microalgal biomass is calculated by dividing E_{CARB} (871 MJ/m^2year) and $S_{EarthPAR}$ (4065 MJ/m^2year), assuming that bioconversion of carbohydrates is 100% efficient. This gives a value of 21.4%, which is stated as the overall maximum theoretical photosynthetic efficiency relative to PAR. High lipid productivity depends on both the microalgal biomass areal productivity and the lipid content that can be generated from the microalgal strain. The lipid productivity is the most important factor influencing the cost of biodiesel production. High lipid content microalgal species and strains also favor the efficiency of biomass processing during oil extraction.

3.3 Technological innovations in illumination sources
3.3.1 Light Emitting Diodes (LEDs)
A light source with narrow spectral output that overlaps the photosynthetic absorption spectrum improves the energy conversion as the emission of light at unusable frequencies is eliminated. Light-emitting diodes (LEDs) are the only light source that currently meet this criterion. LEDs have the ability to produce high light levels with low radiant heat output and maintain useful light output for years. Thus, LEDs can have a very significantly longer life of 100,000 h as compared to 8000 h of fluorescent lights. These advantages make LEDs ideal for microalgal growth in controlled environments of growth chambers. The optimal wavelength conditions will vary from species to species of microalgae (Chen et al 2011). For example, the highest specific growth rate and biomass production from the photosynthetic cultivation of *Spirulina platensis* was obtained using red LED. The superimposed pattern of luminescence spectrum of blue LED (450-470 nm) and that of red LED (650-665 nm) corresponds to the light absorption spectrum of carotenoids and chlorophyll (Yeh & Chung, 2009). Therefore, the red LED favors microalgal growth but switching to illumination with blue LED improved the rate of astaxanthin production by *Haematococcus pluvialis* (Katsuda et al. 2004). Flashing light from blue LEDs is also a promising illumination method for *H. pluvialis* growth and astaxanthin production (Katsuda et al 2006). The use of flashing LED as sources of intermittent light in indoor algal culture can yield a major gain in energy economy comparing to fluorescent light sources (Matthijs et al., 1996). The research results by Nedbal et al. (1996) also suggest that algal growth rates in intermittent light can be higher than those in equivalent continuous light.

Red LEDs were found to reduce the average cell volume of *Chlorella vulgaris* without affecting the total biomass production (Lee and Palsson, 1994). However, under the exposure of fluorescent light, cells regained their normal size.

3.3.2 Optical fiber technology

The use of optical fibers as internal light sources could increase light efficiency whilst simultaneously reducing electricity consumption. For example, solar energy excited optical fiber requires only 1.0 kW-h of electricity (Chen et al., 2011). Spatial (i.e. orientation and dimensions of the photobioreactor) and temporal variations (i.e. due to weather conditions) greatly affect the availability of sunlight. Therefore, internal illumination by optical fiber is unstable. To circumvent this problem, Chen et al. (2011) have conceptualized a photobioreactor that combines optical fiber and multi-LED light sources with both solar panel and wind power generators. This has a potential to be developed into a commercially viable microalgae cultivation system with significantly reduced electricity consumption.

4. Limitations and improvements of photosynthetic biomass production of microalgae

4.1 Low light intensity and distribution

There are several key parameters which determine the microalgal productivity in a photobioreactor. These are lighting, mixing, water, CO_2 pressure, O_2 removal, nutrient supply, temperature, and pH (Kunjapur and Eldrige, 2010). Under nutrient-sufficient and optimal temperature conditions, the maximal culture productivity of photoautotrophic microorganisms is solely limited by the light (Richmond, 2004). The penetration of visible spectrum of light in the microalgal cultures decreases as the cell density increases (Figure 1). The appropriate intensity, duration, and wavelength of light must be provided to enhance the microalgal growth in photobioreactors. Supra-optimal light conditions lead to photoinhibition and sub-optimal light becomes a growth limiting factor. In both conditions, microalgal productivity will be lowered. The photosynthetic conversion efficiency of microalgae will generally be lower than theoretically expected under optimal conditions due to insufficient capacity to utilize the incident radiation (Zhu et al., 2008). The distribution of solar radiation over a greater photosynthetic area can spatially dilute the light in the light saturation zone, thereby reducing the mutual shading of the cells in the culture resulting in higher growth rate and lower accessory pigments content. The distribution of solar radiation can be increased by maintaining the surface to volume ratio as high as possible. The temporal and spectral distribution of irradiation and photon flux density is the main physical parameter that determines the photosynthetic productivity of microalgae. The solar conversion efficiency of microalgal mass culture grown under full sunlight is limited because of two reasons: 1) the photon absorption rate of the chlorophyll antennae of upper layers of cells far exceeds the rate of their utilization hence there is a loss of excess photons as fluorescence and heat leading to photoinhibition; 2) the deprivation of functional photons in the deeper cell layers, which is strongly attenuated due to the filtering effect of upper cells (Naus & Melis, 1991, Neidhardt et al., 1998). Gordon and Polle (2007) argued that a microalgal biomass productivity of 100 g m^{-2} h^{-1} could be obtained solely by improving the flux tolerance rather than by raising intrinsic photosynthetic efficiency.

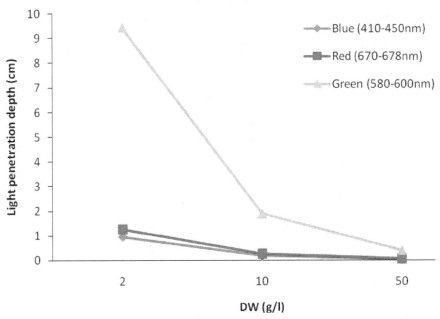

Fig. 1. Effect of biomass concentration on the penetration of incident light into cultures of *Nannochloropsis* sp. (from Richmond 2004).

4.2 Improvements of intrinsic photosynthetic efficiency

The photosynthetic efficiency of microalgae can potentially reach its theoretical maximum, which is calculated to be about 9-10% of total incident solar energy or 20-22% of PAR, being converted into biomass (Beilen, 2010). Such projected ultrahigh microalgal biomass yields of 100 g dry weight m-2h-1 can be realized in photobioreactors with sufficiently thin channels, ultradense cultures, and rapid light/dark cycles wherein optimal synchronization of photonic input with rate limiting dark reaction times is exploited (Gordon and Polle, 2007). However, this does not take into account the intrinsic conversion efficiency of photosynthesis, which is only likely to be improved upon through genetic engineering or synthetic biology. The integration of molecular and photobioreactor engineering is likely the only possible way of obtaining near-theoretical levels of algal biomass productivity while simultaneously augmenting lipid content. At the unicellular level, genetic modification of microalgal photophysiology could decrease light absorption, leading to enhanced availability of functional irradiance at the population level. The PSII and PSI in the light harnessing complex of green algae are associated with large numbers of chlorophyll a and chlorophyll b molecules, which are called antenna molecules. During photosynthetic biomass production in photobioreactors, high photon flux densities saturate the antenna molecules of upper cell layers with excessive photons which do not participate in the photosynthetic biomass production. These excess photons dissipate their energy as heat or fluorescence (photoinhibition) and reduce the overall solar to biomass conversion efficiency of the microalgal culture. Moreover, the lower layers do not receive appropriate amount of photons because of the filtration of light by the cells of upper layer, which accounts for a further loss in the overall biomass productivities (Figure 2). Genetic modifications resulting in truncated chlorophyll antennae size could restrict the high photon absorption by the light harvesting complex. In this context, Polle et al. (2003) have cloned and functionally characterized the Chl antenna size regulatory *Tla1* gene in *Chlamydomonas reinhardtii*. The partially truncated chlorophyll antenna size of the *tla1* mutant prevents the over-absorption of irradiance by cells, thus avoiding wasteful heat losses (Polle et al., 2003). In *Dunaliella salina*, a highly truncated light-harvesting Chl antenna size resulted in aggravated photosynthetic productivity and greater oxygen production under mass microalgal culture (Melis et al., 1999). The *Stm3LR3* mutants of *C. reinhardtii* generated by RNAi technology demonstrated down-regulation of the entire LHC antenna system. The *Stm3LR3* mutant showed reduced fluorescence, increased photosynthetic quantum yield, increased resistance to photoinhibition and faster growth rate under high light levels (Mussgnug et al., 2007).

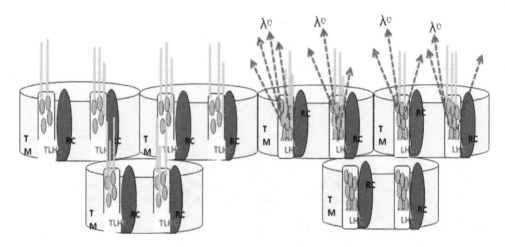

Fig. 2. A diagrammatic representation of wild-type and genetically truncated light harvesting complexes of microalgae. The incident light falling on the antenna molecules in the LHC are wasted as heat and fluorescence, while the lower layer cells are deprived of light. The modified TM has fewer antenna molecules in the TLHC that allows the absorption of light by the cells in the deeper layers. TM (thylakoid membrane), RC (reaction center), LHC (light harvesting complex), TLHC (truncated light harvesting complex).

5. Microalgal biomass harvesting

Conventional harvesting processes for microalgae include dewatering, extraction and purification of biomass. Bulk harvesting of microalgal biomass can be performed by centrifugation, flocculation, gravity sedimentation and/or filtration. Biomass harvesting is one of the most energy intensive processes, and can require high capital investments. Harvesting techniques tend to vary from species to species as various factors (namely density, size and the value of the microalgal end product) will typically inform the most appropriate method. Acoustic focusing, hybrid capacitive deionization or electrophoresis and use of novel materials for conventional membranes and flocculent systems are amongst the range of new innovative strategies that are currently under investigation (Cheng & Ogden, 2011).

6. Biomass conversion to biodiesel

6.1 Lipid extraction and biodiesel formation from microalgal biomass

Conventional methodologies for lipid extraction involve the use of toxic organic solvents such as chloroform, methanol and hexane. While the solvent extraction process is effective it is difficult to adopt on a large scale. Novel methods for lipid extraction involve technologies such as acoustics, sonication, the use of mesoporous nanomaterials, and amphiphilic solvents (Cheng & Ogden 2011). Super critical fluid extraction has been reported to be safer and faster than the conventional solvent extractions (Andrich et al., 2005). Another technique of "milking" microalgae manipulates the hydrophobicity of the solvent system, which allows the extraction of lipids from living algal cells. A flat panel two-phase bioreactor designed by Hejazi & Wijffels (2004) was used in the milking process for *Dunaliella salina* production. In this process, microalgal cells grown under optimal growth conditions are stressed by excess light to stimulate the production of β-carotene, which is then extracted from the cells using lipophilic compounds. Important considerations for application of this "milking" process includes: a) cell wall and membrane properties of the microalgal strain; b) location and accumulation of the product inside the cell; and c) biocompatibility and chemical properties of the solvent used for the "milking" process (Hejazi & Wijffels 2004).

Lipid conversion to biodiesel can easily be achieved by chemical trans-esterification, enzymatic conversion, and catalytic cracking. Chemically, biodiesel is comprised of monoalkyl esters of fatty acids that are derived from triacylglycerols. These triacylglycerides can be produced from crop or microalgae oils, animal fats, and waste cooking oils. Such biodiesel is miscible with petroleum diesel and thus suitable blends of biodiesel-diesel can be obtained. These are denoted as BXX, where XX is the percent of biodiesel in the blend. For example, B40 is 40% biodiesel in a diesel-biodiesel blend (Tat et al., 2007).

6.2 Fuel properties of microalgal biodiesel

There are several properties which determine the suitability of biodiesel as a biofuel including cetane number, kinematic viscosity, cold flow and oxidative stability (Ramos et al, 2009). These properties are greatly influenced by the fatty acid compositions of the feedstock oils (Figure 3, 4). Therefore, to determine the best composition of biodiesel, it is necessary to study the lipid profile of potential biomass feedstocks. The most common fatty acid methyl esters present in most biodiesel are palmitic acid (16:0), stearic acid (18:0), oleic acid (18:1), linoleic acid (18:2) and linolenic acid (18:3) (Knothe, 2008). Biodiesel obtained upon trans-esterification of these common fatty acids has many advantages over petroleum-derived diesel fuel. However, there are several performance problems with biodiesel, notably poor cold flow properties, lower cetane number and insufficient oxidative stability (Knothe, 2009). Ignition delay time and combustion quality of a diesel fuel is determined by the cetane number. An adequate cetane number is required for better engine performance and a high cetane number is also associated with biodiesel with good cold start properties and reduced NO_x exhaust emissions (Ramos et al., 2009; Knothe, 2009; Ladommatos et al., 1996).

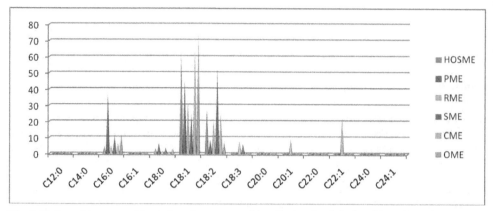

Fig. 3. Fatty acid composition of various vegetable oil crop feedstocks. (Data from Ramos et al 2009). HOSME (high oleic sunflower methyl ester), PME (palm methyl ester), RME (rape methyl ester), SME (soy methyl ester), CME (corn methyl ester), OME (olive methyl ester).

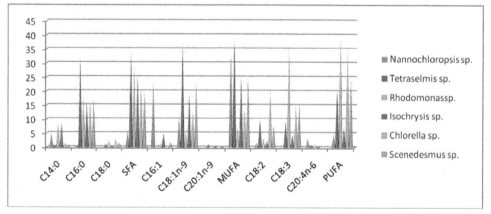

Fig. 4. Fatty acid composition of different microalgal species (Huerlimann et al., 2010). Data for *Chlorella* sp and *Scenedesmus* sp obtained from the author's investigations.

The cetane number of biodiesel can be determined using established standards such as ASTM D975 in the USA. Cetane is the common name for a long straight-chain hydrocarbon, hexadecane ($C_{16}H_{34}$), which is the high quality standard on the cetane scale with an assigned Cetane number of 100. In contrast, the low quality standard is a highly branched compound heptamethylnonane ($C_{16}H_{34}$) which exhibits poor ignition quality and has an assigned cetane number of 15 (Gopinath et al 2010). According to ASTM D975, conventional diesel fuel requires a minimum cetane number of 40 whereas a minimum of 47 has been prescribed for biodiesel (ASTM D6751) or (EN14214). For biodiesel, the cetane number depends on the microalgal feedstock biomass from which the oil is derived and the alcohol that is used during the trans-esterification process. The cetane number decreases with increasing unsaturation and increases with increasing chain length without branching of the CH_2 moities. However, the straight chain fatty acid methyl esters with carbon numbers of 6, 10, 12, 14, 16 and 18 show a non-linear increase with carbon number. Esters of saturated fatty acids (such as palmitic and stearic acids) give high cetane numbers. In summary, cetane number increases with chain length, decreases with unsaturation (or the number of double bonds) and decreases as double bonds and carbonyl group move toward the centre of the chain (Graboski & McCormik, 1998). Oils from different microalgal feedstocks will have different fatty acid compositions and such fatty acids are different with respect to the chain length, degree of unsaturation, or presence of other chemical moieties. While esters of long chain saturated fatty acid show higher cetane number, they can have a high cloud point that results in nozzle clogging (see below). On the other hand, esters of unsaturated fatty acid show low cetane number and are prone to oxidation. Among the non-algal biodiesel feedstocks, palm oil is reported to have the highest cetane number (61), whereas peanut, sunflower and corn oils typically have a cetane values of 53 (Ramos et al., 2009).

To improve the properties of biodiesel, it is necessary to enrich the biodiesel with fatty esters with desirable properties. Genetic engineering can provide important opportunities in this regard. For example, a soybean transgenic crop variety designated 335-13 has been genetically altered to increase concentration of oleic acid by more than 85%, with a corresponding reduction of palmitic acid content to less than 4% (Tat et al., 2007). Saturated and long chain fatty acids gave a high cetane number, which increases with increasing saturation and the chain length (Knothe et al., 2003; Bajpai & Tyagi, 2006; Knothe, 2009). Esters of highly unsaturated fatty acids such as linoleic (18:2) and linolenic (18:3) acids lower the cetane number (Knothe et al., 2003). According to Knothe et al. (2003) high cetane numbers were observed for esters of saturated fatty acids such as palmitic (C16:0) and stearic (C18:0) acids. Feedstock oils rich in these fatty acid compounds would have higher cetane values.

A higher content of polyunsaturated fatty esters in oil derived from the feedstock can reduce the quality of biodiesel upon storage due to oxidation in the presence of air, light, heat, peroxides, trace metals, or even the structural features of the fatty acids themselves. Thus, oxidation stability is another major issue affecting the use of biodiesel fuels (Ramos et al., 2009; Knothe, 2009). The number and position of double bonds in the chains of unsaturated fatty esters are both factors affecting the susceptible to autoxidation reaction (Bajpai & Tyagi, 2006; Frankel, 2005). Most biodiesel fuels contain significant amounts of alkyl esters of oleic, linoleic and linolenic acids, which influence the oxidation stability of the fuels. The relative rates of autoxidation for oleic acid methyl ester, linoleic acid methyl ester and linolenic acid methyl ester have been reported to be 1, 41 and 98 respectively (Knothe et al., 2005; Frankel, 2005; Ramos et al., 2009; Knothe, 2009).

The high viscosity of non-esterified vegetable oils leads to operational problems in diesel engines by increasing the level of engine deposits. Reduction of the high viscosity of vegetable oils is facilitated by the production of alkyl esters from the oil by transesterification. Since viscosity increases with decreasing temperature, handling of fuels in lower temperatures is facilitated by this lower viscosity as well (Knothe, 2009). This can be achieved by increasing the length and degree of saturation of the carbon chains (Knothe et al., 2005). Wax settling and plugging of filters and fuel lines are typical problems associated with biodiesel fuels at low temperatures. The maximum temperature at which the first solids appear for a particular fuel is known as the cloud point (CP) and such solids can lead to fuel filter plugging. The pour point (PP) is typically a few degrees below the cloud point and represents the temperature at which the fuel can no longer be poured (Dunn & Bagby, 1996). Key properties of biodiesel fuels at low-temperature are determined by the cold filter plugging point (CFPP) and low-temperature flow test (LTFT) (Dunn & Bagby, 1996; Knothe, 2009). The cloud point (CP) and CFPP are included in the biodiesel standards but as "soft" specifications (Knothe, 2009). For instance, The cloud point in ATSM D6751 requires a report, while the CFPP in UNE-EN 14214 varies with time of year and geographic location.

The properties of biodiesel at low-temperature are correlated with the properties of individual fatty acids, which mostly depend on the saturated ester content. In contrast, the effect of unsaturated fatty acids is considered negligible (Imahara et al., 2006; Ramos et al., 2009). Saturated fatty acids have significantly higher melting points than unsaturated fatty acids and in a mixture saturated fatty acids will crystallize at higher temperature. Therefore, biodiesel fuels derived from fats or oils with significant amounts of saturated fatty compounds display higher values of CP and CFPP (Knothe, 2003).

7. Summary

Biofuels can broadly be classified as oxygenated (ethanol, biodiesel) and hydrocarbon biofuels (diesel, jet fuel and gasoline). Based on this classification, the different generations of biofuels are - 1st generation, where biofuels are obtained from natural vegetable oils and greases; 2nd generation of lignocellulosic biomass and algal derived fuels. Two biomass crops, *Jatropha* and *Camelina* bridged, the 1st and 2nd generations of biofuels. The next generations of biofuels will be based on the innovative technologies that improve the processing of biomass into various other types of biofuels and improving the existing feedstock species of biofuel using metabolic/ genetic engineering. For example, application of heat and pressure on algae/biomass/waste using innovative approaches like hydrothermal, catalytic and biological biomass conversions for the creation of cost-effective biofuels as a replacement for fossil fuels. Moreover, the next generation fuels are direct replacements for petroleum and are compatible with the existing infrastructure of the petrochemical industry. Genetic modification of microalgae improves their photosynthetic biomass conversion efficiency and hence can lead to higher biomass productivities, which is necessary for economic scalability. The improvements in the existing infrastructure for microalgae biomass production by photo engineering approaches will also play a key role towards the commercial application of next generation microalgae biofuels. In summary, the replacement of petroleum based fuels by bio-based products depends on several key factors, which include selection of the right bio-based product, process modification or product improvement for indirect substitutions, technological interventions to lower the cost of individual processing steps, scalability of biomass production and bioproduct delivery, and availability of sufficient and productive land.

8. Acknowledgements

Simrat Kaur and Charles Spillane acknowledge funding support from Science Foundation Ireland, the Irish Environmental Protection Agency (Fellowship Grant REP957, EPA STRIVE

2009 PHD ET8) and from the Competence Centre for Bio-refining and Bio-energy, Enterprise Ireland (www.ccbb.ie).

9. References

Albarrán-Zavala E & Angulo-Brown F. (2007) A simple thermodynamic analysis of photosynthesis. *Entropy*, 9, 152-168.

Andersen R. A. (2005) Algal Culturing Techniques. Elsevier Academic Press, London, U K.

Andrich G., Nesti U., Venturi F., Zinnai A. & Fiorentini R. (2005) Superctitical fluid extraction of bioactive lipids from microalga *Nannochloropsis* sp. *European J Lipid Science Tech.*, 107, 381-386.

Apt K. E. & Behrens P. W. (1999) Commercial developments in microalgal biotechnology. *J. Phycol.*, 35, 215-226.

Bajpai D. & Tyagi V.K. (2006) Biodiesel: source, production, composition, properties and its benefits. *J. Oleoginous Science* 55, 487–502.

Beilen J. B. V. (2010) Why microalgal biofuels won't save the internal combustion machine. *Biofuels, Bioprod. Bioref.*, 4, 41-52.

Béchet Q., Shilton A., Park J.B.K., Craggs R.J. & Guieysse B. (2011) Universal temperature model for shallow algal ponds provides improved accuracy. *Environmental Science Technology*, 45, 3702-3709.

Borowitzka M. A. (1999) Pharmaceuticals and agrochemicals from microalgae. In: Chemicals from microalgae. Ed. Cohen Z, Taylor & Francis. pp, 313-52.

Chen C.Y., Yeh K. Aisyah R., Lee D. & Chang J. (2011) Cultivation, photobioreactor design and harvesting of microalgae for biodiesel production: A critical review. *Bioresource Technology*, 102, 71-81.

Cheng Kuan-Chen & Ogden K. L. (2011) Algal biofuels: the research. *SBE Supplement: Algal Biofuels, AIChE*, 42-47, www.aiche.org/cep

Chisti Y. (2007) Biodiesel from microalgae. *Biotech. Advances.*, 25,294-306.

Cooney M. J., Young G. & Pate R. (2011) Bio-oil from photosynthetic microalgae: case study. *Bioresource Technology*, 102, 166-177.

Cuaresma M., Janssen M., Vilchez C., & Wijffels R. H. (2011) Horizontal or vertical photobioreactors? How to improve microalgae photosynthetic efficiency. *Bioresource Technology*, 102, 5219-5137.

Dunn R.O. & Bagby M.O. (1996) Low-temperature filterability properties of alternative diesel fuels from vegetable oils. Liquid fuel and industrial product from renewable resources. In: *Proc.Third Liquid Fuel Conference*, ASAE 95–103.

Frankel E.N.(2005) Lipid Oxidation. The Oily Press. U.K: PJ Barnes and Associates: Bridgwater.

Gopinath A., Puhan S. & Nagarajan G. (2010) Effect of biodiesel structural configuration on its ignition quality. *International J. Energy Environment*, 1 (2), 295-306.

Gordon J. F. & Polle J. E. W. (2007) Ultrahigh bioproductivity from algae. *Appl Microbiol Biotechnol.*, 76,969-975.

Graboski M.S. & McCormik R.L. (1998) Combustion of fat and vegetable oil derived fuels in diesel engines. *Prog. Energy Combus. Sci*, 24, 125-164.

Hase R, Oikawa H, Sasao C, Morita M & Watanabe Y (2000) Photosynthetic production of microalgal biomass in a raceway system under greenhouse conditions in Sendai city. *Journal of Bioscience and Bioengineering,*89 (2), 157-163.

Hejazi M. A. & Wijffels R. H. (2004) Milking of microalgae. *Trends Biotech.,* 22 (4), 189-194.

Huerlimann R, de Nys R, Heimann K (2010) Growth, lipid content, productivity, and fatty acid composition of tropical microalgae for scale-up production. *Biotechnol. Bioeng.* 107(2), 245-257.

IEA (2011) Technology Report: Biofuels for transport. International energy Agency, 9 rue de la Fédération, 75739 Pris Cedex 15, France. www.iea.org

Imahara H, Minami E, Saka S. (2006) Thermodynamic study on cloud point of biodiesel with its fatty acid composition. *Fuel,* 85, 1666-70.

Katsuda, T. Lababpour, A. Shimahara, K. & Katoh, S. (2004) Astaxanthin production by *Haematococcus pluvialis* under illumination with LEDs. *Enzyme Microb.Technol.,* 35, 81-86.

Katsuda, T. Shimahara, K. Yamagami , K. Ranjbar, R. & Katoh, S. (2006) Effect of flashing light from blue light emitting diodes on cell growth and astaxanthin production of *Haematococcus pluvialis. J. Biosci. Bioeng.,*102, 442-446.

Kaur S., Gogoi H.K., Srivastava R. B. & Kalita M.C. (2010) Algal Biodiesel: Procedures and Resources for laboratory study. In: *Algal Biotechnology: New Vistas.* Ed. M.K. Das, Daya Publishing House, Delhi. ISBN-13. 9788170356479, 978-8170356479

Knothe G., Matheaus,A.C. & Ryan III T.W. (2003) Cetane numbers of branched and straight chain fatty esters determined in an ignition quality tester. *Fuel* 82, 971-975.

Knothe G. 2009. Improving biodiesel fuel properties by modifying fatty ester composition. *Energy Environment Science* 2, 759-66.

Knothe G. (2008) "Designer" biodiesel: optimizing fatty ester composition to improve fuel properties. *Energy & Fuels,* 22,1358-1364.

Knothe, G., Gerpen, J.V. & Krahl, J. 2005. The Biodiesel Handbook. AOCS Press, Champaign, Illinois.

Kunjapur A. M. & Eldridge R. B. (2010) Photobioreactor design for commercial biofuel production from microalgae. *Ind. Eng. Chem. Res.*49, 3516-3526.

Ladommatos N., Parsi M. & Knowles A. (1996) The effect of fuel cetane improver on diesel pollutant emissions. *Fuel* 75, 8-14.

Lee, C.G. & Palsson, B. O. (1994) High-density algal photobioreactors using light-emitting diodes. *Biotechnol Bioeng.,* 44, 1161-1167.

Matthijs, H.C.P. Balke, H. Van Hes, U.M. Kroon, B.M.A Mur, L.R. & Binot, R.A. (1996) Application of light-emitting diodes in bioreactors: flashing light effects and energy economy in algal culture. *Biotechnol Bioeng,* 50, 98-107.

Melis Anastasios, N. J. & Benemann J. R. (1999). "*Dunaliella salina* (Chlorophyta) with small chlorophyll antenna sizes exhibit higher photosynthetic productivities and photon use efficiencies than normally pigmented cells. *Journal of Phycology,* 10, 515-525.

Molina Grima E., Sanchez Perez J. A., Garcia Camacho F., Fernandez Sevilla J. M., Acien Fernandez F. G. & Urda Cardona J. (1995) Biomass and icosapentanoic acid productivities from an outdoor batch culture of *Phaeodactylum tricornatum* UTEX 640 in an airlift tubular photobioreactor. *Appl. Microbiol. Biotechnol.,* 42,658-63.

Mussgnug J. H., S. T.-H., Jens Rupprecht, Foo A., Klassen V., McDowall A., Schenk P.M., Kruse O. & Hankamer B. (2007). Engineering photosynthetic light capture: impacts on improved solar energy to biomass conversion. *Plant Biotechnology Journal*, 5, 802-814.

Naus J & Melis A. (1991). Changes of photosystem stoichiometry during cell growth in *Dunaliella salina* cultures.*Plant Cell Physiol.*, 32, 569-575.

Nedbal, L. Tichy, V. Xiong, F. & Grobbelaar J.U. (1996) Microscopic green algae and cyanobacteria in high-frequency intermittent light. *J Appl. Phys*, 8, 325-333.

Neidhardt J., Zhang L. & Melis A. (1998) Photosystem II repair and chloroplast recovery from irradiance stress: relationship between chronic photoinhibition, light-harvesting chlorophyll antenna size and photosynthetic productivity in Dunaliella salina (green algae).*Photosynth.Res.*, 56, 175-184.

Nuffied Council on Bioethics (2011). Biofuels: ethical issues. ISBN. 978-1-904384-22-9, www.nuffieldbioethics.org

Polle Juergen E.W., S.-D. K. & Melis A. (2003). tla1, a DNA insertional transformant of the green alga Chlamydomonas reinhardtii with a truncated light-harvesting chlorophyll antenna size.*Planta*, 217, 49-59.

Ramos, M.J., Fernández, C. M., Casas, A., Rodríguez, L. & Pérez, A. 2009. Influence of fatty acid composition of raw materials on biodiesel properties. *Bioresource Technology* 100, 261–68.

Raven J.A. (1988) Temperature and algal growth. *New Phytol.*, 110 (4), 441-461.

Richmond A. (2004) Principles for attaining maximal microalgal productivity in photobioreactors: an overview. *Hydrobiologia*, 512, 33-37.

Richmond A., Boussiba S., Vonshak A. & Kopel R (1993) A new tubular reactor for mass production of microalgae outdoors. *J. Appl. Phycol.*, 5, 327-32.

Sheehan J., Dunahay T., Benemann J. & Roessler P. (1998) A look back at the US Department of Energy's Aquatic Species Program- Biodiesel from Algae. National Renewable Laboratory, Golden, CO, Report NREL/TP-580-24190.

Spektorova L., Creswell R. L. &Vaughan D. (1997) Closed tubular cultivators: an innovative system for commercial culture of microalgae. *World Aquaculture*, 28, 39-48.

Spolaore P., Joannis-Cassan C., Duran E. & Isambert A. (2006) Commercial applications of microalgae. J. *Bioscience and Bioengineering*, 101, 87-96.

Tat M.E., Wang P.S., Gerpen J.H.V. & Clemente T.E. (2007) Exhaust emissions from an engine fueled with biodiesel from high-oleic soybeans. *J Am Oil Chem Soc*, 84, 865-869.

Ugwu C.U., Aoyagi H. & Uchiyama H. (2008) Photobioreactors for mass cultivation of algae. *Bioresource Technology*, 99, 4021-4028.

Weyer K., Daniel B., Darzins A. & Willson B. (2009) Theoretical maximum algal oil production. *Bioenergy Research*, 3, 204-213.

Wigmosta M. S., Coleman A. M., Skaggs R. J., Huesemann M. H. & Lane L. J. (2011) National microalgae biofuel production potential and resource demand. *Water Resource Research*, 47, W00H04.

Zhu Xin-Ghang., Stephen P. L. & Donald R. O. (2008) What is the maximum efficiency with which photosynthesis can convert solar energy into biomass? *Current Opinion in Biotechnology*, 19, 153-159.

Yeh N. & Chung Jen-Ping (2009) High-brightness LEDs- energy efficient lighting sources and their potential in indoor plant cultivation. *Renewable Sustainable Energy Reveiws*, 13, 2175-2180.

Zemke P. E., Wood B. D. & Dye D. J. (2010) Considerations for the maximum production rates of triacylglycerol from microalgae. *Biomass Bioenergy*, 34, 145-151.

9

An Integrated Waste-Free Biomass Utilization System for an Increased Productivity of Biofuel and Bioenergy

László Kótai et al.[*]
Institute of Materials and Environmental Chemistry, Chemical Research Center,
Hungarian Academy of Sciences,
Hungary

1. Introduction

The increase in production and utilization of biomass and other renewable sources of energy are important challenges of the energy industry. It generates, however, demands for ecologically and economically acceptable production systems. Here we report an integrated system of known and new technologies developed for biomass conversion to biofuels. This includes classical and biobutanol based new biodiesels, biogas and electricity production, and an agricultural production system involving fertilization with the ash of the biomass power plants. Basically, three types of agricultural production system are needed for the agricultural segment of the integrated system, namely:

A – plants for combustion in biomass power plants (energy grass)
B – plants for production of vegetable oils for biodiesel production
C – plants for conversion of sugar derivatives to price alcohols, mainly butanol as a diesel fuel source

Depending on the climate, the soil type, the agricultural experiences, and the type of the plants (A,B,C), the produced biomass materials can fulfill more than one requirement as it can be seen in Fig. 1. Depending on the constituents of the biomass (cellulose, starch, lignin, oil, proteins), the energy production can be performed via direct combustion or, after digestion in biogas systems, by using the biogas. The biomass power plants, biogas combustion plants/engines produce hot water, steam and electricity. In plants type B soybean, rape, sunflower or likes are pressed to obtain the oil, while the pressing cake can be used as optimal raw material for biogas plants due to its high protein content, while the

---

[*] János Szépvölgyi[1,6], János Bozi[1], István Gács[1], Szabolcs Bálint[2], Ágnes Gömöry[2], András Angyal[3],
János Balogh[4], Zhibin Li[5], Moutong Chen[5], Chen Wang[5] and Baiquan Chen[5]
1 *Institute of Materials and Environmental Chemistry, Chemical Research Center,*
Hungarian Academy of Sciences, Hungary,
2 *Institute of Structural Chemistry, Chemical Research Center, Hungarian Academy of Sciences, Hungary,*
3 *Axial-Chem Ltd., Hungary,*
4 *Kemobil Co., Hungary,*
5 *China New Energy Co., China,*
6 *Research Institute of Chemical and Process Engineering, University of Pannonia, Hungary.*

stalk can also be used as solid fuel (after drying with the low heating value warm water) in biomass power plants. Generally, the green biomass can be utilized in biogas plant and the dried ones as solid combustion fuel in power plants. The residues of the sugar derivatives producing plants (sugar sorghum, corn, etc.) can supply a biomass power plant with their dried stalk. The complete waste processing in these energy producing units and recirculation of other wastes (potassium sulfate, calcium sulfate or biomass power plant ash) of the integrated system as fertilizers into the agriculture contribute to a sustainable biomass production and fuel production, as well.

2. Energy aspects of biomass utilization

The intensive production of biomass as raw material for fuel or bioenergy production would lead to fast exhausting of the soil and dramatic increasing the production costs without intensive fertilization. Except nitrogen, all of the nutrients (P, K) and microelements can be recycled by reprocessing the residues of the biomass work up or biomass utilizing energy producing technologies. The nitrogen fertilization, however, always requires fossil energy source, since the base material of the two most typical nitrogen fertilizer (ammonia and urea) is the natural gas. It is one of the main reasons of the opposite statements about energy intake and output balance of the biomass based fuel and energy productions. Otherwise, the conversion of the waste to fertilizers (to supply other elements like P, K, S and microelements) should also be the integrated part of the sustainable and economic biomass production system.

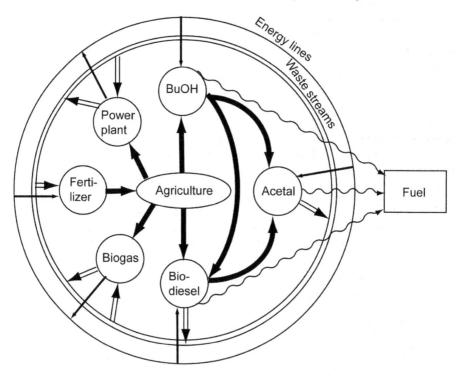

Fig. 1. Waste free biofuel production cycle

The review of the energy inputs required for the production of the raw materials like corn, switchgrass or sunflower, or the cost of the biofuel production form these agricultural materials (Pimental and Petzek, 2005) unambiguously show, that the climate (crop amount), type of the agricultural plant and the type of the processing technologies basically determines the feasibility of an energy positive biofuel production. The energy saving in the production of the various biofuel components from various sources is compared to the production costs related to petrochemical raw materials (Arlie, 1983) is given in Table 1. Selection of the biomass produced and the fuel type prepared from this biomass requires that environmental, economic and energetic viewpoints (Hill et al., 2006) should also be taken into consideration. The social-economic viewpoints play also key role in this decision. That is obvious, that there is not any type of biomass plant which could completely be turned into biofuel, only integrated systems can solve this problem, when more than one type of biomass plants are in synchronized operation. More than one energy producing system is used to utilize the various type of green biomass, and more than one type of biofuel are produced from the various parts of each type of biomass plants. At the same time the energy producing plants or the waste producing and nutrient reprocessing plant (fertilizer plant) can use the wastes of each technological step as raw material to ensure the recyclability.

Product	Petrochemical source	Biomass	Energy gain
MeOH	Natural gas	Wood	0.60
EtOH	Ethylene	Sugarcane	0.73
		Artichoke	0.88
		Corn	0.65
n-BuOH	Propylene	Sugarcane	0.10
		Artichoke	0.75
		Wheat straw	0.80
Acetone	Propylene	Sugarcane	0.80
		Artichoke	1.45
		Wheat straw	1.51

Table 1. Energy gain of sugar-based biofuel components (ton of oil equivalent/ton)

Since the energetically favorable components as BuOH, EtOH and acetone can be produced from biomass the use of these in biofuel including biodiesel production seems to be essential and unavoidable step. The EU biofuel standards now declare that rapeseed oil and ethanol as raw materials are permitted and standardized as biofuels within European Community. Due to the limits of the agricultural area and the productivity of rape in this climate and low energy content of ethanol, however, requires changing this statement and the use of other type of biofuels should be also permitted. Thus, in our integrated system we incorporate new type of blend materials, mainly butanol for replacing the ethanol, and new kind of blends are produced from the wastes of biodiesel and biobutanol production as well. In order to increase the amount of biodiesel produced from one unit of vegetable oil, butyl ester production is suggested instead of the methyl esters, and butoxylation or blending with pure butanol are also possible increments in increasing the efficiency of the biofuel production related to one hectare of the agricultural area.

3. New trends in biodiesel technologies

The new developments in biodiesel production technologies are focused on simplification of the existing technologies by using co-solvents (Guan et al., 2009b), supercritical solvents (Geuens et al., 2008), new catalysts (including heterogeneous and biocatalysts) (Fu and Vasudevan, 2009; Soriano et al., 2009; Leung et al., 2010) and the efforts are directed on founding new blends and raw materials (including the waste glycerol (Guerrero-Perez, et. al., 2009)) to increase the amount of biofuel produced on a given agricultural area. Some of the improved methods are discussed below.

3.1 Trans-esterification with phase mixing materials

Base catalyzed trans-esterification of the vegetable oils with methanol is a slow process in the two-phase system due to the mass transfer limitation. The butanolysis is much faster because it takes place in homogeneous phase. Therefore, the possibilities to create a homogeneous mixture by using an appropriate co-solvent which homogenizes the vegetable oil and methanol has extensively been studied (Guan et al., 2009a; 2009b; Leung et al., 2010; Meng et al., 2009; Soiano et al., 2009). The best solvents are the ether-type compounds like THF, methyl tertiary butyl ether, Me_2O or diethers. Phase diagrams of different vegetable oils - ethereal solvents (1,4-dioxane, THF, Et_2O, diisopropyl ether and MTBE)-methanol have been studied (Boocok et al., 1996a) and THF was found to be the preferred solvent on the basis of the volume required by the miscibility and boiling point considerations. Initially the reactions are very fast , but they slow down drastically due to the polarity changes in the mixed phase containing esters and decreasing amount of alcohol [Boocock et al., 1998). The decreasing polarity of the intermediate mixture could be avoided only by using large excess of methanol (methanol/oil molar ratio 27) when the methyl ester production exceed 94.4 % in 7 min even at room temperature. The separation of the glycerol phase in the presence of THF was much more rapid as in the case of two-phase trans-esterification reaction mixtures (Boocock et al., 1996a). The use of cyclic THF or 1,4-dioxane provides a possibility to decrease the amount of the co-solvent and perform the trans-esterification at higher oil : co-solvent ratios. The THF and the dioxane are miscible with vegetable oils and methanol in any proportion and they have hydrogen bonding ability (Boocock et al., 1996b). In spite of that 5 % dioxane ensures almost complete reaction within 30 min at room temperature at a 6:1 methanol : oil molar ratio, the 1,4-dioxane ring opens up during the trans-esterification reaction at the presence of 1 % KOH catalyst, thus it cannot be recycled at all.

Dimethyl ether can be separated easily with depressurizing the reaction system and Me_2O as a polar compound ensures a sufficiently high trans-esterification reaction rate at room temperature. The effects of the reaction conditions on the reaction time indicated that under the usual conditions, such as 1 % of KOH and 2-fold molar excess of methanol (6:1 methanol : oil molar ratio), the reactions can be performed within 5-10 min which would provide a good technological base for the continuous production of biodiesels (Guan et al, 2009a and 2009b). The gaseous state of form of the Me_2O requires pressurized reactors and the risk of explosion is very high.

Recently a new technology has been developed (Kótai et al., 2008) for the phase-mixing of methanol (or other alcohols) and vegetable oils in trans-esterification reactions with alkoxyalkanol phase transfer agents. These are hemi-ethers of glycols. They are polar end-group, but they have oil-soluble alkyl chain as well. The best candidate is the butylglycol (2-butoxy-ethanol), which can act at even 1 % concentration, and at the 6:1 MeOH : rapeseed oil molar ratio with 1 % KOH a catalyst. The reaction is almost completed within 30 min even at room temperature. By using 5 % of butylglycol, the reaction time is 5-10 min. Since the butyl glycol acts also as an alcohol (not only as an ether), thus not only methyl esters but 2-butoxyethyl esters – $R-C(O)-O-C_2H_4-OC_4H_9$ - are also formed. These esters are formed in an amount of 2-3 %w/w and act as fuel components. In this way, the phase mixing agents built into the ester phase and they contribute to the mass of the biodiesel and do not need to recover it which simplifies the production technology. The catalyst solution prepared from KOH, butylglycol and methanol was used in our plant scale experiment performed in Hungary in 2009, when a continuous trans-esterification process with continuous separator was put into operation. The method could be combined with the ion exchange type removal and recycling of the neutralization agent ($KHSO_4$, chapter 3. 2), because the reaction takes place at room temperature and the amount of soaps formed and appeared in the ester phase was very small. The catalyst distribution between the ester and glycerol phase is around 2:98. The ester phase has been neutralized with $KHSO_4$, when the potassium content is decreased with the continuous operation mode below 50 ppm without any further washing. The further purification steps, washing with water and removing the residual MeOH in vacuum, are the same as in the classical biodiesel technologies, however, the amount of the dissolved MeOH, due to the lack of soaps and residual catalyst is much lower than in the usual technologies. The flow-sheet of the technology is shown in Fig 2.

In order to decrease the length of the tube reactor and the residence time of the mixture in the apparatus, a two-stage trans-esterification seems to be the most reliable, when after decreasing the rate of the reaction, after the first separator a further amount of methanol and catalyst are added, when the reaction rate is suddenly increases: it is attributed to the extra methanol ensuring a large excess for the residual triglyceride, thus the conversion reaches 98 % in 20-30 min reaction time. We have used 50 m^3 tanks for the esterification with intensive stirring. Separators for removing the glycerol phase, and the same volume of the separator was used to separate the neutralization agent and the ester phase which was mixed in a tube after exit of the first separator and before entering into the second one.

3.2 Removal of the residual catalyst from the biodiesel (decontamination)

In spite of the efforts to produce solid phase non-soluble alkaline catalysts or highly active acidic (super-acidic) catalysts (Di Serio et al, 2008; Leung et al., 2010; Soriano et al., 2009), the homogeneous catalytic (KOH or NaOMe catalyzed) trans-esterification of vegetable oils have been the most commonly used method in the biodiesel industry (Huber et al., 2006). However, soap formation during the alkaline-catalyzed trans-esterification is the most problematic by-reaction. In case of low water and carboxylic acid containing vegetable oils the main source of the soap formation is the hydrolysis of the formed methyl esters during washing, which is a strongly pH dependent process. Thus the neutralization preceding the washing is an essential step to minimize the saponification by-reactions. The amount and type of the formed soap is strongly affected by the separation characteristics of the glycerol and the ester phase. The acid treatment is generally needed to start or quicken the phase separation process producing aqueous glycerol solution. It is well known that the distribution of the catalyst between the glycerol and the methyl ester phase is strongly depends on the temperature, type of the catalyst, excess of methanol and the composition of the two separated phases [Chiu et al., 2005; Di Felice et al., 2008; Zhou and Boocock, 2006). Table 2 shows the distribution of 1 % KOH and 0.5 % of H_2SO_4 distribution at different temperatures depending on the amount of methanol at biodiesel : glycerol molar ratio of 1:3.

Catalyst	T, °C	MeOH, mol	K	Catalyst	T, °C	MeOH, mol	K
KOH	25	0	98	H_2SO_4	25	0	60
	25	3	95		25	3	53
	25	6	77		25	6	46
	75	0	47		75	0	31
	75	3	45		75	3	28
	75	6	35		75	6	24

Table 2. Distribution of the KOH catalyst between the ester and glycerol phase in the function of temperature and MeOH excess ($K= C_{glycerol} / C_{ester\ phase}$)

It can be seen that increasing temperature increases the amount of the catalyst in the ester phase. The amount of dissolved methanol also increases the amount of dissolved catalyst. The K values of the distribution of methanol, however, 10.9 and 7.5 at 3 and 8.5 and 4.8 at 6 moles of methanol towards 3 mol of biodiesel and 1 moles of glycerol at 25 and 75 °C, respectively. In the presence of 1 % KOH, these values changed to 2.29, 1.64, 1.90 and 1.48, respectively.

Due to the room-temperature reaction, there is low catalyst and low methanol concentration in the ester phase, thus the amount of the required neutralization agent is also low. Considering the use of a continuous separator, the relative volumes of the separated phases should be adjusted between 10-1:1, so it is possible to use dilute (2 %) $KHSO_4$ solution. The neutralization acts as the first washing step, when the potassium content in the biodiesel phase was proved to be up to 8 ppm without further aqueous washing. Instead of mineral acids we used an acidic salt as hydrogen ion sources, namely, potassium hydrogen sulfate ($KHSO_4$). Potassium hydrogen sulfate has as strong acidic function as the pure sulfuric acid without the disadvantages of the sulfuric acid, e.g. it is a solid crystalline mass can be stored without risk and can be dissolved in water without extreme heat generation, and no acidic vapors as in the case of hydrochloric acid can be felt. Due to the low amount of soaps in the ester phase, the dissolved methanol content of the ester phase is lower than in case of the classical biodiesel technologies. Potassium hydrogen sulfate as a strong "acid" reacts with KOH catalyst and soaps easily and spontaneously as

$$KOR + KHSO_4 = K_2SO_4 + HOR \tag{1}$$

where R = alkylcarbonyl radical (potassium soaps) or H (KOH). The aqeous $KHSO_4$ solution decomposes soaps immediately without emulsion formation. The formed potassium sulfate is neutral, soluble in water, insoluble in the ester phase and has no phase transfer property at all. The concentration of the formed potassium sulfate is low (~2-3 %), since dilute $KHSO_4$ solutions is used for neutralization of the potassium compounds in the ester phase. The aqueous phase will contain potassium sulfate, glycerol, methanol, and other water soluble components. The dilute solution can easily be ion-exchanged with strong acidic cationic exchangers as Varion KSM resin. The glycerol and the methanol containing aqueous phase is a strong polar solution, thus regeneration and recycling of the $KHSO_4$ should be taken place. However, stopping and controlling the operation of the ion exchanger at the stage of the hydrogen sulfate formation is difficult. Therefore, the K_2SO_4 containing material stream is divided into two equal parts. One part of the K_2SO_4 is ion-exchanged in a common way with the formation of H_2SO_4 solution (the

K-ions are bound by the resin phase). The dilute sulfuric acid is combined with the other stream of the K_2SO_4, when $KHSO_4$ forms which can be recycled without further treatment (Kótai et al., 2008a).

$$K_2SO_4 + H_2SO_4 = 2KHSO_4 \tag{2}$$

The exhausted ion exchanger can be regenerated with the 20 % sulfuric acid solution consumed also for the formal neutralization. The potassium sulfate solution obtained during the regeneration process of the ion exchanger resin can be used as a fertilizer component. The ion exchanger can be regenerated and used again several thousand times.

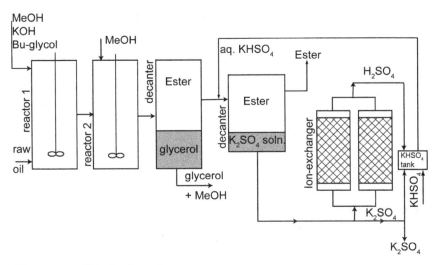

Fig. 2. Flow-sheet of biodiesel synthesis

However, the recycling of the $KHSO_4$ solutions is recommended only 2-20 times due to the accumulation of glycerol and methanol in the aqueous $KHSO_4$ solutions. The process flow is shown in Fig. 2.

The completely neutralized $KHSO_4$ solution turns into K_2SO_4, therefore it can be used as a fertilizer component together with the K_2SO_4 originated from the ion exchange regeneration cycle. Similarly, the potassium hydroxide content of the glycerol phase may be utilized in the same way after processing its glycerol and methanol content with H_2SO_4 catalyst into fuel (see in Chapter 4.2). In this way, the starting potassium based catalyst (KOH) is used not only as a catalyst but as a fertilizer source in an advantageous sulfate form, which helps to avoid the chloride accumulation in the soil and supplies the sulfur deficiency of the soil, especially in the case of rape plant cultures of high sulfur demand (Scherer, 2001).

3.3 Trans-esterification of high acid containing vegetable oils

Production of biodiesels from non-food quality vegetable oils with high fatty acid content is an important area of the biodiesel developments. The common alkaline catalysts cannot be used economically, because they reacts with the fatty acids and form soaps causing high catalyst consumption and a decrease in the catalyst activity. The alkali soaps are the cause of problems in the phase separation and the decreasing in the yield of esters. The calcium soaps do not dissolve in water and not surface active agents, do not cause problems in the phase separation and using calcium methanolate as catalyst in the trans-esterification of the vegetable oils is a well-known process (Lindquist, 1991). Due to the insolubility of CaO and Ca(OH)$_2$ in methanol, the CaO and the Ca(OH)$_2$ are known as less active or inactive compounds for the vegetable oil trans-esterification (Di Serio, et al., 2008), but the activation of calcium oxide to act as a catalysts in this reaction leads to good results (Kótai et al., 2006, 2008b). The free acid content of the vegetable oils and CaO were immediately reacted with formation of calcium soaps, and the Ca(OH)$_2$ formed from the excess of CaO played role in the trans-esterification. The best results (quantitative conversion of triglycerides into methyl esters) was obtained by using calcium hydroxide (5.6 %) prepared in situ by slaking of CaO at a 3:1 CaO to H$_2$O mass ratio and at a 10:4 oil: methanol ratio for 1 h under reflux. Excess of CaO is essential in this reaction. Using half amount of catalyst and methanol, the conversion was 90 % only, and 9% triglyceride and 1 % diglyceride could be detected in the ester phase. In case of high fatty acid containing starting vegetable oils the calcium soaps formed is partially soluble in the methanol containing ester phase, thus after separation of the soap containing glycerol the ester phase was treated with a slight excess of concentrated sulfuric acid. The sulfuric acid has multiple roles in the system,

1.   it forms calcium sulfate and releases the bound fatty acids
2.   catalyzes the reaction of the liberated fatty acids and the methanol excess used in the trans-esterification
3.   destroys the instable polyunsaturated compounds which could cause deposition in the engines
4.   decreases the iodine number and the ash content in the ester phase

Since the optimal trans-esterification can be performed at 2-4 fold excess of methanol, the residual methanol content means a very high (30-60 fold) excess toward the liberated fatty acids taking part in the acid catalyzed trans-esterification (Khelevina 1968). The anhydrous CaSO$_4$ formed in the neutralization step promotes the esterification reaction of the liberated fatty acids due to binding of the water formed in the equilibrium esterification reactions:

$$R\text{-}COOH + MeOH = RCOOMe + H_2O \tag{3}$$

$$CaSO_4 + 0.5H_2O = CaSO_4.0.5H_2O \tag{4}$$

Although the calcium sulfate has di-hydrate, in this system only the hemihydrate has formed (Kótai et al., 2008b). The solid CaSO$_4$.0.5H$_2$O can be used as sulfur fertilizer and improve the properties of unbound (sandy like) soils due to its binding properties with water (plaster of paris). During the removal of the polyunsaturated (Benjumea et al., 2011) labile compounds by sulfuric acid various compounds having acidic character form. The acid number of the ester phase is around 3.2 mg KOH/g oil. It is essential to remove these acids from the ester phase. It has been tried with filtering this ester phase through calcium or iron carbonate containing mineral granulates. This method, however, have not been effective, and in case of calcium carbonate some oil soluble products were also formed which increased the ash content of the ester. By using Varion ADA type resin (alkyl-ammonium type ion exchanger in OH form) the acidic constituent could easily be removed. The acid number was found to be 1.25 mg KOH/g oil, practically the same as was found for this oil after complete vacuum distillation (1.24 mg KOH/g oil) (Kótai et al., 2008b).

4. Butanol as raw material in biodiesel production

By using of butanol as blending material for gasoline and diesel fuels has many advantages towards fuel ethanol (Andersen et al., 2010; Bruno et al., 2009, Duck and Bruce, 1945, Workman et al., 1983), additionally, it can also be used as reactive component in biodiesel production. Butanol can substitute methanol as an alcohol for esterification (Wahlen et al., 2008; Nimcevic et al., 2000, Stoldt and Dave, 1998), it can be an acetal forming compounds for transforming of acetone or other ketones into acetals, or can easily be converted to butyraldehyde for the conversion of glycerol formed during trans-esterification of oils into 1,3-dioxane- or dioxolane type fuel additives (Silva et al., 2010)

4.1 Butanol as blending or reactive component in biodiesel production

The butanol or the butanol - acetone - ethanol mixture produced in the ABE fermentation have been tested already during the World War II as a blending components for gasoline powered engines (Duck and Bruce, 1945). The engines can be powered with gasoline containing butanol up to 40 % without any technical modification of the engine. The characteristics of pure butanol and ABE solvent mixtures as gasoline or diesel fuel additives have been tested in detail. Generally, the blended mixtures produce almost the same power and thermal efficiency as the gasoline (Schrock and Clark, 1983). The same time the blending has a positive effect via substantially decreasing the NO_x content of the exhaust gases.

Since the modern mobile agricultural equipment used in the production of biomass is mostly diesel powered, both ethanol and butanol have been tested for using them as blending components of diesel fuels. By the addition of butanol or ABE blends to diesel fuel the thermal efficiency could be increased, the exhaust gas temperatures are lowered and the soot formation is decreased. The operational parameters of the engine have been studied in detail (Workman et al., 1983). Butanol proved to be an alternative diesel fuel blend. By using butanol as a reactive component in vegetable oil butyl ester preparation or preparation of butanol based acetals, the residual butanol content does not has to be removed from the reaction mixture because it can act as blending component. A special case of the extraction of butanol from aqueous solutions during ABE fermentation is when the extractant is vegatable oil (Welsh and Williams, 1989). The butanol to be extracted can be reacted easily with the extractant. Since the vegetable oil alkyl esters are also extractants of the butanol (Crabbe et al., 2001, Ishizaki et al., 1999), this method can easily be integrated into a catalyst free supercritical trans-esterification technology, when the partially trans-esterified vegetable oil is recycled into the extraction process until its complete transformation into butylester. Since the butanol content extracted in the last step does not need to be separated from the butyl ester product, the method is advantageously integrated into the waste free biodiesel production system.

The use of butanol as a reactive component in biodiesel production has a lot of advantage. First of all, the convenient base catalyzed reaction takes place in homogeneous phase, thus the reaction is faster, the separation of the glycerol is better and the temperature of the reaction can be lowered. The excess of the butanol does not need to be removed from the ester phase. The viscosity of the butylester mixture prepared from soya oil was found to be 4.50 mm^2/s at 40 °C, the cetane index was 69. The cloud point of butyl-biodiesel is -3 °C. The flash point and pour point were found to be 44 and –13 °C, respectively (Wahlen et al., 2008). Since butanol reacts with free acids faster than methanol, the high acid containing vegetable oils (free fatty acid content of vegetable oils varies from 7 to 40 %), the waste cooked oils or other high free acid containing oils can also be used as raw materials. These high free fatty acid containing oils could not be trans-esterified economically with methanol and basic catalysts, and in the presence of acidic catalysts the reactions are very slow.

Solid phase heterogeneous catalysts have not widely been available for use them in industrial scale for these type of oils (Di Serio et al., 2008). The acid catalysts simultaneously catalyze the esterification of the free acids and the trans-esterification of the glycerides however, butyl esters are formed more easier than methyl esters (Wahlen et al., 2008). Methanol, ethanol, n-propanol and n-butanol have been reacted with oleic acid as a model for free fatty acids at 4:1 alcohol/acid molar ratio in the presence of 5 % sulfuric acid (as catalyst) at 80 °C for 16 min. The conversion was best (90%) for n-butanol, and the worst in the case of methanol (85%). The difference in trans-esterification activity of C_{1-4} alcohols in the presence of H_2SO_4 catalyst is more significant. The methanol can react with the soybean oil (10:1 methanol/bound fatty acid ratio) at 60 °C with 5 % sulfuric acid as catalyst only with 2 % conversion within 32 min. By using 12:1 alcohol/soybean oil ratio and 80 °C temperature, the methanol and ethanol gave 18 % conversion in 16 min, while the propanol and butanol showed 50 % conversion during the same time. The reaction of butanol with vegetable oils at a mixed feedstock containing oleic acid and soya oil with a ratio of 5:1-1:5 required minimum 2:1 butanol/fatty acid (free and bound) molar ratio at 110 °C in the presence of 5% H_2SO_4 catalyst . Using microwave heating at 6:1 butanol/soybean oil ratio in the presence of 3 % H_2SO_4, a 98 % conversion was achieved within 50 min. By using microwave heating the trans-esterification reaction of the vegetable oils with butanol can be performed without any catalyst under supercritical conditions (Geuens et al., 2008). Since the butanol boiling point is higher than methanol boiling point, the reactions takes place at higher temperatures without using extremely large pressures. The best results were achieved at 310 °C and 80 bar pressure in SiC coated tube reactor. The lack of the catalyst results very a small amount of glycerol without soap formation. The excess of butanol does not need to be separated from the ester phase or can be flashed out from the glycerol phase for recycling.

By using n-butanol instead of methanol and butoxylation of the unsaturated alkyl chain improve the ratio of the fossil energy used to produce a unit of renewable energy source. The highly unsaturated oils cause gum and deposit formation, but their epoxidation with peroxy-acetic acid and contacting the epoxides with n-butanol in the presence of 2 % sulfuric acid as catalyst at 80 C°, results 100% conversion of the epoxides. The selectivity is 87 %, and the 46 % conversion of the unsaturated alkyl chains does not cause an increase in the cloud point (Smith et al., 2009). As it can be seen, the butanol increases the amount of the biofuel produced from raw vegetable oil, by molar weight increasing referring to methyl esters (Table 3., Nimcevic et al., 2000), by incorporating butoxy groups into the unsaturated alkyl chains and by mixing the excess butanol with the formed fuel.

Ester	Combustion value		Alcohol molar fraction in the ester molecule
	MJ/kg	MJ/kmol	
Methyl	39.83	14156	8.7
Ethyl	40.03	14787	12.2
Butyl	40.52	16103	18.4

Table 3. Alcohol inputs in the production of biodiesel

In addition to these possibilities, in case of high acid containing raw vegetable oils, the acid content can also be transformed into butyl-type biofuel and does not need to be recovered as soaps. The abovementioned possibilities can be applied as parts of an integrated system together with other techniques to improve biofuel production, e.g. during the conversion of vegetable oils into methyl esters, the free fatty acids liberated from calcium soaps can be esterified with butanol instead of methanol as well (see Chapter 3.3).

4.2 Transformation of the wastes of butanol and biodiesel production into fuel

Glycerol is a very hygroscopic material and its combustion heat is low due to its high oxygen content. Neither its viscosity nor its hygroscopic nature or the miscibility properties indicate direct applicability as a fuel component. However, the glycerol is a reactive compound, thus the glycerol formed in the biodiesel synthesis can be transformed into lower oxygen containing compounds or to their mixture by various reactions (Guerro-Perez et al., 2009, Mota et al., 2009). In order to decrease the oxygen content of the products formed, the most reliable way is water elimination. It can be performed by reduction or by condensation reactions performed with reactants containing O= or HO-functions. The structure and reactivity ensure a series of water elimination (intra or intermolecular) reactions and formation of a variety of compounds.

The glycerol can act as a multifunctional primary and secondary alcohol and can easily be dimerized or polymerised into compounds with residual alcohol functions and alcoholic type reactivity. The glycerol can also be reacted with various other alcohol derivatives (with methanol residue from trans-esterification or with ethanol or butanol from ABE fermentation) into ethers. Transformation into cyclic acetals by using oxo-compounds e.g. acetone from ABE fermentation or acetone – acetaldehyde - butyraldehyde mixtures from the oxidation of not separated ABE products can also be performed. The formed acetals are cyclic dioxolane and dioxane type primary or secondary alcohols, or their stereoisomers (if R_1 and R_2 are not the same), respectively (Ferreira, et. al., 2010; Kótai and Angyal, 2011).

By partial oxidation of the mixture of primary alcohols from the first ABE fraction containing acetone, ethanol and butanol to aldehydes, a mixture of alcohols and oxo-compounds can be prepared. By using the waste glycerol containing methanol from the biodiesel production (or formaldehyde from the methanol oxidation) can provide a complex reaction mixture which can be condensed into an un-separated multicomponent mixture of various oxygenates with lower oxygen content than the starting glycerol. This mixture does not require complete separation into components or individual compounds to use it as a fuel. The reaction has been studied in the presence of various acidic catalysts as sulfuric acid, sulfonated styrene-divinyl-benzene copolymers and p-toluene-sulfonic acid. All of the catalyst gave similar results, the main product have been the 1,3-dioxolane derivatives. Various other components have also been formed in 1-2 % amount of each. The low-boiling fractions contain mainly the starting alcohols, acetone and dialkoxypropane derivatives, the higher fractions contain mainly 2,2-dimethyl-4-hydroxymethyl-1,3-dioxolane, its mixture with the starting alcohols and the formed dialkoxypropanes. The 2,2-dimethyl-5-hydroxy-1,3-dioxane has appeared only in the distillation residue because its boiling point is higher than 120 °C. In the acetalization of acetone with glycerol the two possible isomers 1,3-dioxolane or 1,3-dioxane ring containing products can also be formed in the 1,2- or 1,3-type cyclization reactions. The molar ratio of the dioxolane /dioxane and the yields slightly depend on the type of the acidic catalyst. The composition of a typical reaction mixture is illustrated in Table 4. The main product is the 2,2-dimethyl-4-hydroxymethyl-1,3-dioxolane, a smaller amount of the 2,2-dimethyl-5-hydroxy-1,3-dioxane and dialkoxy-propanes are also formed. Two mixed methoxy-group containing acetals are formed, as well. Thus, it seems to be probably that the primarily formed 2,2-dimethoxypropane has reacted with the higher primary alcohols. In order to increase the complexity of the mixture, which is an optimal situation for fuels, the glycerol - MeOH mixture was mixed with the first un-separated fraction of the butanol production which contains EtOH, acetone and BuOH, and reacted with various oxo- compounds prepared from the abovementioned alcohols by oxidation (CH_2O, CH_3CHO and butyraldehyde).

Compound	Alcohol	Fraction	Peak area	B.p. range
MeC(OMe)2Me	MeOH	I-IV	1	58-99 °C
MeC(OMe)(OEt)Me	MeOH, EtOH	I-IV	2	58-99 °C
MeC(OEt)2Me	EtOH	III,IV	1	71-99 °C
MeC(OMe)(OBu)Me	MeOH, BuOH	I-IV	1	58-99 °C
2,2-Me$_2$-4-CH$_2$OH-1,3-dioxolane	glycerol	III-VI	88	71-120 °C<
2,2-dimethyl-5-OH-1,3-dioxane	glycerol	VI	2	120 °C<

Table 4. The acetals formed in the reaction of ABE solvents and glycerol containing methanol with Varion KSM acidic ion exchanger catalyst at 3 h reflux

It can be seen that from the same molar amounts of the alcohols the acetone prefers the reaction with the glycerol, or the dialkoxy-propanes formed reacts with the glycerol via re-formation of the alcohols.

In this way, the waste stream from ABE and biodiesel production with or without oxidative treatment results an un-separated mixture containing various alcoholic and oxo-components which can react with each other in various water elimination reactions to from a variety of lower oxygen containing acetal/ether type compounds. The formed mixture contains components with a wide boiling range. Table 5 contains the product distribution in a mixture formed in the reaction of 1-1 equivalents of acetone, acetaldehyde, n-butyaldehyde and formaldehyde by 1 equivalent of glycerol and 2-2 equivalents of MeOH, EtOH and BuOH with Varion KSM sulfonated ion exchanger as catalyst under 3 h reflux. The reaction mixture has been separated into five fractions to study the distribution of each component formed and the starting material in the fractions. Depending on the reaction conditions, molar ratios of each reactant and the catalyst, the product distribution can be varied. Two isomers of 2-alkyl-4-hydroxymethyl dioxolanes are formed which have different boiling points. As an example, the effect of glycerol formal on the properties of the biodiesels can be seen in Table. 6. (Puche, 2009)

Compounds	Alcohol	Oxo-reactants	Fraction	Peak area
Dialkoxi-methanes				
$(MeO)_2CH_2$	MeOH	CH_2O	I-II	2
$(EtO)_2CH_2$	EtOH	CH_2O	I-IV	5
$BuOCH_2OMe$	BuOH,MeOH	CH_2O	I-III	15
$BuOCH_2OEt$	BuOH,EtOH	CH_2O	I-IV	14
$(BuO)_2CH_2$	BuOH	CH_2O	II-V	5
Dialkoxyethanes				
$(BuO)(MeO)CHCH_3$	BuOH,MeOH	CH_3CHO	I-IV	1
$(BuO)_2CHCH_3$	BuOH	CH_3CHO	IV-V	2
Dialkoxybutanes				
$(BuO)_2CHCH_2CH_2CH_3$	BuOH	PrCHO	IV-V	2
1,3-Dioxolanes (2 isomers)				
2-Me-4-CH_2OH-1,3-dioxolane	glycerol	CH_3CHO	I-V	6
2-Me-4-CH_2OH-1,3-dioxolane	glycerol	CH_3CHO	V	4
2-Pr-4-CH_2OH-1,3-dioxolane	glycerol	PrCHO	V	12
2-Pr-4-CH_2OH-1,3-dioxolane	glycerol	acetone	IV-V	11
1,3-Dioxanes				
2-Me-5-OH-1,3-dioxane	glycerol	CH_3CHO	IV-V	2
2-Pr-5-OH-1,3-dioxane	glycerol	PrCHO	V	3

Table 5. The identified components of the reaction between biodiesel waste and partially oxidized ABE production waste streams in the presence of Varion KSM catalyst

	RME + glycerol formal				
Glycerol formal content	0	0.5 %	1%	5%	10%
Density, g/cm³	0.8592	0.8620	0.8631	0.8711	0.8802
Freezing point, °C	-7	-16	-21	-21	-21
Viscosity at -10 C°, cSt	Solid	No data	548.2	343.3	No data

Table 6. Effect of glycerol formal on properties of methyl ester of rapeseed oil

Not only acetals, but other ether type components can also be used as fuel blends. The condensation products formed with alcoholic functions can be used for further acetal formation. The dioxolane and dioxane type compounds with alcoholic function groups can be esterified or etherified in a further reaction into other valuable products (Jalinski, 2006).

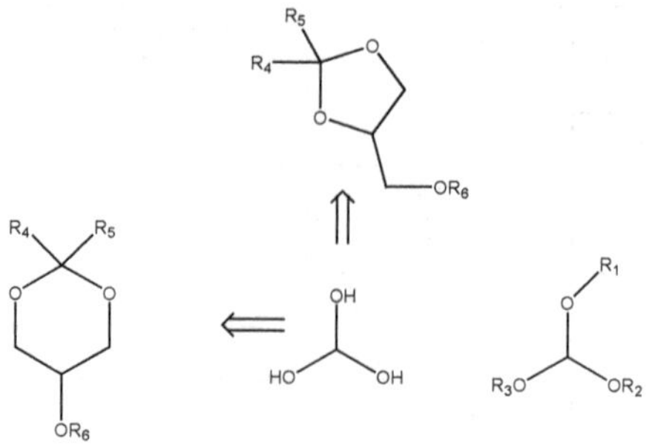

(6)

The general scheme for transformation of glycerol into fuel components with ABE components is given by eqn. (6), where R_1, R_2 and R_3 are Me, Et, Bu, $CH_3C(O)$-, $C_3H_7C(O)$-, R_4 and R_5 are H, Me, Pr, and R_6 means Me, Et, Bu, $CH_3C(O)$-, $C_3H_7C(O)$ or other groups derived from the alcohol-type glycerol condensation products. Transformation of all three hydroxyl groups of the glycerol into alkoxy groups (methoxy, ethoxy or butoxy), or esterifying them with low carbon chain carboxylic acids (acetic acid, butyric acid) decrease the hydrophil nature and oxygen content and increase the combustion heat, the miscibility with fuel. Thus, these compounds are advantageous fuel additives (Mota et al., 2009). Since ethanol, butanol, methanol, acetic and butyric acid are products/by-products and intermediates of the ABE fermentation or biodiesel production, these reactions are candidates for integration into a complex biomass utilization system. The intermediate acetic and butyric acid can also be used as acylation agents for the cyclic acetals, and in this way all product of the ABE fermentation become fuel component. Not only these organic acids but carbonic acid can also acts as acid residue in the esterified products. The carbonate compounds prepared form acetals formed from n-butyraldehyde or acetone and glycerol lowering the soot and the particulate formation during ignition of the diesel fuels (Delfort, 2004). Alkylation or acylation of free hydroxy-groups in 1,3-dioxolane and dioxane type fuel blends increases their solubility with two order of magnitudes (Jalinski, 2006)].

It is obvious, that glycerol which has primary and secondary alcohol functions, and can be condensed with itself to different kind of polyglycerols (Barrault et al., 1998). Polyglycerols can be obtained at high temperature vapor phase reaction over solid catalysts as alkali and alkaline earth metal hydroxides or carbonates, zeolites, La-ion-exchanged zeolites and ion-exchanger resins (Barrault et al., 1998). In the presence of resins, the main product is the diglycerol.

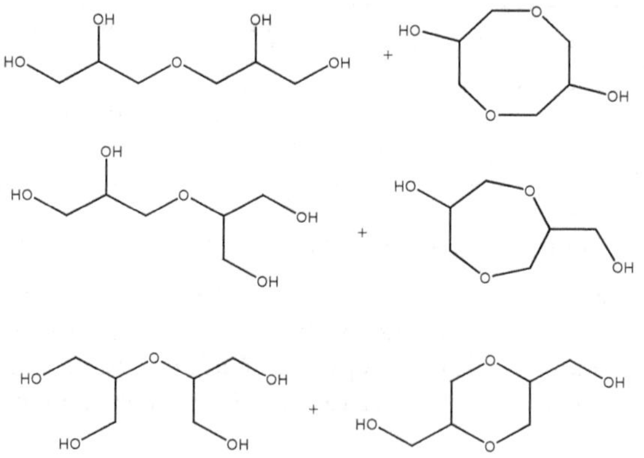

(7)

Since the glycerol has hydroxyl groups with various reactivity, depending on the catalyst and the reaction conditions, various dimers and even more type of oligomers and polymers can be formed. By using these dimers (oligomers) in acetal forming reactions, the complexity of fuel mixture can be further increased.

Not only water elimination, but increasing carbon chain length can decrease the relative oxygen content and increase the combustion heat and improve the fuel properties. Selective etherification of glycerol or the free alcoholic function groups of the condensates formed from the glycerol. The alcohol functions of glycerol or other alcohols formed during polymerization of glycerol or acetal production can easily be alkylated by reaction with isoalkenes (Klepacova et al., 2003). Trans-esterification of crude soya oil with methanol in the presence of NaOH catalyst, then separating the glycerol phase reacted with the mixture in the presence of Amberlyte-15 acidic ion exchanger catalysts for 2 when isobutylene converts the glycerol into ethers. The mixture formed contains 9 % triether, 47 % diether, 21 % mono-ether, 5 % unreacted glycerol, 14 % isobutylene and 4 % methyl esters. By separating and recycling the starting materials and the mono-ethers the residue can be mixed with the ester phase formed in the trans-esterification when a mixture is formed containing 12 % ethers and 88 % methyl esters. Its clouding point is below 0 °C and having a viscosity of 5.94 cSt which is lower with 9 °C and 0.5 cSt, respectively, if this parameters are compared to the ester phase without the addition of glycerol ethers (Barrault et al., 1998).

The oxygenate mixtures produced in the abovementioned ways ensures that a very complex mixture of compounds could be manufactured, in which all components of the ABE fermentation and biodiesel production turn into fuel component. These blending materials have very advantageous properties, decrease the viscosity, decrease the pouring point and soot formation and improve the cetane number. In this way, vegetable oil ester (mainly butyl ester), butanol and acetal or other oxygenate mixture containing biodiesels are formed with much higher production efficiency compared to the classical vegetable oil methyl esters. Thus, our technology can provide an aromatic hydrocarbon-free fuel which can be used even in highly populated large cities. Since biodiesels, fossil diesels and the gasolines can be mixed with pure butanol up to an amount of 40% without influencing the fuel properties, and these oxygenates can also be used around in an amount of 20 %, these new kind of fuel mixtures can provide a solution for the EU demand (incorporation of 20 % biocomponent into fuels until 2020).

5. Other aspects of the integrated biomass utilization system

It is an obvious question that which bioalcohol should be used for the replacement of methanol in biodiesel production, or it is worth to change the ethanol blends of fuels to butanol which has much better fuel properties and energy content than the ethanol.

Comparison of technical and economical assessment for corn and switch grass fermented by yeast into ethanol and C. acetobutylicum into butanol showed (Pfromm et al., 2010) that biobutanol production is not competitive with ethanol production. As an example, the carbon balances for corn are illustrated in Fig. 3. However, involving new technologies, new raw materials (e.g. sugar sorghum) and the extractive fermentation processes combined with immobilized cell techniques, and decrease the production cost by means of the new separation technologies, the butanol becomes competitive as blending or reactive component in biofuel production.

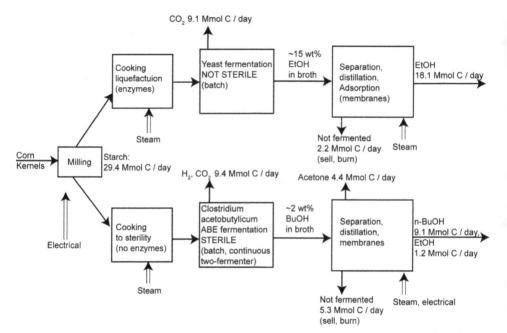

Fig. 3. Comparison of the carbon balances in the fermentation of corn into ethanol (yeast) and ABE (bacteria)

5.1 New trends in biobutanol technology

The industrial production of biobutanol has roughly a 100 years history, here only some new trends are reviewed in order to indicate the new perspectives of the biobutanol technology. The so called ABE fermentation produces acetone, butanol and ethanol in a ratio of about 2:1:7 or 3:1:6 depending on the bacteria and the fermentation conditions. A small amount of acetic acid and butyric acid are also formed. The starting materials are C_5 and C_6 sugars, e.g. starch or cellulose hydrolizates, and there are some bacteria strain as well which can utilize cellulose directly. The main problem of the biotechnological butanol production is the toxicity of the solvents formed (mainly the butanol) towards the microorganism (Costa, 1981). The most intensively studied area of the developments is the genetic engineering to produce butanol tolerant bacterium strains or produce less sensitive genetically modified yeast and saving the microorganism from the toxic effects, e.g. by immobilization and capsulation of the bacteria (Park et al., 1989), or by the removal of the accumulated solvents before reaching the toxicity level (Papadopoulos and Linke, 2009; Schmidt et al., 1988;). Combination of the methods provides a good chance to start a continuous ABE fermentation (Hartmeier et al., 1991, Ishii et al., 1985, Kótai and Balogh, 2011). Since the energy demand of the butanol recovery from the dilute solution is one of the main cost factor extraction with a suitable solvent or adsorption on a cheap heterogeneous carrier can be candidates for the development of energy efficient butanol production. Due to the low adsorption capacity of known adsorbents like activated carbon, or the affinity of solid sorbents towards water allowed only utilization at low level. The extraction seemed to be more effective, but the solvents have to meet serious requirements like:

- Non-toxic to the microorganism and high stability,
- High distribution coefficient and selectivity with respect to the product,
- Low viscosity and solubility in the aqueous phase,
- Gravitation separation by density difference,
- Large interfacial tension and low tendency to emulsify the broth,
- High boiling point difference with respect to ABE solvents and low price

By using immobilized microorganism and in situ extraction the continuous production of butanol and ABE solvents can be performed easily, especially, if the integrated biomass utilization system ensures the sugar solution from sorghum processing. Thus, a large amount of ballast materials from corn or starch hydrolysis which increase the dry material content of the mash can be avoided The main problems are;

1. that the solvents which have high selectivity to the ABE solvents are very toxic to the microorganisms,
2. the best distribution coefficients for butanol found among the non-toxic solvents is only 3.5 (oleyl alcohol). In order to apply the high distribution coefficient of a toxic solvent associated with the requirements of the continuous production of butanol, a special system to produce ABE solvents has been developed.

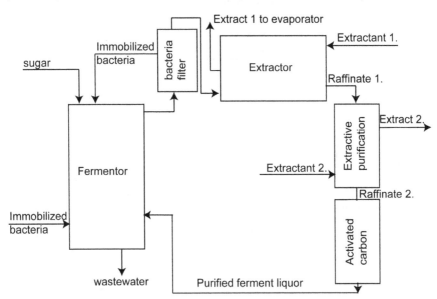

Fig. 4. Flow-sheet of continuous butanol fermentation with outer extraction

Heptanal, which has distribution coefficient for butanol of $K_B=11.3$ has been selected for the new technique. The toxicity of the heptanal is avoided via a by-pass extraction and a post-extractive purification for removal of traces of extractant from the ferment liquor before recycling it to the fermentation. The flow-sheet of particular method can be seen in Fig. 4. The technology using immobilized bacteria and heptanal as solvent is under development now. After starting the fermentation and before reaching the toxic level of butanol, a part of the ferment liquor is pumped out into a counter-current extractor filled with heptanal. The extraction is a continuous process. It means that the raffinate phase is contacted with another non-toxic solvent and/or a sorbent which removes the toxic solvent from the raffinate before recycling it into the fermentation. It should be noted here, that in case of heptanal the distribution of the ABE solvents can be considered to be advantageous.

Solvent	Heptanal	Oleyl alcohol
Acetone	1.65	0.40
Ethanol	1.01	0.10
Butanol	11.13	3.75

Table 7. Distribution coefficients of ABE solvents

The boiling point of heptanal is higher than the ABE solvents, thus the evaporation of the heptanal is avoided and does not require energy, By using a special fermentor - extractor system (Fig. 5) the extract phase containing the ABE solvents are contacted with another heptanal phase having a smaller volume (1/10-1/5) than the volume of the primary extract. The contact takes place through a special porous wall based on a pumicite - cement composite (Kótai and Balogh, 2011; Kótai et al., 2011.).

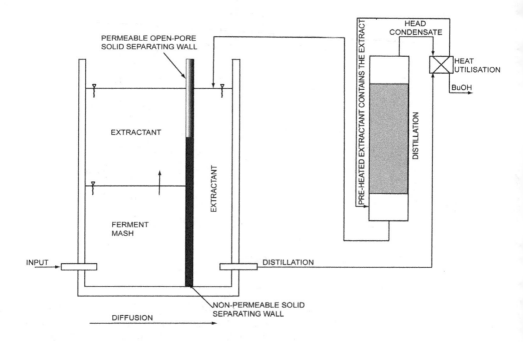

Fig. 5. Fermentor with composite wall

The porous material absorbs the solvents as a sponge, and it acts as a liquid transmitting media. There is no physical mixing only the diffusion controls the distribution of the solvents at the two sides of the wall. In order to keep the concentration difference as driving force, the ABE solvents is produced continuously at the A side (including the extraction step) and the extracted solvents are continuously removed at the B side of the wall, e.g. vacuum distillation, sorption by solid solvents as sulfonated styrene-divinyl-benzene copolymers (Kótai and Balogh, 2011), or with other methods. Due to their higher affinity towards water than the ABE solvents these copolymers ca not remove the AB solvents from the water directly. The Varion KSM sulfonated styrene-divinylbenzene copolymer can absorb the following amount of solvents (Kótai and Balogh, 2011)

Solvent	Amount, g/100g	Solvent	Amount, g/100g
Water	87	acetone	40
MeOH	75	heptanal	35
EtOH	73	Oleyl alcohol	35
BuOH	63	Hexane	30

Table 8. The adsorbed amount of solvents by solid sorbent based on sulfonated polymer

By using waste ion-exchangers, metal-containing activated carbons or metal-free carbons were prepared by leaching of metals from these metal-containing carbons [Kótai and Angyal, 2011], which are candidates as selective spherical sorbents with low hydrodynamic resistance for removing residual extractants and absorb the contaminants occurring in small amount in the ferment liquor. The Fe-containing sorbent are magnetic and could be separated very easily (without filtering) with an electromagnet (Kótai and Angyal, 2011). Elastic waste tires were also sulfonated for the preparation of sulfonated organic polymers to absorb butanol from organic media (Kótai and Balogh, 2011).

The supplement of the continuous reactor with sugar solution from the sorghum processing results a much cleaner technology than the process using the corn mash containing a lot of solid ballast material. Additional advantage of the method is the small volume of the B extract, because the small volume of invaluable solvents (water) needs small volume and investment of the distillation unit. A rough comparison of the energy and material balance shows, that selecting the appropriate bacteria and conditions, it is possible to reach the almost maximal theoretical utilization of glucose (0.37 g butanol/g glucose), so the amount of waste water and the energy to produce a unit of butanol in principle can be decreased to 25-50 %.

Combination of these techniques with new developments in membrane technologies, mainly with selective membrane separation and pervaporation (Liu et al., 2005; Thongsukmak & Sirkar), can provide a feasible butanol production technology.

5.2 Recycling of ash from biomass power plants

Combustion of biomass to produce energy (heat or electricity) always leaves back ash in various amount and composition depending on the biomass used as raw materials in the furnaces. This ash highly alkaline due to its potassium carbonate content (pH >10). Therefore it cannot be used directly as a fertilizer; only very small amount can be applied even for acidic soils. Since the ash contains all microelements and non-volatile elements in the amount absorbed and used by the plants harvested from the given area, their amount and ratio is optimal for the plant. Its insoluble content (e.g. phosphates) can be converted via digestion to utilizable material. In addition, it can be also supplemented with nitrogen fertilizers or other additives. In this way the processed ash becomes a useful material called eco-fertilizer (Angyal et al., 2006). In order to ensure the waste free technological viewpoints, the biomass power plant ash has to transform into fertilizer. It has been known for a long time that the alkaline ash can be neutralized with mineral acids, when potassium containing fertilizers are formed. However, these methods are difficult to perform technically, due to the large volume of gas formed in these reactions. Roughly 150 normal m^3 of carbon dioxide evolves in the reaction of 1 ton of ash with intensive foaming. A biomass power plant with electric capacity of 50 MW produces roughly 50-60 tons ash/day which means producing of > 300 normal m^3 of CO_2 gas/h. Furthermore, a very large volume of dilute fertilizer solution is produced, thus the cost of transportation is very high. The evaporation of the solution is not feasible. In order to solve these problems and ensure recycling the ash components for sustainable biomass production, a new technology has been developed for the neutralization of the biomass power plant ashes. This method ensures complete digestion of potassium and phosphorus content of the ash and decreases the K and P fertilization costs which are essential in the case of plants with fast metabolism such as sugar sorghum or energy grass. The nitrogen supplement is the only one which should be ensured in a usual way with the addition of N fertilizers. In this way the sulfur deficiency of the soils, or extreme sulfur demand as in case of oily plants like rape can also be satisfied (Scherer, 2001).

The new method based on the reaction of sulfuric acid and the biomass ash in a long quasi-closed tube reactor equipped with a screw for moving the reaction mixture. The reaction substantially proceeds during the mixture motion from the one end of the reactor to the other. This technology ensures not only continuous production of neutralized ash, but other important changes also appear in the chemical constitution of the starting ash. Normally, concentrated sulfuric acid does not wet the ash and does not react with it at all, their mixing can be proceed without CO_2 evolution. The reaction starts only in the presence of water. The mortar like mass prepared from the ash/limestone mixture and cc. sulfuric acid are mixed with water and reacted in the tube reactor. The reaction starts only after the dilution of the sulfuric acid. The most advantageous concentration is ~50 % (of sulfuric acid) when the carbonates reacts in a self-sustainable way due to producing water in the neutralization reactions. The key element in the reaction is the quasi-closed tube reactor. During moving the mortar like mass toward the opened end of the tube reactor, the neutralization reaction takes place and CO_2 gas evolves as usual. However, the reaction mass acts as a plug which ensures that the formed CO_2 gas cannot released from the reactor. The in situ evolved CO_2 gas makes micro-bubbles in the material, because of the overpressure of the quasi-closed equipment, and the swelled mass fills out the tube reactor. However, the wall of the reactor and the reaction mass as a plug ensures an overpressure within the reactor. Due to the overpressure no fizzing out occurs within the reactor and the micro-bubbles of the carbon dioxide gas are kept in the mortar-like mass. If the amount of the mass is adjusted to be less in its volume than the volume of the reactor, the mass is blown up due to the evolved gas and fills completely the space within the reactor. The carbon dioxide micro-bubbles have a slight overpressure, thus leaving the tube reactor at the opened end, the gas leaves the semi-solidified mass and the places of the micro-bubbles becomes opened pores. Technologically this method ensures processing of the ash with sulfuric acid in a small volume of continuously operated tube reactor having only a volume of ~3 times larger than the volume of the ash that can be processed (~ 60-80 times larger reactors as used in the classical neutralization methods do not needed (Angyal et al, 2006). Since the formed mortar like mass dries and solidifies easily, after granulating or pelletizing into the usual shape of solid fertilizers, the formed eco-fertilizer can be spread as solid by common facilities.

This method of neutralization has numerous advantages towards the classical neutralization technologies, not only the formation of solid fertilizer instead of dilute liquids, and thus avoiding the high volume expensive reactors during manufacturing, but from chemical viewpoints as well. Normally, the ash formed from straw and energy grass contains a mixed potassium calcium carbonate (Buetschliite), $K_2Ca(CO_3)_2$ as main components, the second most important phase is the KCl, and K_2CO_3 and K_2SO_4 can also be detected by powder X-ray diffraction. Similar amount of magnesium hydroxide and sodium carbonate can also be detected. The ratio of potassium chloride and sulfate depends on the soil composition, and the fertilization and type of fertilizer used (KCl or K_2SO_4) during the production of the wheat of course. Expressing the important metal content in the form oxides are as it follows: ~40 % of K_2O, ~10% CaO, 3.5 % MgO and 2.5 % of Na_2O. The straw contains a lot of chlorides (~7 %), the other anions as sulfate and carbonate expressed in SO_3 and CO_2 are ~10 % and ~20 %, respectively. The potassium-calcium carbonate (or potassium and calcium carbonate as well) easily reacts with diluted (~50 %) sulfuric acid, but not only the expected K_2SO_4 and $CaSO_4$ but their double salts as syngenite ($K_2Ca(SO_4)_2.H_2O$) and polyhalite ($K_2Ca_2Mg(SO_4)_4.2H_2O$) are formed as main products. The syngenite is less soluble (but not completely insoluble) in water and has ion-exchange properties toward ammonium ion, because due to their similar sizes of potassium and ammonium ions they can substitute each other in the structure of this compound.

$$K_2Ca(CO_3)_2 + 2H_2SO_4 = K_2Ca(SO_4)_2.H_2O + H_2O + 2CO_2 \qquad (8)$$

The excess of sulfuric acid is neutralized to pH=6 with limestone powder and can be used as simple and general potassium and sulfate fertilizer which has opened pore structures which can absorb water and keep it in the pores, this way increasing the water retaining capacity of the soil. This fertilizer contains soluble phosphates and microelements previously digested by the sulfuric acid treatment. Furthermore, via controlling the amount of the calcium carbonate powder in the last step of the manufacturing, acidic (sub-neutralized), neutral or alkaline (over-neutralized) fertilizers can also be produced. No liquid or solid waste form in this technology, all the used components are built into the structure of the product. Since the water absorbing capacity of these materials, due to the porosity controlled by the synthesis conditions is very high (50-120 % of its mass), it is obvious, that not only water but various aqueous solutions can also be absorbed in these pores. This behavior opens new perspectives, namely absorbing different other fertilizers, e.g. nitrogen fertilizers, insecticides, or any other solutions of important compounds which should be injected into the soils.

The most common nitrogen fertilizer is the ammonium nitrate, however, the metal-doped NH_4NO_3 production has serious problems because the ammonium nitrate is prilled from the melt, and the melted ammonium nitrate is easily exploded due to the catalytic effect of metal compounds. Thus, these metal microelements cannot be added to the melted NH_4NO_3 before prilling. Using ammonium nitrate solutions, the metals compounds can be added in the required amounts without any risk to the porous granulated eco-fertilizer. Although drying after the absorption of the aqueous solutions (e.g NH_4NO_3 solutions) requires extra energy, but the complete process energetically is still more advantageous, due to the following considerations. Ammonium nitrate solution is prepared in an exothermic reaction of a ~65 % of nitric acid with ammonia gas, when an aqueous solution of NH_4NO_3 solution (roughly 80%) is formed. Evaporation of this solution at high temperature leads to the melt of ammonium nitrate which is prilled in the next step of manufacturing. Since we use the NH_4NO_3 solution, which is contacted with the granules, the water removal, without melting of the NH_4NO_3, requires less energy than the final step of the solid NH_4NO_3 manufacturing. This concentrated NH_4NO_3 solution has acidic character, and easily react with the syngenite and other components of the eco-fertilizer. The concentrated NH_4NO_3 solutions are not only physically absorbed and imbibed in the pores of the eco-fertilizer, but chemically reacts with its components, as well.

$$K_2Ca(SO_4)_2.H_2O \text{ (syngenite)} \quad (NH_4)_2Ca(SO_4)_2.H_2O \text{(koktait)}$$

The potassium ions can be substituted with ammonium ion with the formation of partially or completely ion-exchanged syngenite-like isomorphous compounds. The completely substituted product is called to be koktaite, $(NH_4)_2Ca(SO_4)_2.H_2O$, which is less soluble in water, thus releases nitrogen slowly into the soil (Angyal et al., 2006; Coates and Woodward, 1988; Von Maessenhausen et al., 1988). The formed potassium nitrate transformed into a solid solution with the excess of the ammonium nitrate, the typical composition of this product was the $K_{0.27}(NH_4)_{0.73}NO_3$. The koktaite and NH_4-syngenites are sparingly soluble in water, thus the ammonium ion concentration liberated in the presence of water is constant at a given temperature and ionic strength of sulfate ion. Since not the full amount of the ammonium ion is liberated, no damages to the plant and losses by washing away, respectively occur even if using in high doses. When the plant absorbs the ammonium-ion from the soil, due to the equilibrium conditions, a part of the solid will dissolve and supply the water with a new amount of ammonium ion. Since the equilibrium concentration is closely constant, the amount of water will control the amount of the released ammonium ion, namely, the release of the ammonium ion from this koktaite type compounds is controlled by raining or irrigating. In drought situation, when there is no absorption of ammonium ion from the soil by the plant, there is no dissolution of ammonium syngenites and releasing ammonium ion which would be decomposed by the soil bacteria as it happens in case of water soluble ammonium ion containing fertilizers. Besides ammonium nitrate, other fertilizer components can also be used to adjust the main element concentrations, such as K, P or N, and to change the available form of these elements in various chemical compounds. The K_2HPO_4 does not react at all with other components of the ash. It is interesting, that ammonium salts as NH_4Cl and $(NH_4)_2SO_4$ cannot transform the syngenite completely into $(NH_4)_2SO_4$, even if the ammonium sulfate is in excess, but in the presence of urea, the transformation is complete. Both KCl and K_2SO_4 decompose the ammonium syngenite, but the mixture of the K_2SO_4 and the NH_4Cl produces $(NH_4)_2Ca(SO_4)_2.H_2O$. Thus, the main factor is probably the ammonium to potassium ion ratio. There is an important difference between the behavior of the potassium sulfate and potassium chloride. The latter compound is more reactive, and KCl, KNO_3 and NH_4MgCl_3 are also formed in its presence. Using various additives not only the ratio of the agriculturally important elements (K, P, N, S) are controlled but their chemical forms can also be altered. Using various kind of soil bacteria and supplementary materials to ensure theirs intensive growing is another possibility for nitrogen-fixation in the treated area. By using the eco-fertilizer technology supplemented with absorbing of aqueous liquids ensures the recycling of the by-product of the biomass power plant providing energy and electricity for the biofuel (biodiesel and biobutanol and supplemented) plants. This way we can sustain the production of the renewable energy plants, e.g. sugars sorghum, while the soil quality is maintained and improved, respectively.

6. Conclusion

By proper selection of biomass available from a given area, the sugar and energy sources, and the relative amount of the vegetable oil produced can be adjusted. In order to decrease the processing cost of raw materials into sugar containing mash for fermentation plants, the classical sugar sources as corn can be replaced with sugar sorghum, which can be processed similar to sugarcane. Combustion of the residual biomass in power plants or their digestion into biogas depend on the water and protein content of the residue and the heat or electricity demand of fuel-producing (biodiesel, biobutanol, acetals, etc.) or waste processing (fertilizer production) plants. Generally, it is more advantageous to use biomasses of high protein and water content in biogas plants. Burning the biogas or by using it as fuel in gas-engines the amount of heat and electricity can be controlled. Wastes of high cellulose content can be advantageously burned in power plants, sometimes after drying with the low heat value warm water streams of energy production. Wastes of fuel production can be utilized by combination these two methods of energy production. The ash and the solid residues from biomass power plants can be utilized as fertilizers by mixing them with potassium sulfate or calcium sulfate formed during recovery of the catalyst ($KHSO_4$ or H_2SO_4) in biodiesel or acetal plants. Finally, there are two other wastes. The first is K_2SO_4 from the biodiesel technology, and the other is the ash from the combustion. Beyond the integration of energy producing and consuming plants and controlling the ratio of the raw materials and the type of the energy (heat or electricity), the production technology is also to be changed mainly in biodiesel, biobutanol and fertilizer plants. In this way the energy consumption of each technological step can be decreased.

7. References

Andersen, V. F., Anderson, J. E., Wallington, T. J., Mueller, S. A. & Nielsen, O. J., *Energy & Fuels*, Vol. 24, (2010), pp. 2683-2691. ISSN 0887-0624

Angyal, A., Hujber, O., Kótai, L., Legeza, L. & Sajó, I. E., *HU Patent Appl. 0600390*, 2006.

Arlie, J.-P., *Revue de l' Institute Francais du Petrole*, Vol. 38, No.2. (1983), pp. 251-257. ISSN 0020-2274

Barrault, J., Pouilloux, Y., Vanhove, C., Cottin, K., Abro, S. & Clacens, J. M., *Chemistry and Industry*, Vol. 75, (1998), pp. 13-23. ISSN 0009-3068

Benjumea, P., Agudelo, J. R. & Agudelo, A. F., *Energy & Fuel*, Vol. 25, (2011), pp. 77-85. ISSN 0887-0624

Boocock, D. G. B., Konar, S. K. & Sidi, H., *Journal of the American Oil Chemists Society*, Vol 73 (1996), pp. 1247-1251, ISSN 0003-021X

Boocock, D. G. B., Konar, S. K., Mao, V. & Sidi, H., *Biomass & Bioenergy*, Vol.11, No.1. (1996), pp. 43-50, ISSN: 0961-9534

Boocock, D. G. B., Konar, S. K., Mao, V., Lee, C. & Buligan, S., *Journal of the American Oil Chemists Society*, Vol. 75, (1998), pp. 1167-1172, ISSN 0003-021X

Bruno, T. J., Wolk, A. & Naydich, A., *Energy & Fuel*, 23 (2009), pp. 2295-2306. ISSN 0887-0624

Coates, R. V. & Woodard, G. D., *Journal of Science of Food & Agriculture*, Vol. 14, No.6, (1963), pp. 398-404, ISSN 0022-5142

Costa, J. M., *Proc. Ann. Biochem. Eng. Symp.* 11th, (1981), pp. 83-90.

Crabbe, E., Nolasco-Hipolito, C., Kobayashi, G., Sonomoto, K. & Ishizaki, A., *Process Biochemistry*, Vol. 37, (2001), pp. 65-71, ISSN 0032-9592.

Delfort, B., Durand, I., Jaecker, A., Lacome, T., Montagne, X. & Fabrice, P., US 7097674, 2004.

Di Serio, M., Tesser, R., Pengmei, L. & Santacesaria, E., *Energy & Fuels*, Vol. 22, (2008), pp. 207-217. ISSN 0887-0624

Duck, J. T. & Bruce, C. S., *Journal of Research of the National Bureau of Standards*, (1945), pp.439-465, ISSN: 0160-1741.

Ferreira, P., Fonseca, I. M., Ramos, A. M., Vital, J. & Castanheiro, J. E., , *Applied Catalysis B. Environmental*, Vol.98, (2010), pp. 94-99, ISSN 0926-3373

Fu B., & Vasudevan, P. T., *Energy & Fuels*, vol. 23, (2009), pp. 4105-4111. ISSN 0887-0624

Geuens, J., Kremsner, J. M., Nebel, B. A., Schober, S., Dommisse, R. A., Mittelbach, M., Tavernier, S., Kappe, C. O. & Maes, B. U. W., *Energy & Fuels*, vol. 22, (2008), pp. 643-645. ISSN 0887-0624

Guan, G., Kusakabe, K., Sakurai, N. & Moriyama, K., *Fuel*, Vol. 88, (2009), pp. 81-6, ISSN 0016-2361

Guan, G., Sakurai, N. & Kusakabe, K., *Chemical Engineering Journal*, Vol.146, (2009), pp. 302-6, ISSN 1385-8947.

Guerro-Perez, M. O., Rosas, J. M., Bedia, J., Rodriguez-Mirasol, J. & Cordero, T., *Recent Patents on Chemical Engineering*, Vol. 2, (2009), pp. 11-21. ISSN 2211-3347

Hartmeier, W., Buecker, C. & Wallrath, J., *Biochem. Bioeng., 2nd Int. Symp., Stuttgart*, (1991), pp. 236-241.

Hill, J., Nelson, E., Tilman, D., Polasky, S. & Tiffany, D., *Proceedings of the National Academy of Sciences* Vol. 103, No. 30, (2006), pp. 11206-11210, ISSN 1091-6490.

Huber, G. W., Iborra, S. & Corma, A., *Chemical Reviews*, Vol. 106, (2006), pp. 4044-4098, ISSN 0009-2665

Ishii, S., Taya, M., Kobayashi, T., *Journal of Chemical Engineering of Japan*, Vol. 18, No.2, (1985), pp. 125-130, ISSN 0021-9592

Ishizaki, A., Michiwaki, S., Crabbe, E., Kobayashi, G., Sonomoto, K. & Yoshino, S., *Journal of Bioscience and Bioengineering*, Vol. 87, No. 3, (1999), pp. 352-356 ISSN 1389-1723

Jalinski, T. J., *WO 2006/084048 A1*, 2006

Khelevina, O. G. & Kanyaev, N. P., *Izvestiya Vysshikh Uchebnii Zavedenii, Khimiya i Khimicheskaya Tekhnologiya*, Vol. 11, No.2, (1968), pp. 166-170, ISSN 0579-2991.

Klepacova, K., Mravec, D., Hajekova, E. & Bajus, M., *Petroleum & Coal*, Vol. 45, (2003), pp. 54-57, ISSN 1335-3055

Kótai, L. & Angyal, A., *Research Report, GOP 1.1.1.,* Axial-Chem Ltd., 2011.

Kótai, L., Angyal, A., Somogyi, I., Bihatsi, L., May, Z., Gömöry, Á. & Tamics, E., *HU 08/00437A*, 2008.

Kótai, L. & Balogh, J., *Research Report, GOP 1.1.1.,* Kemobil Co., 2011.

Kótai, L., Gömöry, Á., Gács, I., Holly, S., Sajó, I. E., Tamics, E., Aradi, T. & Bihátsi, L., *Chemistry Letters*, Vol. 37, No. 10, (2008), pp. 1076-7, ISSN 0366-7022

Kótai, L., Szépvölgyi, J., Tamics, E., *HU 11/00286* A. 2011.

Kótai, L., Tamics, E., Sas, J., Deme, P., Gömöry, Á. & Aradi, T., *HU 06/00886A*, 2006.

Leung, D. Y. C., Wu, X. & Leung, M. K. H., *Applied Energy*, Vol. 87, (2010), pp. 1083-1085, ISSN 0306-2619

Lindquist, C.-J., *WO91/15452*, 1991.

Liu, F., Liu, L., Feng, X., *Separation & Purification Technology*, Vol. 42, (2005), pp. 273-282. ISSN 1383-5866

Maeda, K., Kuramochi, H., Asakuma, Y., Fukui, K., Tsuji,. T., Osako, M. & Sakai, S., *Chemical Engineering Journal*, Vol. 169, (2011), pp. 226-30, ISSN 1385-8947.

Meng, Z., Jiang, J. & Li, X., *Taiyangneng Xuabao*,Vol.30, No.3, (2009), pp. 385-389. ISSN 02540096

Mota, C. J. A., da Silva, C. X. A. & Goncalves, V. L. C., *Quimica Nova* Vol.32, No.3, (2009), pp. 639-648, ISSN 0100-4042

Nimcevic, D., Puntigam, R., Woergetter, M. & Gapes, J. R., *Journal of the American Oil Chemists Society*, Vol. 77, No. 3, (2000), pp. 275-280. ISSN 0003-021X

Papadopoulos, A. I. & Linke, P., *Chemical Engineering and Processing: Process Intensification*, Vol. 48, (2009), pp. 1047-1060, ISSN 0255-2701.

Park, C.-H., Okos, M. R. & Wankat, P. C., *Biotechnology and Bioengineering*, Vol. 34, (1989), pp. 18-29, ISSN 0006-3592

Pimentel, D. & Petzek, T. W., *Natural Resources Research*, Vol. 14, No.1, (2005), pp. 65-76, ISSN 1520-7439.

Pfromm, P. H., Amanor-Boadu, V., Nelson, R., Vadlani, P. & Madl, R., *Biomass and Bioenergy*, Vol. 34, (2010), pp. 515-524, ISSN 0961-9534

Puche, J. D., *US 7637969B2*, 2009.

Scherer, H. W., *European Journal of Agronomy*, Vol. 14, (2001), pp. 81-111, ISSN 1161-0301

Schmidt, A., Windsperger, A. & Friedl, A., *US 4749495*, 1988.

Schrock, M. D. & Clark, S. J., *Transactions of the ASAE* (1983), pp. 723-727, ISSN 0001-2351

Silva, P. H. R., Goncalves, V. L. C. & Mota, C. J. A., *Bioresource Technology*, Vol. 101, (2010), 101, pp. 6225-6229, ISSN 0960-8524.

Smith, P. C., O'Neill, B. K., Ngothai, Y. & Nguyen, Q. D, *Energy & Fuels*, Vol. 23, (2009), pp. 3798-3803, ISSN 0887-0624

Soriano, N. U., Jr., Venditti, R. & Argyropoulos, D. S., *Fuel*, Vol. 88, (2009), pp. 560-565, ISSN 0016-2361

Stoldt, S. H. & Dave, H., *EP 0860494* A1, 1998.

Thongsukmak, A. & Sirkar, K. K., *Journal of Membrane Science*, Vol. 302, (2007), pp. 45-58, ISSN 0376-7388.

Von Maessenhausen, W., Czikkely, V. & Jung, J., *US4883530* B2, 1988.

Wahlen, B. D., Barney, B. M. & Seefeldt, L. C., *Energy & Fuels*, Vol. 22 (2008), pp. 4223-28. ISSN 0887-0624.

Welsh, F. W. & Williams, R. E., *Journal of Chemical Technology & Biotechnology*, Vol. 46, No. 3, (1989), pp. 169-78, ISSN 0268-2575

Workman, J.P., Miller, G. L. & Smith, J. L., *Transactions of ASAE*, 1983, 642-645, ISSN 0001-2351

10

Advantages and Challenges of Microalgae as a Source of Oil for Biodiesel

Melinda J. Griffiths, Reay G. Dicks,
Christine Richardson and Susan T. L. Harrison
*Centre for Bioprocess Engineering Research (CeBER), University of Cape Town,
South Africa*

1. Introduction

Microalgal oil is currently being considered as a promising alternative feedstock for biodiesel. The present demand for oil for biofuel production greatly exceeds the supply, hence alternative sources of biomass are required. Microalgae have several advantages over land-based crops in terms of oil production. Their simple unicellular structure and high photosynthetic efficiency allow for a potentially higher oil yield per area than that of the best oilseed crops. Algae can be grown on marginal land using brackish or salt water and hence do not compete for resources with conventional agriculture. They do not require herbicides or pesticides and their cultivation could be coupled with the uptake of CO_2 from industrial waste streams, and the removal of excess nutrients from wastewater (Hodaifa et al., 2008; An et al., 2003). In addition to oil production, potentially valuable co-products such as pigments, antioxidants, nutraceuticals, fertilizer or feeds could be produced (Mata et al., 2010; Rodolfi et al., 2009).

Despite these advantages, algal fuel is not currently in widespread use, largely due to its high cost of production (Chisti, 2007; Miao & Wu, 2006). Despite strong interest from the commercial and scientific sectors, there are currently no industrial facilities producing biodiesel from algae (Lardon et al., 2009). One of the major economic and technological bottlenecks in the process is biomass and lipid production by the algae (Borowitzka, 1992; Sheehan et al. 1998; Tsukahara & Sawayama, 2005). Productive strains and optimized culture conditions able to produce cells with a simultaneously high growth rate and lipid content are required. The high cost and energy demand of harvesting unicellular algae also remains a major challenge. The small cell size (often < 10 μm in diameter) and dilute biomass produced requires innovative solutions to minimize the consumption of water and energy as well as processing costs (Rodolfi et al., 2009).

This chapter provides an overview of microalgae as a source of oil for biodiesel, focusing on:

- A description of algae and their properties with regards to oil production
- Requirements and key factors in microalgal cultivation
- Methods and challenges in harvesting and processing of algal biomass
- Economic and environmental feasibility of microalgal biodiesel
- Mechanisms to enhance lipid productivity of microalgae and future research directions.

2. Microalgae

The term 'algae' is used to describe a huge variety of prokaryotic (strictly termed Cyanobacteria) and eukaryotic organisms with a range of morphologies and phylogenies. They represent a wide array of species, inhabiting environments from deserts to the Arctic Ocean, including both salt and fresh water. They vary in colour, shape and size, from picoplankton (0.2 to 2 µm) to giant kelp fronds up to 60 m in length (Barsanti & Gualtieri, 2006). Macroalgae (e.g. seaweeds) are generally large (can be seen without the aid of a microscope), multicellular and often show some form of cellular specialisation. Microalgae are usually less than 2 mm in diameter and unicellular or colonial. Microalgae have been investigated for a variety of commercial applications. Annual global microalgal production is currently estimated at about 10 000 metric tons, with the main algae cultivated being *Spirulina* (accounting for roughly half of the worldwide algal production), *Chlorella*, *Dunaliella* and *Haematococcus*.

Algae have been investigated as a source of energy in many different contexts, from direct combustion to the production of hydrogen gas. Anaerobic digestion can be applied for the generation of methane or biogas (Golueke et al., 1957). Algal species with high oil content are particularly attractive as a feedstock for biodiesel production. Research into algae for the mass-production of oil has focused on the microalgae due to their high lipid content compared to macroalgae. Most algal species considered for biodiesel production are either green algae (Chlorophyta) or diatoms (Bacillariophyta) (Sheehan et al., 1998). They are generally photosynthetic, but several species are able to grow heterotrophically or mixotrophically (Barsanti & Gualtieri, 2006).

Microalgae have higher growth rates than land-based plants. Due to their simple cellular structure and existence in an aqueous environment, the entire cell surface is available for light capture and mass transfer, leading to high rates of substrate uptake and photosynthetic efficiency (Miao & Wu, 2006; Sheehan et al., 1998). In contrast to land-based oil crops, where only the seeds are harvested, each algal cell contains lipid and hence the yield of product from biomass is much higher (Becker, 1994). Due to these differences, the oil yield per area of microalgal cultures potentially exceeds that of the best oilseed crops (Table 1).

Oil source	Yield (L.m^{-2}.yr^{-1})	Reference
Algae	4.7 to 14	Sheehan et al., 1998
Palm	0.54	Mata et al., 2010
Jatropha	0.19	Sazdanoff, 2006
Rapeseed	0.12	Sazdanoff, 2006
Sunflower	0.09	Sazdanoff, 2006
Soya	0.04	Sazdanoff, 2006

Table 1. Average productivities of some common oil seed crops compared to algae

3. Biodiesel from microalgae

Microalgal lipids can be extracted to yield oil similar to that from land-based oilseed crops. The amount and composition of the oil varies between algal species. Algal oil can be converted to biodiesel through the same methods applied to vegetable oil. The idea of using microalgae as a source of transportation fuel is not new. Research in this field has been conducted since the 1950s (Oswald & Golueke, 1960). In the 1970s, several large, publicly funded research programs were set up in the USA, Australia and Japan (Regan & Gartside, 1983; Sheehan et al., 1998). The US Department of Energy invested more than US$ 25 million between 1978 and 1996 in the Aquatic Species Program to develop biodiesel production from algae (Sheehan et al., 1998). The main focus of the program was the production of biodiesel from high lipid-content algae grown in open ponds, utilizing waste CO_2 from coal fired power plants. Over 3000 species were collected and many of them screened for lipid content.

Early in the program, it was observed that environmental stress, particularly nutrient limitation (nitrogen for green algae and silicon for diatoms) led to an increase in accumulation of lipids. Promising species were investigated to determine the mechanism of this 'lipid trigger'. Researchers in the program were the first to isolate the enzyme Acetyl CoA Carboxylase from a diatom. This enzyme catalyzes the first committed step in the lipid synthesis pathway. Acetyl CoA Carboxylase was over-expressed successfully in algae; however, the anticipated increase in oil production was not demonstrated. The program close out report (Sheehan et al., 1998) concluded that, although algae used significantly less land and water than traditional crops, and sufficient resources did exist for algal fuel to completely replace conventional diesel, the high cost of microalgae production remained an obstacle. Even with the most optimistic lipid yields, production would only have become cost effective if petro-diesel had risen to twice its 1998 price.

The last decade has seen a renewal of interest in biofuels and microalgae as a feedstock source. An increase in oil prices, additional pressure to find alternatives to dwindling oil supplies and an urgent need to cut carbon emissions contributing to global warming has led to a renewed interest in algae as a source of energy, particularly lipid producing algae as a source of biodiesel.

4. Microalgal lipids

The main components of algae cells are proteins, carbohydrates and lipids (Becker, 1994). Microalgae naturally produce lipids as part of the structure of the cell (e.g. in cell membranes and as signalling molecules), and as a storage compound, similar to fat stores in animals and humans (Tsukahara & Sawayama, 2005). The term lipid encompasses a variety of compounds with different chemical structures (e.g. esters, waxes, cholesterol). The most common lipids are composed of a glycerol molecule bound to three fatty acids, known as triacylglycerol or TAG, or to two fatty acids with the third position taken up by a phosphate (phospholipids) or carbohydrate (glycolipids) group. Fatty acids consist of a long unbranched carbon chain. They are classified according to the number of carbon atoms in the chain and the number of double bonds, for example saturated (no double bonds), monounsaturated (one double bond) or polyunsaturated (more than one double bond). Microalgae commonly contain fatty acids ranging from C12 to C24, often with C16 and C18 unsaturates. Certain species contain significant amounts of polyunsaturated fatty acids.

Storage lipids, generally in the form of TAG, accumulate in lipid vesicles called oil bodies in the cytoplasm. Most fast-growing species have relatively low lipid content during normal growth, with these lipids mainly consisting of phospho- or glycolipids associated with cell membranes. Under certain conditions, generally triggered by stress or the cessation of growth, lipid content can increase to over 60% of cell dry weight (DW), mostly composed of

TAG (Shifrin & Chisholm, 1981; Piorreck et al., 1984; Spoehr & Milner, 1949; De la Pena, 2007; Becker, 1994).

TAGs are the most suitable class of lipids for biodiesel production. Phospholipids are particularly undesirable as they increase consumption of catalyst and act as emulsifiers, impeding phase separation during transesterification (Mittelbach & Remschmidt, 2004; Van Gerpen, 2005). Phospholipids, and some sulphur-containing glycolipids, also increase the phosphorous and sulphur content of the fuel respectively, which must both be below 10 mg.L^{-1} to meet the European biodiesel standard EN 14214. The type of fatty acids found in the oil can have a profound effect on the biodiesel quality. The fatty acid chain length and degree of saturation (determined by the number of double bonds) affects properties such as the viscosity, cold flow plug point, iodine number and cetane number of the fuel (Ramos et al., 2009). For biodiesel production, it is therefore important to maximize not only total lipid production, but also TAG content and appropriate fatty acid profile.

Lipid synthesis relies on carbon compounds generated from CO_2 by photosynthesis, as well as energy and reducing power (in the form of ATP and NAD(P)H respectively). The latter are produced during the light reactions of photosynthesis, while CO_2 uptake is mediated by the Calvin cycle during the dark reactions of photosynthesis. The output of the Calvin cycle is a three-carbon compound (glyceraldehyde 3-phosphate), which is converted through glycolysis into acetyl CoA. The conversion of acetyl CoA to malonyl CoA is the first committed step in lipid biosynthesis (Livne & Sukenik, 1992). Throughout metabolism there are a number of branch points at which metabolic intermediates are partitioned between the synthesis of lipids and other products such as carbohydrates and proteins (Lv et al., 2010). For example, acetyl CoA is a substrate for lipid synthesis as well as entry into the TCA cycle, which generates energy and biosynthetic precursors for proteins and nucleic acids. Both external and internal constraints, such as the availability of nutrients and the enzymatic reaction rates, limit the supply of metabolic intermediates. The production of storage lipids is particularly energy and resource intensive (Dennis et al., 1998; Roessler, 1990) and therefore usually occurs at conditions of reduced growth.

5. Cultivation of microalgae

The use of microalgae for energy generation requires large-scale, low-cost production. This demands cheap, scalable reactor design with efficient provision of the requirements for high algal productivity. Design considerations include optimum surface area to volume ratio for light provision, optimal mixing to keep cells in suspension and for distribution of nutrients, control over water balance and sterility, as well as maintenance of favorable temperature. A wide variety of reactor designs have been proposed, each with advantages and drawbacks.

5.1 Reactor systems

Microalgal production is a technology halfway between agriculture, which requires large areas for sunlight capture, and fermentation, which involves liquid culture of microorganisms (Becker, 1994). As light does not penetrate more than a few centimetres through a dense algal culture, scale-up is based on surface area rather than volume (Scott et al., 2010). Many different types of algal cultivation systems have been developed, but they can be divided into two main categories: open and closed.

Open systems consist of natural waters such as lakes, ponds and lagoons, or artificial ponds and containers that are open to the atmosphere. Most commercial production to date has taken place in open ponds as these systems are easy and cheap to construct (Pulz, 2001). The most common technical design is the raceway pond: an oblong, looped pond mixed by a paddlewheel, with water depths of 15 to 20 cm (Becker, 1994). Biomass concentrations of between 0.1 and 1 g.L-1 and biomass productivities of between 50 and 100 mg.L-1.day-1 are possible (Chisti, 2007; Pulz, 2001). The main advantages of open systems are their low cost and ease of construction and operation. They also offer the potential for integration with wastewater treatment processes or aquaculture systems (Chen, 1996).

Disadvantages of open systems include contamination with unwanted species such as foreign algae, yeast, bacteria and predators, evaporation of water, diffusion of CO_2 to the atmosphere and low control over environmental conditions, particularly temperature and solar irradiation (Becker, 1994; Pulz, 2001). In addition, the relatively low cell densities achieved can lead to higher cost of cell recovery (Chen, 1996). Only a few microalgal species have been successfully mass cultivated in open ponds. These tend to be either fast-growers that naturally outcompete contaminating algae (e.g. *Chlorella* and *Scenedesmus*), or species that grow in a specialised environment such as high salt (e.g. *Dunaliella salina*) or high pH (*Spirulina platensis*), which limits growth of competitors and predators (Chen, 1996). Due to the lack of control over cultivation conditions resulting in low productivity, and the fact that many desirable species cannot be effectively maintained in open systems, attempts have been made to overcome some of these limitations through the use of enclosed reactor systems.

Closed systems, or photobioreactors, consist of containers, tubes or clear plastic bags of various sizes, lengths and orientations (Pulz, 2001). Commonly used designs include vertical flat-plate reactors and tubular reactors, either pumped mechanically or by airlift (Scott et al., 2010). Closed reactors offer a much higher degree of control over process parameters, leading to improved heat and mass transfer, and thus higher biomass yields. They can also offer a much higher surface area to volume ratio for light provision, better control of gas transfer, reduction of evaporation and easier installation in any open space (Chen, 1996). Additionally, the risk of contamination is reduced, CO_2 can be contained, production conditions can be reproduced and temperature can be controlled.

Productivity in closed systems can be much higher than open systems, with biomass concentrations of up to 8 g.L-1 and productivities of between 800 and 1300 mg.L-1.day-1 (Pulz, 2001). However, they are generally much more costly to build and more energy demanding to operate than open systems (Table 2). Closed systems can also have problems with fouling and oxygen build-up. Large systems can be difficult to clean and sterilize and long sections of enclosed tubing may require oxygen purging. High oxygen concentrations cause the key enzyme Rubisco to bind oxygen instead of carbon dioxide, leading to photorespiration instead of photosynthesis (Dennis et al., 1998). Although closed bioreactors offer a much higher degree of control over process parameters and can have higher yields, it is uncertain whether the increased productivity can offset the higher cost and energy requirements. For a commodity product such as vegetable oil for biodiesel, low cost, high volume production is demanded, while quality is less critical (Pulz, 2001). In this case, the more favourable economics and energy requirements of open ponds may well outweigh the advantages of closed reactors.

A hybrid system combining the cost effectiveness of open ponds with the controlled environment of closed systems is appealing and has been tested in a few cases. Generally production is divided into an initial growth or inoculum production stage in closed reactors, followed by a stress or scaling up stage in open ponds (Huntley & Redalje, 2006).

Parameter	Open	Closed
Control over process parameters	Low	High
Contamination risk	High	Low
Water loss due to evaporation	High	Low
CO_2 loss	High	Low
O_2 build-up	Low	High
Area required	High	Low
Productivity	Low	High
Consistency and reproducibility	Low	High
Weather dependence	High	Low
Cost	Low	High
Energy required	Low	High

Table 2. Comparison of open ponds and closed photobioreactors. Adapted from Pulz (2001).

5.2 Cultivation parameters

Several factors need to be considered in the cultivation of algal biomass. These include the provision of light, carbon and nutrients such as nitrate, phosphate and trace metals, the mixing regime, maintenance of optimal temperature, removal of O_2 and control of pH and salinity (Becker, 1994; Grobbelaar, 2000; Mata et al., 2010). The optimal and tolerated ranges tend to be species specific, and may vary according to the desired product.

5.2.1 Temperature

Light and temperature are among the most difficult parameters to optimise in large-scale outdoor culture systems. Daily and annual fluctuations in temperature can lead to significant decreases in productivity. Optimal growth temperatures are generally between 20 and 30°C (Chisti, 2008). Many algal species can tolerate temperatures of up to 15°C lower than their optimum, with reduced growth rates, but a temperature of only a few degrees higher than optimal can lead to cell death (Mata et al., 2010). Closed systems in particular often suffer from overheating during hot days, when temperatures inside the reactor can reach in excess of 50°C. Heat exchangers or evaporative water-cooling systems may be employed to counteract this (Mata et al., 2010). Low seasonal and evening temperatures can also lead to significant losses in productivity.

5.2.2 Light and mixing

The efficient production of algal biomass relies on the optimal provision of light energy to all cells within the culture. Most algal growth systems become light limited at high cell densities. Due to absorption and shading by the cells, light only penetrates a few centimetres into a dense algal culture (Richmond, 2004). The average provision of light is linked to reactor depth or diameter, cell concentration and mixing. A larger surface area to volume ratio, usually achieved through areas of thin panelling or narrow tubing, results in higher light provision.

Photosynthetic efficiency is highest at low light intensities. At high light levels, although photosynthetic rate may be faster, there is less efficient use of absorbed light energy.

Above the saturation point, damage to photosynthetic machinery can occur in a process known as photoinhibition (Scott et al., 2010). In a dense culture exposed to direct sunlight, cells at the surface are likely to be photoinhibited, while those at the centre of the reactor are in the dark. Mixing is therefore important not only in preventing cell settling and improving mass transfer, but also exposing cells from within a dense culture to light at the surface.

The frequency of light-dark cycling has been reported to affect algal productivity (Grobbelaar, 1994; Grobbelaar, 2000). Algae are less likely to become photoinhibited when the light is supplied in short bursts because the photosystems have time to recover during the dark period (Nedbal et al., 1996). While high rates of mixing facilitate rapid circulation of cells between light and dark zones in the reactor, high liquid velocities can damage algal cells due to increased shear stress (Mata et al., 2010). High rates of mechanical mixing or gas sparging also have large energy requirements, jeopardizing the process energy balance and increasing costs (Richardson, 2011).

5.2.3 Gas exchange

In order to maintain a high photosynthetic rate, the influx of carbon and energy must be non-limiting. In photoautotrophic growth, energy is provided by light and carbon in the form of CO_2. In order to be taken up by cells, the CO_2 must dissolve in the water. The rate of dissolution is determined by the CO_2 concentration gradient as well as by the temperature, rate of gas sparging and surface area of contact between the liquid and gas (a function of agitation and bubble size). Reactor geometry, methods of gas introduction and reactor mixing can all influence the rate of CO_2 delivery (Bailey & Ollis, 1977). Certain strains of microalgae can tolerate up to 12% CO_2 (Pulz, 2001). The 0.03% CO_2 content of ambient air is suboptimal for photosynthesis (Pulz, 2001), hence for optimal microalgal growth, additional CO_2 must be provided. This is usually done by direct injection of a CO_2 enriched air stream. As the addition of CO_2 acidifies the medium, care must be taken not to adversely decrease the pH (Anderson, 2005). It is debatable whether direct gas injection is the optimal method of CO_2 delivery. Efficiencies of carbon uptake are very low at high CO_2 concentrations, as most CO_2 exits the top of the reactor. Novel strategies of CO_2 provision include microporous hollow fibre membranes and separate gas exchanger systems (Carvalho et al., 2006).

5.2.4 Salinity, nutrients and pH

The major nutrient requirements for microalgal growth are nitrogen and phosphorous, with certain diatoms, silicoflagellates and chrysophytes also requiring silicon (Anderson, 2005). Requirements of nutrients, pH and osmolarity are species dependent. Deviation from optimal levels may cause a decrease in biomass productivity, but can have other advantages, for example, high salinity may limit contamination. Sufficient supply of all essential nutrients is a prerequisite for efficient photosynthesis and growth, but limitation of key nutrients (e.g. nitrate, phosphate or silica) may cause accumulation of desired products such as lipid.

5.2.5 Nutritional mode

Most microalgae are photoautotrophs (utilizing sunlight as their source of energy and CO_2 as a carbon source). This is the most common growth mode employed in algal cultivation

(Chen, 1996). However, several species (e.g. *Chlorella*, *Chlamydomonas*, *Phaeodactylum*, *Nitzschia*, *Tetraselmis* and *Crypthecodinium*) are also capable of heterotrophic growth (utilizing organic carbon such as glucose, acetate or glycerol as the sole source or carbon and energy) or mixotrophic growth (photoautotrophic growth supplemented by an organic carbon source).

The advantages of using an organic carbon substrate are that it decreases dependence on light provision, allowing growth in conventional fermenters in the dark. Optimal growth conditions can be maintained, allowing higher cell concentrations and hence increased volumetric productivities to be reached (Chen, 1996). Higher productivities of both biomass and lipid have been reported under heterotrophic growth compared to autotrophic (Ceron Garcia, 2000; Miao & Wu, 2006). Disadvantages of feeding an organic carbon source include the fact that there are a limited number of algal species that can utilize organic carbon sources, the risk of bacterial contamination is greatly increased and the carbon substrate adds an additional cost, along with the environmental burden of its production. The use of a substrate such as glucose, commonly sourced from crop plants, adds a trophic level to the process, thereby removing the simplicity of the concept of microalgae as cellular factories producing liquid fuel from pure sunlight and CO_2.

5.2.6 Cultivation strategy

The optimal cultivation strategy (e.g. batch, fed-batch or continuous cultivation mode) is determined by the kinetics of growth, product accumulation and substrate uptake (Shuler & Kargi, 2005). For production of a primary product such as protein or biomass for food or feed, optimisation of biomass productivity is the main objective. In this case, batch or continuous systems are generally used. For production of a secondary product such as carotenoids or storage lipids, the use of two or more production stages to enhance yield has been proposed (Ben-Amotz, 1995; Huntley & Redalje, 2006; Richmond, 2004). The first stage is designed to optimize growth, while the second stage provides conditions that retard growth and encourage product synthesis, usually by applying some form of stress, e.g. nutrient deprivation in the case of lipid accumulation. Another potential two-stage strategy that could enhance lipid productivity is an initial photosynthetic stage, followed by a second heterotrophic phase, where feeding with an organic carbon source such as glucose may boost lipid content.

6. Harvesting and processing

The economic recovery of microalgal biomass remains a major challenge. Microalgae for biofuel are a low value product suspended in large volumes of water. Harvesting contributes 20 to 40% of the total cost of biomass production (Gudin & Therpenier, 1986; Molina Grima et al., 2003). The difficulty in separation can be attributed to the small size of the cells (3 to 300 μm, Henderson et al., 2008), their neutral buoyancy and the fact that photoautotrophic microalgal cultures are relatively dilute, achieving concentrations in the order of 1 to 8 g.L^{-1} (Pulz, 2001). Each algal species presents unique challenges due to the array of sizes, shapes, densities and cell surface properties encountered. A low-cost, energy efficient method with a high recovery efficiency and concentration is required, minimizing cell damage and allowing for water and nutrient recycle (Fig. 1).

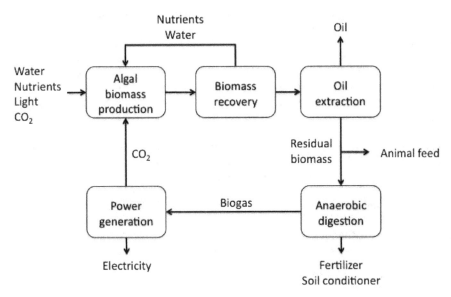

Fig. 1. Conceptual overview of microalgae process options (adapted from Chisti, 2008)

6.1 Factors affecting separation

Several natural properties of microalgal cells affect the choice and efficiency of harvesting methods. Factors relevant to separation include density, surface charge, size, shape, hydrophobicity, salinity of the medium, adhesion and cohesion properties and settling or floating velocities. Table 3 highlights the variability in some of these parameters between species, indicating that species-specific solutions may be required. Algal cell characteristics can vary with culture age and growth conditions. For example, changes in biochemical composition such as lipid content could affect the buoyancy of algal cells. Surface charge, the chemical structure of the cell wall and the amount and composition of the extracellular organic matrix (EOM) can vary with growth phase and greatly influence the degree to which cells repel or stick to one another (Bernhardt & Clasen, 1994; Henderson et al., 2008). Danquah et al. (2009) found a strong correlation between growth phase and settling efficiency, with improved filtration, flocculation and sedimentation rates during the stationary phase.

Species	Density (kg.m⁻³)	Zeta potential (mV)	Culturing pH	Morphology	Diameter, length (µm)
Microcystis	1200	-7.5 to -26	5.6 - 9.5	Globular sphere	3-7
Chlorella vulgaris	1070	-17.4	7	Single cell spherical	3.5
Cyclotella sp.	1140	-19.8 to -22.3	4 - 10	Chains of spheres	6.1
Syendra acus	1100	-30 to -40	7.6	Needles	4.5-6, 100-300

Table 3. Characteristics relevant to harvesting of some microalgae (Henderson et al., 2008). Zeta potential is a measure of the degree of repulsion between adjacent particles due to surface charge.

Morphological characteristics that influence harvesting include cell motility, size, shape, cell wall elongations such as spines and flagella, colony formation, and the presence of extracellular mucilage layers or capsules (Petrusevski et al., 1995; Jarvis et al., 2009). Larger particles allow for easier separation due to increased surface area and mass. Filamentous morphology or appendages also allow for easier filtration as cells cannot pass through filter pores. Density affects sedimentation and flotation. Most algae have a specific gravity close to that of water, rendering them with a neutral buoyancy. Some cyanobacteria can adjust their density through gas vacuoles (Anderson, 2005), rendering sedimentation difficult but enabling potential surface collection.

Cell surface charges influence the electrostatic interactions between cells and between cells and surfaces or bubbles. This directly affects the adhesion, adsorption, flotation and flocculation properties of algal cells. Most algae have a negatively charged cell surface, leading to electrostatic repulsion between cell walls. Addition of positively charged ions to the solution can help to neutralize the negative surface charge and aid cell flocculation. Changing the pH of the solution can also cause flocculation (Chen et al., 1998). Hydrophobicity is another, non-electric property affecting interaction of algal cells with each other and external surfaces. Most algae are naturally hydrophilic (Fattom & Shilo, 1984), but this can be altered by surfactants and pH (Jameson, 1999). Increasing the hydrophobicity of cells could cause them to adhere to bubbles, filters or other separation catalysts.

6.2 Harvesting methods

Harvesting requires one or more solid-liquid separation techniques (Molina Grima et al., 2003). In order to achieve the levels of concentration required, various chemical, biological and physical separation steps may be necessary. Common methods of cell harvesting include flocculation, filtration, sedimentation, centrifugation and flotation (Mata et al., 2010). The small cell size of microalgae makes them difficult to dewater. Flocculation is used to 'clump' the cells, grouping them together to form larger particle sizes. This is often suggested as a pretreatment step prior to filtration, sedimentation or flotation. Flocculation occurs when the repulsion between cells is reduced, allowing them to either aggregate directly onto each other or through an intermediate bridging surface. The extent of flocculation is dependent on pH, temperature, density, hydrophobicity, surface charge and culture age (Lee et al., 1998). Flocculation can be induced by addition of positively charged ions or polymers, e.g. minerals such as lime, calcium and salts, metal salts such as aluminium sulphate and ferric chloride, and naturally occurring flocculants such as starch derivatives and tannins. Drawbacks to the use of chemical flocculants are the high dosages required, the need for pH correction (Pushparaj et al., 1993) and the contamination of the biomass and media with the flocculant, meaning that media cannot be recycled without removal of the chemical. Autoflocculation can be induced through pH change (Csordas & Wang, 2004), nutrient limitation (Schenk et al., 2008), excretion of macromolecules (Benemann et al., 1980) or aggregation between microalgae and bacteria (Lee et al., 2009).

Conventional filtration is only effective for larger (> 70 μm) or filamentous species such as *Coelastrum* and *Spirulina* (Brennan & Owende, 2010, Lee et al., 2009). For smaller cells, micro-filtration, ultra-filtration and membrane-filtration can be used, though usually only for small volumes (Brennan & Owende, 2010, Petrusevski et al., 1995, Borowitzka, 1997). Fouling (accumulation of material on the surface of the membrane, slowing filtration) is a major problem. If filtration were to be considered for mass production, a high driving force for separation (high pressure or suction) would be required, which necessitates a high energy input. Microstraining (filtration by natural gravity using low speed rotating drum filters) is a promising method due to ease of operation and low energy consumption (Mohn, 1980). Another option is cross-flow membrane filtration (Zhang et al., 2010). Using a tangential, turbulent flow of liquid across the membrane prevents clogging of the filter with cells. The efficiency of the process is very dependent on cell morphology and the transmembrane pressure (Petrusevski et al., 1995). A more unconventional approach is magnetic filtration. Here addition of magnetic metals, either taken up by algal cells, or used to flocculate them, could allow capture using a magnetic field (Bitton et al., 1975).

Sedimentation is the process whereby solid particles suspended in a fluid are settled under the influence of gravity or some other force. In microalgae, it depends on coagulation or flocculation of cells to produce flocs with a large enough size (> 70 µm) or high enough density to induce settling (Vlaski et al., 1997). Sedimentation is typically used in wastewater treatment. It is suitable for large throughput volumes and has low operational costs. Flocculation, using a dense substance such as calcium carbonate, can greatly reduce settling time. Ultrasound (acoustic energy) can be used to induce aggregation and facilitate sedimentation (Bosma et al., 2003), however the energy requirement may be too high for large-scale use.

Centrifugation is essentially sedimentation under a rotational force rather than gravity. The efficiency of centrifugation depends on the size and density of the particles, the speed of the rotor, the time of centrifugation and the volume and density of the liquid. Almost all microalgae can be harvested by centrifugation. It is a highly efficient and reliable method, can separate a mixture of cells of different densities and does not require the addition of chemicals, but has a high energy consumption (Chisti, 2007). It is routinely used for recovery of high value products, or for small scale research operations, although large, flow-though centrifuges can be used to process large volumes. Many algae require speeds of up to 13 000 g which results in high shear forces (Harun et al., 2010; Knuckey et al., 2006) and can damage sensitive cells.

Flotation operates by passing bubbles through a solid-liquid mixture. The particles become attached to the bubble surface and are carried to the top of the liquid where they accumulate. The concentrated biomass can be skimmed off (Uduman et al., 2010). Flotation is considered to be faster and more efficient than sedimentation (Henderson et al., 2008). It is associated with low space requirements and moderate cost. Addition of chemical coagulants or flotation agents is often required to overcome the natural repulsion between the negatively charged algal particles and air bubbles. The pH and ionic strength of the medium are important factors to optimize this recovery technique.

6.3 Processing

After harvesting, the major challenge is in releasing the lipids from their intracellular location in the most energy efficient and economical way possible. Algal lipids must be separated from the rest of the biomass (carbohydrates, proteins, nucleic acids, pigments) and water. Common harvesting methods generally produce a slurry or paste containing between 5 and 25% solids (Shelef et al., 1984). Removing the rest of the water is thought to be one of the most expensive steps with literature values ranging from 20 to 75% of the total processing cost (Uduman et al., 2010, Molina Grima et al., 2003). Shelef et al. (1984) highlight a number of possible techniques for drying biomass: flash drying, rotary driers, toroidal driers, spray drying, freeze-drying and sun drying. Because of the high water content, sun-drying is not an effective method and spray-drying is not economically feasible for low value products (Mata et al., 2010). The selection of drying technique is dependent on the scale of operation, the speed required and the downstream extraction process (Mohn, 1980).

Lipid extraction can be done in a number of ways. Solvent extraction techniques are popular, but the cost and toxicity of the solvent (e.g. hexane) is of concern and solvent recovery requires significant energy input. Other methods involve disruption of the cell wall, usually by enzymatic, chemical or physical means (e.g. homogenization, bead milling, sonication (Mata et al., 2010)), allowing the released oil to float to the top of the solution. Ultrasound and microwave assisted extraction methods have been investigated (Cravotto et al., 2008). Supercritical CO_2 extraction is an efficient process, but is too expensive and energy intensive for anything but lab-scale production. Direct transesterification (production of biodiesel directly from algal biomass) is also possible. Some of these techniques do not require dry biomass, but the larger the water content of the algal slurry, the greater the energy and solvent input required.

Once the algal oil is extracted, it can be treated as conventional vegetable oil in biodiesel production. Direct pyrolysis, liquefaction or gasification of algal biomass have also been suggested as means of producing fuel molecules. One of the concerns for biodiesel production through transesterification, shared with any biodiesel feedstock, is the quality of the biodiesel produced. Biodiesel must meet certain international regulations, for example, the ASTM international standards or the EN14214 in Europe. It has been calculated that the fatty acid profile of certain microalgal species will produce biodiesel that does not meet these specification, therefore blending or additives may be required (Stansell, 2011).

7. Economic and environmental feasibility

In order to be economically feasible, microalgal biodiesel must be cost competitive with petroleum-based fuels. We have investigated the relationship between algal lipid productivity and cost in order to determine the range of productivities that need to be achieved for economic viability. Based on values from Chisti (2007), a model was set up to estimate cost per litre of algal oil as a function of algal biomass productivity and lipid content. Where the cost of producing a litre of algal biodiesel was below the price of a litre of fossil-fuel derived diesel, it was considered economically viable (i.e. no profit margin was introduced). The price of fossil-fuel derived diesel is partly dependent on the price of crude oil, which has varied widely in the last few years, hence several scenarios were evaluated.

Assumptions made in the execution of the model were:
1. Cost per kg algal biomass: US$ 0.6 for raceway ponds, and US$ 0.47 for photobioreactors (Chisti, 2007)
2. In order to be economically viable, the cost of algal oil per litre must be less than 6.9 x 10^{-3} times the cost of crude oil in US$ per barrel (Chisti, 2007)
3. Density of algal oil: 0.86 $g.cm^{-3}$ (Barsanti & Gualtieri, 2007)

The economic model was run for three prices of crude oil, based on fluctuations over the last few years. These scenarios of 'high' ($ 130), 'medium' ($ 90) and 'low' ($ 50) cost of crude oil per barrel gave the price limits for algal oil of 0.90, 0.62 and 0.35 US$ per L respectively. The results of the model are shown in Fig. 2a (raceway ponds) and 2b (closed photobioreactors).

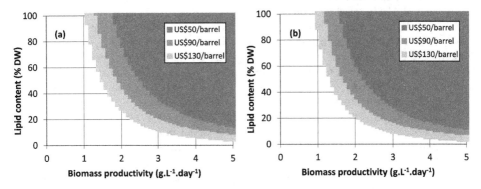

Fig. 2. Lipid contents and biomass productivities required for economic feasibility in (a) large-scale, outdoor raceway ponds and (b) large-scale, outdoor photobioreactors. Dark grey region: productivities economically feasible at US$ 50 per barrel crude oil (cost of algal oil per L lower than cost of regular diesel per L). Additional region for crude oil price US$ 90 per barrel = mid-grey and US$ 130 = light grey

Based on this model, the results for raceway ponds show that algal biodiesel will not be economically feasible, either in ponds or photobioreactors, at current costs below a biomass productivity of 1 g.L^{-1}.day^{-1}. Assuming a maximum realistically achievable lipid content of 50% DW, algal biodiesel becomes economically feasible at biomass productivities of 1.5 g.L^{-1}.day^{-1} (US$ 130 per barrel crude oil), close to 2 g.L^{-1}.day^{-1} (US$ 90), and 2.5 g.L^{-1}.day^{-1} (US$ 50) in raceway ponds. At lower lipid contents, higher biomass productivity is required, e.g. at a lipid content of 25% DW, algal biodiesel only becomes cost effective at 2 g.L^{-1}.day^{-1} for US$ 130 per barrel. The model for photobioreactors is based on a lower cost per kg algal biomass than raceway ponds, hence economic feasibility is reached at slightly lower biomass productivities and lipid contents, e.g. at a biomass productivity of 2 g.L^{-1}.day^{-1}, a lipid content of only 20% DW is required to be viable at US$ 130 per barrel crude oil.

Currently reported biomass productivities in outdoor raceway ponds average around 0.17 g.L^{-1}.day^{-1}, with a lipid content of 26% DW (Griffiths and Harrison, 2009), which is far from being economically feasible. Biomass productivities for closed photobioreactors (1.33 g.L^{-1}.day^{-1}) are closer to being within the economically viable range, if they can be maintained in the long term, concurrent with sufficiently high lipid content. As a reflection of this, there are currently no industrial facilities producing biodiesel from microalgae (Lardon, 2009). For cultivation to be economically viable, productivities must be increased, costs lowered, or additional income streams developed. The economics of algal biofuel production could be greatly improved through the production of co-products. For example, high value compounds such as pigments could be produced along with lipid. The residual biomass after lipid extraction could be sold as animal feed, fertilizer or soil conditioner, anaerobically digested to produce biogas, gasified or merely burned to provide some of the heat or electricity required in the process.

In addition to economic feasibility, algal biodiesel must be environmentally desirable. It is critical that the energy embodied in the fuel produced is greater than the energy input required to produce it. Net energy analysis and life cycle analysis (LCA) are tools used to quantify the environmental burdens at every stage of production, from growth of the biomass to combustion of the fuel. Lardon et al. (2009) conducted a life-cycle analysis of a hypothetical algal biodiesel production facility. Two different culture conditions: fertilizer feeding and nitrogen starvation, as well as two different extraction options: dry or wet, were investigated. The study confirmed the potential of microalgae as an energy source, but highlighted the necessity of decreasing energy and fertilizer consumption. Energy inputs, such as the energy required for mixing and pumping, the embodied energy in the materials used and the energy cost of harvesting and processing must be minimized. Recycling of material and energy from waste streams is also important wherever feasible (Scott et al., 2010). The use of nitrogen stress, as well as the optimization of wet extraction were indicated as desirable options. The anaerobic digestion of residual biomass was also suggested as a way of reducing external energy usage and recycling of nutrients.

We conducted a LCA on a hypothetical algal biodiesel process. Biomass production in three different reactor types (open ponds and two types of closed reactor: horizontal tubular and vertical tubular) was evaluated. In all cases, harvesting was modeled as an initial settling step followed by centrifugation. Hexane extraction was used to recover the oil, with the residual biomass sent for anaerobic digestion and the resulting energy from biogas production recycled to the process. The hexane was recovered and the oil converted to biodiesel using an enzymatic process. The basis chosen was production of 1000 kg of biodiesel from *Phaeodactylum tricornutum*. The net energy return (the energy embodied in the biodiesel produced divided by the energy input required) was positive (1.5) for the open pond, neutral (0.97) for the horizontal tubular reactor and negative (0.12) for the vertical tubular reactor. In this model, open ponds were the most energetically favorable reactor type, yielding 50% more energy than was put in. Horizontal tubular reactors required an energy input equivalent to the output, and vertical tubular reactors were the most unfavorable, requiring several times the energy input as that in the product, where system optimization was not conducted.

The overriding energy input in the process was found to be that required to run the reactor. Reactor energy was by far the most dominant determinant of the overall process energy requirement. This was largest in the vertical tubular reactor as these were continually mixed by gas sparging. Energy required for pumping between unit processes was also significant, particularly at lower biomass concentrations due to the larger volume of culture to be processed. The major energy inputs in downstream processing were that embodied in the lime used as a flocculation agent, and the energy required for solvent recovery. Lipid productivity and species choice had a significant impact on the energy balance.

8. Optimizing lipid productivity

Increasing microalgal lipid productivity improves both the economics and energy balance of the process. The land area and size of culture vessels required, as well as the energy and water requirements for large-scale algal culture are strongly dependent on algal productivity. With a higher productivity, lower cultivation, mixing, pumping and harvesting volumes would be required to yield the same amount of product, resulting in lower cost and energy requirements. More concentrated cell suspensions could also make downstream processing more efficient. The genetic characteristics of an algal species determine the range of its productivity. The levels reached in practice within this range are determined by the culture conditions. The two main approaches to enhancing productivity are: 1. selection of highly productive algal species and 2. designing and maintaining optimal conditions for productivity.

The choice of algal strain is a key consideration. The diversity of algal species is much greater than that of land plants (Scott et al., 2010) allowing selection of species best suited to the local environment and goals of the project. Although there have been several screening programs, building on the work of the Aquatic Species Program (Sheehan et al., 1998), the majority of strains remain untested, few species have been studied in depth and the data reported in the literature is often not comparable due to the different experimental procedures used. We conducted a broad literature review of the growth rates and lipid contents of 55 promising microalgal species under both nutrient replete and limited conditions. The original study (Griffiths & Harrison, 2009) has been extended here through the use of two key assumptions to convert data into common units of biomass and lipid productivity.

Lipid productivity is determined by both growth rate and lipid content. Lipid content (P) was typically reported as percentage dry weight (% DW). Data presented in pg lipid.cell^{-1} was discarded if no cell weight was available for conversion. Growth rates were reported as doubling time (T_d) or specific growth rate (μ). These were inter-converted according to Equation 1.

$$T_d = \frac{\ln 2}{\mu} \tag{1}$$

Standard units of g.L^{-1}.day^{-1} were chosen for biomass productivity. Specific growth rate (μ, in units of day^{-1}) can be converted to volumetric biomass productivity (Q_V, in g.L^{-1}.day^{-1}) where the biomass concentration (X, in g.L^{-1}) is known (Equation 2). Biomass productivity is often reported on the basis of surface area (Q_A), in units of g.m^{-2}.day^{-1}. This can be converted to Q_V using Equation 3 where the depth (D, in m) of the culture vessel can be calculated from the reactor geometry.

$$Q_V = \mu \times X \tag{2}$$

$$Q_V = \frac{Q_A}{D \times 1000} \tag{3}$$

Lipid productivity (Q_P) was infrequently reported in the literature, and was generally reported in g.L^{-1}.day^{-1} or mg.L^{-1}.day^{-1}. This parameter could be calculated from volumetric biomass productivity (Q_V, in g.L^{-1}.day^{-1}) and lipid content (P in % DW) where appropriate data were available (Equation 4).

$$Q_P = Q_V \times P \tag{4}$$

The calculation of lipid productivity for the majority of species necessitated two assumptions:
1. Conversion of areal productivities (in g.m^{-2}.day^{-1}) to volumetric productivities (g.L^{-1}.day^{-1}), using an average depth of 0.1 m, based on best fit of the data
2. Conversion of specific growth rate to biomass productivity using an average biomass concentration of 0.15 g.L^{-1}, based on typical experimental results.

The average literature values for the 55 species are shown in Table 4. Among the species with the highest reported lipid productivity were *Neochloris oleoabundans*, *Navicula pelliculosa*, *Amphora*, *Cylindrotheca* and *Chlorella sorokiniana* (Fig. 3). Other findings were that green algae (Chlorophyta) generally showed an increase in lipid content when nitrogen deficient, whereas

Species	Taxa[a]	Media[b]	Lipid content N replete (% dw)	Lipid content N defficient (% dw)	T_d (days)	Q_A (g.m^{-2}.day^{-1})	Q_V (g.L^{-1}.day^{-1})	Ave Q_V (g.L^{-1}.day^{-1})	Lipid productivity Calculated (mg.L^{-1}.day^{-1})	Lipid productivity Literature (mg.L^{-1}.day^{-1})
Amphiprora hyalina	B	M	22	28	0.41			0.30	67	
Amphora	B	M	51		0.83	40.0		0.23	117	160
Anabaena cylindrica	Cy	F	5	5	1.00			0.10	5	
Ankistrodesmus falcatus	C	F	24	32	0.33	31.6	0.46	0.36	85	
Chaetoceros calcitrans	O	M	40				0.04	0.04	16	18
Chaetoceros muelleri	O	M	19	27	0.46		0.07	0.26	50	22
Chlamydomonas applanata	C	F	18	33						
Chlamydomonas reinhardtii	C	F	21		0.26			0.40	83	
Chlorella emersonii	C	F	29	63	0.80		0.03	0.08	23	
Chlorella minutissima	C	M	31	57	1.60		0.03	0.05	15	
Chlorella protothecoides	C	F	13	23	1.68			0.07	8	
Chlorella pyrenoidosa	C	F	16	64	0.28			0.47	76	
Chlorella sorokiniana	C	F	18	18	0.35		0.55	0.62	110	45
Chlorella vulgaris	C	F	24	42	0.70	10.7	0.11	0.16	40	30
Crypthecodinium cohnii	D	M	25		0.38			0.28	70	
Cyclotella cryptica	O	M	18	34	0.56			0.20	36	
Cylindrotheca	B	M	27	27	0.30			0.43	114	
Dunaliella primolecta	Pr	S	23	14		9.1		0.09	21	
Dunaliella salina	Pr	S	19	10	0.44			0.27	53	
Dunaliella tertiolecta	Pr	S	15	18	0.48			0.22	35	
Euglena gracilis	Eg	F	20	35	0.60			0.18	37	
Hymenomonas carterae	H	M	20	14	1.71			0.06	12	
Isochrysis galbana	H	M	25	29	0.89	11.5	0.16	0.15	37	38
Monodopsis subterranea	E	F	25	13			0.19	0.19	48	30
Monoraphidium minutum	C	F	22	52	0.35			0.30	65	
Nannochloris	C	M/F	28	30	0.49	31.9	0.23	0.27	74	77
Nannochloropsis	E	M	31	41	1.20		0.27	0.24	72	52
Nannochloropsis salina	E	M	27	46		13.9		0.14	38	
Navicula acceptata	B	F	33	35	0.42			0.29	96	
Navicula pelliculosa	B	F	27	45	0.23			0.46	124	
Navicula saprophila	B	F	24	51	0.38			0.28	68	
Neochloris oleoabundans	C	F	36	42			0.46	0.46	164	136
Nitzschia communis	B	M			0.96			0.18		
Nitzschia dissipata	B	M	28	46	0.39			0.27	73	
Nitzschia frustulum	B	M	26							
Nitzschia palea	B	M	47	40						48
Oscillatoria	Cy	F	7	13	0.28			0.37	27	
Ourococcus	C	F	27	50	3.01			0.03	9	
Pavlova lutheri	H	M	36				0.21	0.21	75	50
Pavlova salina	H	M	31				0.16	0.16	49	49
Phaeodactylum tricornutum	B	M	21	26	1.02	20.0	0.34	0.18	38	45
Porphyridium purpureum	R	M	11		0.45		0.23	0.23	24	35
Prymnesium parvum	H	M	30		0.74			0.14	42	
Scenedesmus dimorphus	C	F	26		0.46			0.23	57	
Scenedesmus obliquus	C	F	21	42	2.74		0.12	0.10	22	
Scenedesmus quadricauda	C	F	18				0.19	0.19	35	35
Selenastrum gracile	C	F	21	28						
Skeletonema costatum	O	M	16	25	0.66		0.08	0.15	24	17
Spirulina maxima	Cy	S	7		1.34			0.16	11	
Spirulina platensis	Cy	S	13	10	0.60	25.0		0.23	29	
Synechococcus	Cy	M	11		0.36			0.29	32	75
Tetraselmis suecica	P	M	17	26	1.51	28.1	0.59	0.39	65	32
Thalassiosira pseudonana	O	M	16	26	0.49		0.08	0.26	43	17
Thalassiosira weissflogii	O	M	22	24	0.58			0.18	41	
Tribonema	O	M	12	16	1.82		0.51	0.33	33	
Total (Average)			23	32	0.80	22.2	0.23	0.23	52	51
Freshwater			21	36	0.82	21.2	0.24	0.26	54	35
Marine			25	31	0.82	22.7	0.21	0.21	49	47
Chlorophyta			23	41	1.01	24.7	0.24	0.25	58	65
Other taxa			25	30	0.72	20.4	0.24	0.23	57	50
Cyanobacteria			8	9	0.72	25.0		0.23	21	75

[a] Key to taxa: C = Chlorophyta, Cy = Cyanobacteria, D = Dinophyta, E = Eustigmatophyta, Eg = Euglenozoa, H = Haptophyta, O = Ochrophyta, Pr = Prasinophyta, [b] Key to media: F = Freshwater, M = Marine, S = Saline

Table 4. Growth and lipid parameters of 55 species of microalgae, along with their taxonomy and media type (adapted from Griffiths and Harrison, 2009). The average of literature values for lipid content under nitrogen (N) replete and deficient growth conditions, doubling time (T_d), and areal (Q_A) and volumetric (Q_V) biomass productivities are shown in columns 4 to 8. Average biomass productivity calculated from T_d, μ, Q_A and Q_V is shown in column 9, and calculated and literature lipid productivity in columns 10 and 11 respectively. Blanks represent no data available

diatoms and other taxa were more variable in their response, although all those subjected to silicon deprivation showed an increase in lipid content. This increase in lipid content, however, does not necessarily translate into increased lipid productivity due to decreased growth rates under nutrient stress conditions. Response of biomass productivity to nutrient deprivation is variable between species and further investigation is necessary.

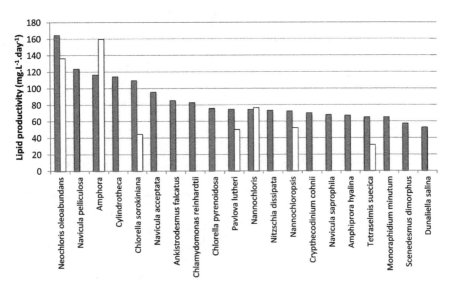

Fig. 3. Average calculated (grey bars) and literature (empty bars) biomass productivity for the 20 most productive species investigated (adapted from Griffiths & Harrison, 2009)

In Fig. 4, the impact of biomass productivity and lipid content on calculated lipid productivity is analyzed through correlation. A relationship is demonstrated between lipid productivity and biomass productivity. All species with a high biomass productivity (above 0.4 g.L^{-1}.day^{-1}), and all but one above 0.3 g.L^{-1}.day^{-1}, have a high lipid productivity, greater than 60 mg.L^{-1}.day^{-1}. However, there are a few species with high lipid productivity despite an average biomass productivity, indicating that lipid content is also a factor. Lipid content correlates poorly with lipid productivity, indicating that lipid content alone is not a good indicator of suitability for biodiesel production. There are several species with low lipid productivity despite an above-average lipid content (> 22%). The species with high lipid productivities (> 60 mg.L^{-1}.day^{-1}) range in lipid content from 16% DW to 51%. Further, species with high lipid content (> 30%) vary in lipid productivity between 15 and 164 mg.L^{-1}.day^{-1}.

Once the species has been chosen, the next critical factor is the optimisation of culture conditions. In addition to optimal temperature and pH, conditions that maximize autotrophic growth rate are optimal light, carbon and nutrient supply. Microalgal lipid accumulation is affected by a number of environmental factors (Guschina & Harwood 2006; Roessler 1990), and often enhanced by conditions that apply a 'stress' to the cells. Lipids appear to be synthesised in response to conditions when energy input (rate of photosynthesis) exceeds the capacity for energy use (cell growth and division) (Roessler 1990). Enhanced cell lipid content has been found under conditions of nutrient deprivation (Hsieh & Wu, 2009; Illman et al., 2000; Li et al., 2008; Shifrin & Chisholm, 1981; Takagi et al.,

2000), high light intensity (Rodolfi et al., 2009), high temperature (Converti et al., 2009); high salt concentration (Takagi et al., 2000) and high iron concentration (Liu et al., 2008).

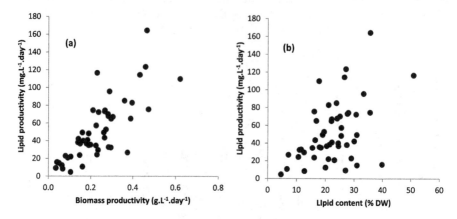

Fig. 4. Correlation of calculated lipid productivity with (a) biomass productivity and (b) lipid content under nutrient replete conditions

Nitrogen (N) deprivation is the most frequently reported method of enhancing lipid content, as it is cheap and easy to manipulate. N deficiency has a reliable and strong influence on lipid content in many species (Chelf, 1990; Rodolfi et al., 2009; Shifrin & Chisholm, 1981). Unfortunately, stress conditions that enhance lipid content, such as nitrogen deprivation, typically also decrease the growth rate, and thus the net effect on lipid productivity must be ascertained (Lardon et al., 2009). Maximum biomass productivity and lipid content in *Chlorella vulgaris* occur under different conditions of nitrogen availability, suggesting that a two-stage cultivation strategy may be advantageous. From studies we have conducted on *C. vulgaris*, it appears that an intermediate level of nitrogen limitation creates the optimum balance between biomass and lipid production. The optimum cultivation strategy tested was batch culture, using a low starting nitrate concentration (between 250 and 300 mg.L^{-1} nitrate), ensuring that nitrogen in the medium was depleted towards the end of exponential growth. Other cultivation strategies (e.g. two-stage batch, fed-batch or continuous) were found not to improve upon the productivity achieved in N limited batch culture.

Although high lipid productivity is a key factor in species selection, other characteristics such as ease of cultivation, tolerance of a range of environmental conditions (particularly temperature and salinity), flue-gas contaminants and high O_2 concentrations, as well as resistance to contaminants and predators are likely to be equally as important.

9. Conclusion and future research directions

Algal biodiesel continues to hold promise as a sustainable, carbon neutral source of transportation fuel. The technical feasibility of algal biodiesel has been demonstrated (Miao & Wu, 2006; Xiong et al., 2008), but the economics and energy demands of production require substantial improvement. The necessary changes appear attainable through the enhancement of productivity, the reduction of cost and energy demand for key processes and the application of the biorefinery concept (co-production of valuable products or processes). Current research is focussed on achieving this through a combination of biological and engineering approaches. The major challenges currently being addressed are:

- Increasing productivity in large-scale outdoor microalgal culture
- Minimizing contamination by predators and other algal species
- Mitigating temperature changes and water loss due to evaporation
- Optimizing supply of light and CO_2
- Developing cheap and efficient reactor designs
- Developing cost and energy-efficient methods of harvesting dilute suspensions of small microalgal cells
- Decreasing the overall energy and cost requirements, particularly for pumping, gas transfer, mixing, harvesting and dewatering
- Improving resource utilization and productivity through a biorefinery approach
- Producing valuable co-products
- Decreasing environmental footprint through recycling of water, energy and nutrients.

These topics have captured the imagination of several researchers and some innovative solutions are being investigated. The overall goal of biofuel production is to optimise the conversion of sunlight energy to liquid fuel. In algal cultivation, techniques to improve light delivery include manipulating the reactor design, the use of optics to deliver light to the centre of the reactor, optimising fluid dynamics to expose all cells to frequent light flashes, increasing the efficiency of photosynthesis and carbon capture (e.g. enhancing the carbon concentrating mechanism), and using mixed-species cultures to utilise different intensities or wavelengths of light (Scott et al., 2010).

One of the major problems with light delivery is poor penetration of light into dense cultures due to mutual shading by the cells. Under high light conditions, microalgal cells absorb more light than they can use, shading those below them and dissipating the excess energy as fluorescence or heat. In nature, this confers individual cells an evolutionary advantage, however, in mass production systems it is undesirable as it decreases overall productivity. It would be advantageous to minimize the size of the chlorophyll antennae in cells at the surface, so as to permit greater light penetration to cells beneath (Melis, 2009). Reducing the size of the light harvesting complexes through genetic modification has been shown to improve productivity (Nakajima et al., 2001). The goal now is to engineer cells that change antennae size according to light intensity.

Although the TAG content of cells can be enhanced by manipulation of the nutrient supply, there is a tradeoff between growth and lipid production. For optimum productivity, cells that can maintain a simultaneously high growth rate and lipid content are required. Strategies to achieve this include screening for novel species, and genetic engineering of well characterised strains. The genes and proteins involved in regulation of lipid production pathways are currently being investigated through synthetic biology and the modelling of carbon flux through metabolism. Key enzymes and branch-points can then be manipulated to improve productivity. For example, carbohydrate and lipid production compete directly for carbon precursors. Shunting carbon away from starch synthesis by downregulation of the enzyme ADP-glucose pyrophosphorylase in *Chlamydomonas* has been shown to enhance TAG content 10-fold (Li et al., 2010).

The challenge of harvesting small algae cells from dilute suspensions has yet to be solved in a cheap, energy efficient manner. Ideally the addition of chemical agents that impede the recycling of the culture medium and nutrients should be avoided. A series of methods is likely to be used e.g. flocculation followed by sedimentation, or settling followed by centrifugation. Promising ideas for harvesting techniques include concentration using sound waves and triggering of autoflocculation on command. Another attractive idea is direct product excretion, where algae secrete fuel molecules into the medium as they are produced, allowing continuous production and harvesting without cell disruption. The Cyanobacterium *Synechocystis* has recently been successfully modified to excrete fatty acids (Liu, 2011).

The use of nutrients from waste sources (e.g. CO_2 from flue-gas and nitrate and phosphate from wastewater) could help to reduce costs and energy input, as well as contributing to environmental remediation. Potential co-products include fine chemicals such as astaxanthin, B-carotene, omega-3 fatty acids, polyunsaturated fatty acids, neutraceuticals, therapeutic proteins, cosmetics, aquafeed and animal feed (Mata et al., 2010). Algae could also potentially be modified to synthesize other types of fuel e.g. ethanol, butanol, isopropanol and hydrocarbons (Radakovits et al., 2010) or downstream processing of algae could be modified to process the entire biomass to energy containing fuels through thermal processes.

10. Acknowledgements

This work is based upon research supported by the South African National Energy Research Institute (SANERI), the South African Research Chairs Initiative (SARChI) of the Department of Science and Technology, the Technology Innovation Agency (TIA) and the National Research Foundation (NRF). The financial assistance of these organizations is hereby acknowledged. Opinions expressed and conclusions arrived at are those of the authors and are not necessarily to be attributed to SANERI, SARChI, TIA or the NRF.

11. References

An, J-Y, Sim, S-J, Lee JS & Kim BW. (2003). Hydrocarbon production from secondarily treated piggery wastewater by the green alga *Botryococcus braunii*. *Journal of Applied Phycology*, 15(2-3): 185-191

Anderson, RA. (2005). *Algal Culturing Techniques*. Elsevier, London

Bailey, J & Ollis, D. (1977). *Biochemical Engineering Fundamentals*. McGraw-Hill, New York

Barsanti, L & Gualtieri, P. (2006). *Algae: Anatomy, Biochemisty and Biotechnology*. CRC Press, Taylor and Francis Group, Florida

Becker, EW. (1994). *Microalgae: Biotechnology and Microbiology*. Cambridge University Press, Cambridge

Ben-Amotz, A. (1995). New Mode of *Dunaliella* Biotechnology: Two-Phase Growth for B-Carotene Productio. *Journal of Applied Phycology*, 7(1):65-68

Benemann, J, Koopman, B, Weissman, J, Eisenberg, D & Gobell R. (1980). Development of Microalgae Wastewater Treatment and Harvesting Technologies in California. In *Algae Biomass: Production and Use*. Edited by Shelef, G & Soeder, C. pp. 457-495 . Elsevier, Amsterdam

Bernhardt, H & Clasen, J. (1994). Investigations into the Flocculation Mechanisms of Small Algal Cells. *Journal of Water Supply: Research and Technology – AQUA*, 43(5): 222-232

Bitton, G, Fox, JL & Strickland, HG. (1975). Removal of Algae from Florida Lakes by Magnetic Filtration. *Applied and Environmental Microbiology*, 30(6):905-908

Borowitzka, MA. (1992). Algal Biotechnology Products and Processes - Matching Science and Economics. *Journal of Applied Phycology*, 4:267-279

Borowitzka, MA. (1997). Microalgae for Aquaculture - Opportunities and Constraints. *Journal of Applied Phycology*, 9(5):393-401

Bosma, R, van Spronsen, WA, Tramper, J & Wijffels, RH. (2003). Ultrasound, a New Separation Technique to Harvest Microalgae. *Journal of Applied Phycology*, 15(2-3):143-153

Brennan, L & Owende, P. (2010). Biofuels from microalgae: A Review of Technologies for Production, Processing, and Extractions of Biofuels and Co-Products. *Renewable and Sustainable Energy Reviews*, 14(2):557-577

Carvalho, AP, Meireles, LA & Malcata, FX. (2006). Microalgal Reactors: A Review of Enclosed System Designs and Performances. *Biotechnology Progress*, 22(6):1490-1506

Ceron Garcia, MC, Fernandez Sevilla, JM, Acien Fernandez, FG, Molina Grima, E & Garcia Camacho, F. (2000). Mixotrophic Growth of *Phaeodactylum tricornutum* on Glycerol: Growth Rate and Fatty Acid Profile. *Journal of Applied Phycology*, 12:239-248

Chelf, P. (1990). Environmental Control of Lipid and Biomass Production in Two Diatom Species. *Journal of Applied Phycology*, 2(2):121-129

Chen, F. (1996). High Cell Density Culture of Microalgae in Heterotrophic Growth. *Trends in Biotechnology*, 14(11):421-426

Chen, YM, Liu, JC & Ju, Y. (1998). Flotation Removal of Algae from Water. *Colloids and Surfaces B: Biointerfaces*, 12(1):49-55

Chisti, Y. (2007). Biodiesel from Microalgae. *Biotechnology Advances*, 26(3):294-306

Chisti, Y. (2008). Biodiesel from Microalgae Beats Bioethanol. *Trends in Biotechnology*, 25(3):126-31

Converti, A, Casazza, A, Ortiz, EY, Perego, P & Del Borghi, M. (2009). Effect of Temperature and Nitrogen Concentration on the Growth and Lipid Content of Nannochloropsis Oculata and *Chlorella vulgaris* for Biodiesel Production. *Chemical Engineering and Processing: Process Intensification*, 48(6):1146-1151

Csordas, A. (2004). An Integrated Photobioreactor and Foam Fractionation Unit for the Growth and Harvest of *Chaetoceros* spp. in Open Systems. *Aquaculture Engineering*, 30(1-2):15-30

Danquah, MK, Gladman, B, Moheimani, N & Forde, GM. (2009). Microalgal Growth Characteristics and Subsequent Influence on Dewatering Efficiency. *Chemical Engineering Journal*, 151(1-3):73-78

De la Pena, M. (2007). Cell Growth and Nutritive Value of the Tropical Benthic Diatom, *Amphora* sp., at Varying Levels of Nutrients and Light Intensity, and Different Culture Locations. *Journal of Applied Phycology*, 19(6):647-655

Dennis, DT, Turpin, DH, Lefebvre, DD & Layzell, DB. (1998). *Plant metabolism*. Longman, Singapore

Fattom, A & Shilo, M. (1984). Hydrophobicity as an Adhesion Mechanism of Benthic Cyanobacteria. *Applied and Environmental Microbiology*, 47(1):135-143

Golueke, CG, Oswald, WJ & Gotaas, HB. (1957). Anaerobic Digestion of Algae. *Applied Microbiology*, 5(1):47-55

Griffiths, MJ & Harrison, STL. (2009). Lipid Productivity as a Key Characteristic for Choosing Algal Species for Biodiesel Production. *Journal of Applied Phycology*, 21(5):493-507

Grobbelaar, JU. (1994). Turbulence in Mass Algal Cultures and the Role of light/dark Fluctuations. *Journal of Applied Phycology*, 6(3):331-335

Grobbelaar, JU. (2000). Physiological and Technological Considerations for Optimising Mass Algal Cultures. *Journal of Applied Phycology*, 12:201-206

Gudin, C & Therpenier, C. (1986). Bioconversion of Solar Energy into Organic Chemicals by Microalgae. *Advanced Biotechnology Processes*, 6:73-110

Harun, R, Singh, M, Forde, GM & Danquah, MK. (2010). Bioprocess Engineering of Microalgae to Produce a Variety of Consumer Products. *Renewable and Sustainable Energy Reviews* 14(3):1037-1047

Henderson, R, Parsons, SA & Jefferson, B. (2008). The Impact of Algal Properties and Pre-Oxidation on Solid/Liquid Separation of Algae. *Water Research*, 42:1827-1845

Hodaifa, G, Martínez, M & Sánchez, S. (2008). Use of industrial wastewater from olive-oil extraction for biomass production of *Scenedesmus obliquus*. *Bioresource Technology*, 99(5):1111-1117

Hsieh, C & Wu, W. (2009). Cultivation of Microalgae for Oil Production with a Cultivation Strategy of Urea Limitation. *Bioresource Technology*, 100(17):3921-3926

Huntley, ME & Redalje, DG. (2006). CO$_2$ Mitigation and Renewable Oil from Photosynthetic Microbes: A New Appraisal. *Mitigation and Adaptation Strategies for Global Change*, 12(4):573-608

Illman, A, Scragg, AH & Shales, SW. (2000). Increase in *Chlorella* Strains Calorific Values when Grown in Low Nitrogen Medium. *Enzyme MicrobialTechnology*, 27(8):631-635

Lee, SJ, Kim, S, Kim, J, Kwon, G, Yoon, B & Oh, H. (1998). Effects of Harvesting Method and Growth Stage on the Flocculation of the Green Alga *Botryococcus braunii*. *Letters in Applied Microbiology*, 27(1):14-18

Jameson, GJ. (1999). Hydrophobicity and Floc Density in Induced-Air Flotation for Water Treatment. *Colloids and Surfaces A: Physicochemical and engineering aspects*, 151:269-281

Jarvis, P, Buckingham, P, Holden, B & Jefferson, B. (2009). Low Energy Ballasted Flotation. *Water Research*, 43:3427-3434

Knuckey, R, Brown, M, Robert, R & Frampton, D. (2006). Production of Microalgal Concentrates by Flocculation and their Assessment as Aquaculture Feeds. *Aquaculture Engineering*, 35(3):300-313

Lardon, L, Heìlias, A, Sialve, B, Steyer, JP & Bernard, O. (2009). Life-Cycle Assessment of Biodiesel Production from Microalgae. *Environmental science & technology*, 43(17):6475-6481

Lee, AK, Lewis, DM & Ashman, PJ. (2008). Microbial Flocculation, a Potentially Low-Cost Harvesting Technique for Marine Microalgae for the Production of Biodiesel. *Journal of Applied Phycology*, 21(5):559-567

Li, Y, Han, D, Hu, G, Dauvillee, D, Sommerfeld, M, Ball, S & Hu, Q. (2010). Chlamydomonas Starchless Mutant Defective in ADP-Glucose Pyrophosphorylase Hyper-Accumulates Triacylglycerol. *Metabolic Engineering*, 12(4):387-91

Li, Y, Horsman, M, Wang, B, Wu, N & Lan, CQ. (2008). Effects of Nitrogen Sources on Cell Growth and Lipid Accumulation of Green Alga Neochloris Oleoabundans. *Applied microbiology and biotechnology*, 81(4):629-36

Liu, X, Sheng, J & Curtiss, R. (2011). Fatty Acid Production in Genetically Modified Cyanobacteria. *Proceedings of the National Academy of Sciences of the United States of America*, 108(17):6899-904

Liu, Z, Wang, G & Zhou, B. (2008). Effect of Iron on Growth and Lipid Accumulation in *Chlorella vulgaris*. *Bioresource Technology*, 99(11):4717-22

Livne, A & Sukenik, A. (1992). Lipid Synthesis and Abundance of Acetyl CoA Carboxylase in *Isochrysis galbana* (Prymnesiophyceae) Following Nitrogen Starvation. *Plant and Cell Physiology*, 33(8):1175-1181

Lv, J, Cheng, L, Xu, X, Zhang, L & Chen, H. (2010). Enhanced Lipid Production of *Chlorella vulgaris* by Adjustment of Cultivation Conditions. *Bioresource Technology*, 101(17):6797-804.

Mata, TM, Martins, AA & Caetano, NS. (2010). Microalgae for Biodiesel Production and Other Applications: A Review. *Renewable and Sustainable Energy Reviews*, 14(1):217-232

Melis, A. (2009). Solar Energy Conversion Efficiencies in Photosynthesis: Minimizing the Chlorophyll Antennae to Maximize Efficiency. *Plant Science*, 177(4):272-280

Miao, X & Wu, Q. (2006). Biodiesel Production from Heterotrophic Microalgal Oil. *Bioresource Technology*, 97(6):841-846

Mittelbach, M & Remschmidt, C. (2004). *Biodiesel - The Comprehensive Handbook*. Martin Mittelbach

Mohn, FH. (1980). Experiences and Strategies in the Recovery of Biomass from Mass Cultures of Microalgae. In *Algae biomass: Production and Use*. Edited by Shelef, G & Soeder, C. pp. 547-571. Elsevier, Amsterdam

Molina Grima, E, Belarbi, E, Acien Fernandez, FG, Robles Medina, A & Chisti, Y. (2003). Recovery of Microalgal Biomass and Metabolites: Process Options and Economics. *Biotechnology Advances*, 20(7-8):491-515

Nakajima, Y, Tsuzuki, M &Ueda, R. (2001). Improved Productivity by Reduction of the Content of Light-Harvesting Pigment in *Chlamydomonas perigranulata*. *Journal of Applied Phycology*, 13(2):95-101

Nedbal, L, Tichy,V, Xiong, F & Grobbelaar, JU. (1996). Microscopic Green Algae and Cyanobacteria in High-Frequency Intermittent Light. *Journal of Applied Phycology*, 8(4-5):325-333

Oswald, WJ & Golueke, C. (1960). Biological Transformation of Solar Energy. *Advanced Applied Microbiology*, 2:223-262

Petrusevski, B, Bolier, G, van Breemen, A & Alaerts, GJ. (1995). Tangential Flow Filtration: A Method to Concentrate Freshwater Algae. *Water Research*, 29:1419-1424

Piorreck, M, Baasch, K & Pohl, P. (1984). Biomass Production, Total Protein, Chlorophylls, Lipids and Fatty Acids of Freshwater Green and Blue-Green Algae Under Different Nitrogen Regimes. *Phytochemistry*, 23(2):207-216

Pulz, O. (2001). Photobioreactors: Production Systems for Phototrophic Microorganisms. *Applied Microbiology and Biotechnology*, 57(3):287-293

Pushparaj, B, Pelosi, E, Torzillo, G & Materassi, R. (1993). Microbial Biomass Recovery using a Synthetic Cationic Polymer. *Bioresource Technology*, 43(1):59-62

Radakovits, R, Jinkerson, RE, Darzins, A & Posewitz, MC. (2010). Genetic Engineering of Algae for Enhanced Biofuel Production. *Eukaryotic Cell*, 9(4):486-501

Ramos, MJ, Fernandez, CM, Casas, A, Rodriguez, L & Perez, A. (2009) Influence of Fatty Acid Composition of Raw Materials on Biodiesel Properties. *Bioresource Technology*, 100(1):261-268

Regan, DL & Gartside, G. (1983). *Liquid Fuels from Microalgae in Australia*. CSIRO, Melbourne

Richardson, C. 2011. *Investigating the role of reactor design to maximise the environmental benefit of algal oil for biodiesel*. Master's thesis. University of Cape Town.

Richmond, A. (2004). *Microalgal culture: biotechnology and applied phycology*. Blackwell Science, Oxford

Rodolfi, L, Chini Zittelli, G, Bassi, N, Padovani, G, Biondi, N, Bonini, G & Tredici, MR. (2009). Microalgae for Oil: Strain Selection, Induction of Lipid Synthesis and Outdoor Mass Cultivation in a Low-Cost Photobioreactor. *Biotechnology and Bioengineering*, 102(1):100-112

Roessler, P. (1990). Environmental Control of Glycerolipid Metabolism in Microalgae: Commercial Implications and Future Research Directions. *Journal of Phycology*, 26:393-399

Sazdanoff, N. (2006). *Modeling and simulation of the algae to biodiesel fuel cycle*. Honour's thesis, Ohio State University

Schenk, PM, Thomas-Hall, S, Stephens, E, Marx, UC, Mussgnug, JH, Posten, C, Kruse, O & Hankamer, B. (2008). Second Generation Biofuels: High-Efficiency Microalgae for Biodiesel Production. *BioEnergy Research* 1(1):20-43

Scott, SA, Davey, MP, Dennis, JS, Horst, I, Howe, CJ, Lea-Smith, D & Smith, AG. (2010). Biodiesel from Algae: Challenges and Prospects. *Current Opinions in Biotechnology*, 21:277-286

Sheehan, J, Dunahay, T, Benemann, J & Roessler, P. (1998). *A Look Back at the U.S. Department of Energy's Aquatic Species Program: Biodiesel from Algae*. Close-Out report. National Renewable Energy Lab, Department of Energy, Golden, Colorado, U.S.A. Report number NREL/TP-580-24190, dated July 1998

Shelef G, Sukenik A, Green M (1984) *Microalgae Harvesting and Processing: a Literature Review*. SERI report.

Shifrin, NS & Chisholm, SW. (1981). Phytoplankton Lipids: Interspecific Differences and Effects of Nitrate, Silicate and Light-Dark Cycles. *Journal of Phycology*, 374-384

Shuler, ML & Kargi, F. (2005). *Bioprocess Engineering: Basic Concepts*. Pearson Education, Singapore

Spoehr, HA & Milner, HW. (1949). The Chemical Composition of *Chlorella*, Effect of Environmental Conditions. *Plant Physiology*, 24:120-149

Stansell, G, Gray, VM & Sym, S. (2011). Microalgal Fatty Acid Composition: Implications for Biodiesel Quality. *Journal of Applied Phycology*, In press. DOI: 10.1007/s10811-011-9696-x

Takagi, M, Watanabe, K, Yamaberi, K & Yoshida, T. (2000). Limited Feeding of Potassium Nitrate for Intracellular Lipid and Triglyceride Accumulation of *Nannochloris* sp. UTEX LB1999. *Applied Microbiology and Biotechnology*, 54(1):112-7

Tsukahara, K & Sawayama, S. (2005). Liquid Fuel Production using Microalgae. *Journal of the Japan Petroleum Institute*, 48(5):251-259

Uduman, N, Qi, Y, Danquah, MK, Forde, GM & Hoadley, A. (2010). Dewatering of Microalgal Cultures: A Major Bottleneck to Algae-Based Fuels. *Journal of Renewable and Sustainable Energy*, 2(1)

Van Gerpen, J. (2005). Biodiesel Processing and Production. *Fuel Processing Technology* 86:1097-1107

Vlaski, A, van Breemen, A & Alaers, G. (1997). The Role of Particle Size and Density in Dissolved Air Flotation and Sedimentation. *Water Science and Technology*, 36(4):177-189

Xiong, W, Li, X, Xiang, J & Wu, Q. (2008). High-Density Fermentation of Microalga *Chlorella prototothecoides* in Bioreactor for Microbio-Diesel Production. *Applied Microbiology and Biotechnology*, 1:29-36

Zhang, X, Hu, Q, Sommerfeld, M, Puruhito, E & Chen, Y. (2010). Harvesting Algal Biomass for Biofuels using Ultrafiltration Membranes. *Bioresource Technology*, 101(14):5297-304

Permissions

The contributors of this book come from diverse backgrounds, making this book a truly international effort. This book will bring forth new frontiers with its revolutionizing research information and detailed analysis of the nascent developments around the world.

We would like to thank Margarita Stoytcheva and Gisela Montero, for lending their expertise to make the book truly unique. They have played a crucial role in the development of this book. Without their invaluable contribution this book wouldn't have been possible. They have made vital efforts to compile up to date information on the varied aspects of this subject to make this book a valuable addition to the collection of many professionals and students.

This book was conceptualized with the vision of imparting up-to-date information and advanced data in this field. To ensure the same, a matchless editorial board was set up. Every individual on the board went through rigorous rounds of assessment to prove their worth. After which they invested a large part of their time researching and compiling the most relevant data for our readers. Conferences and sessions were held from time to time between the editorial board and the contributing authors to present the data in the most comprehensible form. The editorial team has worked tirelessly to provide valuable and valid information to help people across the globe.

Every chapter published in this book has been scrutinized by our experts. Their significance has been extensively debated. The topics covered herein carry significant findings which will fuel the growth of the discipline. They may even be implemented as practical applications or may be referred to as a beginning point for another development. Chapters in this book were first published by InTech; hereby published with permission under the Creative Commons Attribution License or equivalent.

The editorial board has been involved in producing this book since its inception. They have spent rigorous hours researching and exploring the diverse topics which have resulted in the successful publishing of this book. They have passed on their knowledge of decades through this book. To expedite this challenging task, the publisher supported the team at every step. A small team of assistant editors was also appointed to further simplify the editing procedure and attain best results for the readers.

Our editorial team has been hand-picked from every corner of the world. Their multi-ethnicity adds dynamic inputs to the discussions which result in innovative outcomes. These outcomes are then further discussed with the researchers and contributors who give their valuable feedback and opinion regarding the same. The feedback is then collaborated with the researches and they are edited in a comprehensive manner to aid the understanding of the subject.

Apart from the editorial board, the designing team has also invested a significant amount of their time in understanding the subject and creating the most relevant covers. They scrutinized every image to scout for the most suitable representation of the subject and create an appropriate cover for the book.

The publishing team has been involved in this book since its early stages. They were actively engaged in every process, be it collecting the data, connecting with the contributors or procuring relevant information. The team has been an ardent support to the editorial, designing and production team. Their endless efforts to recruit the best for this project, has resulted in the accomplishment of this book. They are a veteran in the field of academics and their pool of knowledge is as vast as their experience in printing. Their expertise and guidance has proved useful at every step. Their uncompromising quality standards have made this book an exceptional effort. Their encouragement from time to time has been an inspiration for everyone.

The publisher and the editorial board hope that this book will prove to be a valuable piece of knowledge for researchers, students, practitioners and scholars across the globe.

List of Contributors

C.L. Bianchi, C. Pirola, D.C. Boffito, A. Di Fronzo and G. Carvoli
Università degli Studi di Milano, Dipartimento di Chimica Fisica ed Elettrochimica, Milano, Italy

D. Barnabè, R. Bucchi and A. Rispoli
Agri2000 Soc. Coop., Bologna, Italy

Vivian Feddern, Anildo Cunha Junior, Marina Celant De Prá, Paulo Giovanni de Abreu, Jonas Irineu dos Santos Filho, Martha Mayumi Higarashi, Mauro Sulenta and Arlei Coldebella
Embrapa Swine and Poultry, Brazil

Carlos A. Guerrero F., Andrés Guerrero-Romero and Fabio E. Sierra
National University of Colombia, Colombia

Jianguo Zhang and Bo Hu
University of Minnesota, USA

Emad A. Shalaby
Biochemistry Dept., Faculty of Agriculture, Cairo University, Egypt

Maddalena Rossi, Alberto Amaretti, Stefano Raimondi and Alan Leonardi
University of Modena and Reggio Emilia, Italy

Jin Liu
Department of Applied Sciences and Mathematics, Arizona State University, Polytechnic Campus, Mesa, USA

Junchao Huang
Kunming Institute of Botany, Chinese Academy of Sciences, China
School of Biological Science, The University of Hong Kong, Hong Kong, China

Feng Chen
School of Biological Science, The University of Hong Kong, Hong Kong, China
Institute for Food & Bioresource Engineering, College of Engineering, Peking University, Beijing, China

Simrat Kaur and Charles Spillane
Genetics and Biotechnology Laboratory, Botany and Plant Science, C306 Aras de Brun, National University of Ireland Galway, Ireland

Mohan C. Kalita
Department of Biotechnology, Gauhati University, Assam, India

Ravi B. Srivastava
Defence Institute of High Altitude Research, Defence Research & Development Organisation, Leh (Jammu and Kashmir), India

László Kótai, János Bozi and István Gács
Institute of Materials and Environmental Chemistry, Chemical Research Center, Hungarian Academy of Sciences, Hungary

János Szépvölgyi
Institute of Materials and Environmental Chemistry, Chemical Research Center, Hungarian Academy of Sciences, Hungary
Research Institute of Chemical and Process Engineering, University of Pannonia, Hungary

Szabolcs Bálint and Ágnes Gömöry
Institute of Structural Chemistry, Chemical Research Center, Hungarian Academy of Sciences, Hungary

András Angyal
Axial-Chem Ltd., Hungary

János Balogh
Kemobil Co., Hungary

Zhibin Li, Moutong Chen, Chen Wang and Baiquan Chen
China New Energy Co., China

Melinda J. Griffiths, Reay G. Dicks, Christine Richardson and Susan T. L. Harrison
Centre for Bioprocess Engineering Research (CeBER), University of Cape Town, South Africa

Printed in the USA
CPSIA information can be obtained
at www.ICGtesting.com
JSHW011422221024
72173JS00004B/636

9 781632 400789